Food Related Diseases: Identification, Prevention and Management

Food Related Diseases: Identification, Prevention and Management

Editor: Hugh Freeman

MURPHY & MOORE
www.murphy-moorepublishing.com

Murphy & Moore Publishing,
1 Rockefeller Plaza,
New York City, NY 10020, USA

Visit us on the World Wide Web at:
www.murphy-moorepublishing.com

ISBN: 978-1-63987-231-2 (Hardback)

Cataloging-in-Publication Data

Food related diseases : identification, prevention and management / edited by Hugh Freeman.
p. cm.
Includes bibliographical references and index.
ISBN 978-1-63987-231-2
1. Foodborne diseases. 2. Foodborne diseases--Diagnosis. 3. Foodborne diseases--Prevention.
4. Foodborne diseases--Treatment. 5. Food poisoning. 6. Communicable diseases.
I. Freeman, Hugh.
RA601.5 .F66 2022
615.954--dc23

Table of Contents

Preface

This book aims to highlight the current researches and provides a platform to further the scope of innovations in this area. This book is a product of the combined efforts of many researchers and scientists, after going through thorough studies and analysis from different parts of the world. The objective of this book is to provide the readers with the latest information of the field.

A large number of diseases have been identified which are caused due to the consumption of foods contaminated with viruses, bacteria and other parasites. Such contaminations can occur at any stage of food delivery. Some common pathogens responsible for spoiling the food are salmonella, campylobacter, Escherichia coli, staph aureus and norovirus. Symptoms of food related diseases include diarrhea, dizziness, fever, nausea, vomiting and loss of appetite. Such infections can be detected by culturing stool samples of the patients and are often resolved in less than seven days. The spread of such diseases can be controlled by properly washing and thoroughly cooking food items. This book includes some of the vital pieces of work being conducted across the world, on various topics related to food related diseases. It aims to shed light on some of the unexplored aspects of food related diseases and the recent researches in this field. This book is appropriate for students seeking detailed information in this area as well as for experts.

I would like to express my sincere thanks to the authors for their dedicated efforts in the completion of this book. I acknowledge the efforts of the publisher for providing constant support. Lastly, I would like to thank my family for their support in all academic endeavors.

Editor

1

Staphylococcal Food Poisoning and Novel Perspectives in Food Safety

Zhenbo Xu, Brian M. Peters, Bing Li, Lin Li and Mark E. Shirtliff

Abstract

Ingestion of food is the major way for human beings to obtain nutrient substances for basic living, and therefore, the quality and safety of food is a major concern. Foodborne illness includes any illness resulting from the consumption of contaminated food that contains pathogenic bacteria, viruses, or chemical or natural toxins. Consequently, food safety is considered to be a globally expanding issue and thus a leading topic in public health, no longer limited to foodborne illnesses but extended to nearly every safety issue regarding "farm to table" food approaches. Bacterial foodborne infections occur by ingestion of food contaminated with growth of pathogenic bacteria, toxin production, and continuous bacterial growth in intestines. In the past decade, a large number of cases or reports have been available on food containing unhealthy, harmful, or toxic substances (other than food poisoning outbreaks) worldwide. Foodborne microorganisms, primarily associated with pathogenic bacteria and toxic substances produced in food, have presented major challenges for food safety. As a global foodborne pathogen, *Staphylococcus aureus* is typically capable of causing a large number of infections, including food poisoning. Thus, this chapter aims to review several factors contributing to the rise of staphylococci as a growing concern for the food safety industry, including the growth of *S. aureus*, production and regulation of staphylococcal enterotoxins, the viable putative but nonculturable (VPNC) state, and antimicrobial resistance of *Staphylococcus* caused by the indiscriminate use of antibiotics in both clinical and veterinary settings.

Keywords: *Staphylococcus aureus*, Enterotoxins, Food poisoning, Biofilm

1. Introduction

Foodborne illness, also known as foodborne infection, foodborne diseases, or food poisoning, is defined as any illness resulting from the consumption of contaminated food that contains pathogenic bacteria, viruses, or chemical or natural toxins (such as poisonous mushrooms) [1,

2]. Remaining as the leading concern in public health and food safety globally, the annual occurrence of foodborne illness has been reported to be 47.8 million foodborne cases (with 128,000 hospitalizations and 3,000 deaths) in the United States, 750,000 cases (with 113,000 hospitalizations and 460 deaths) in France, and 5.4 million cases (with 18,000 hospitalizations and 120 deaths) in Australia [3–5]. Foodborne infections and diseases are caused by a large variety of pathogens that contaminate food and related products, accounting for the major source of all foodborne illnesses, with an estimate of 14 million illnesses, 60,000 hospitalizations, and 1,800 deaths per year [2]. Therefore, bacterial foodborne infections occur when ingested food is contaminated with pathogenic bacteria, toxins produced (even if the growth of host bacteria is terminated), and continuous bacterial growth in intestines (for setting up an infection that causes illness).

1.1. Staphylococci

As a group of Gram-positive, facultative aerobic, frequently unencapsulated, osmotolerant microorganisms, staphylococci are carried, mostly transiently, by approximately 50% of healthy adults on the skin and anterior nares and widespread in untreated water, raw milk, and sewage. Staphylococci are responsible for various tissue infections and a multitude of diseases [6–9]. Up to date, more than 30 distinct types of staphylococci strains have been verified to be infectious for humans, and the associated illnesses and diseases range from mild to severe, from no treatment required to even potentially fatal [6, 7, 10]. Most of these infections are caused by *Staphylococcus aureus*, which has been regarded as a leading human and zoonotic pathogen implicated in both clinical medicine and food safety and typically is capable of causing a large number of infections, including skin infections and sometimes pneumonia, endocarditis, osteomyelitis, gastroenteritis, scalded skin syndrome, and toxic shock syndrome (TSS) [6]. Aside from this bacterium, coagulase-negative staphylococci (CoNS) are regarded as a frequent cause of nosocomial infection and bacteremia, especially in patients with indwelling medical devices [12, 13]. CoNS have also become the most frequently isolated pathogens in intravascular catheter-related infections (CRI), accounting for an estimated 28% of all nosocomial bloodstream infections [14].

1.2. Staphylococcal Food Poisoning (SFP)

Staphylococcal food poisoning (SFP) is a noncontagious gastrointestinal illness caused by the ingestion of food contaminated with preformed staphylococcal enterotoxins (SEs), with low fatality and symptoms commonly including nausea, retching, vomiting, stomach cramps, exhaustion, and diarrhea [1, 15]. The onset of SFP symptoms commonly occurs between 0.5 and 6 h, and the illness typically lasts for 1 day (up to 3 days), with rapid recovery [1]. Serving as one of the most economically important foodborne diseases, SFP is currently a major issue for global public health programs worldwide [16, 17]. In the United States, *Staphylococcus* has been among the top 5 pathogens responsible for foodborne illnesses. From 1983 to 1997, the annual number of SFP cases had been estimated to be 185,000, with 1,750 hospitalizations and 2 deaths, totaling a cost of 1.5 billion dollars [1, 2]. Based on the recent surveillance in 2011, SFP was found to be account for 241,148 foodborne illnesses, 1,064 hospitalizations, and 6 deaths per year [1, 18]. In Europe, SFP ranks as the fourth most common causative agent of foodborne illness, with 926 SFP outbreaks reported in 15 European Union (EU) countries from 1993 to 1998 [2, 19]. In Japan, according to the Ministry of Health and Welfare of Japan, for a period of 20 years (1980–1999), a total of 2,525 outbreaks of SFP were reported, which involved 59,964 persons, resulting in 3 deaths [17]. In addition, an extensive SFP outbreak occurred in Japan in 2000, with a large number of patients (13,420 cases) ingested with dairy products contaminated with SEs [20]. In China, for a majority of regions, *S. aureus* was recovered from more than 15% of food samples, and in occasional outbreak cases, the identification rate approached 90% [2, 8, 21]. However, due to the lack of comprehensive surveillance and investigation and the prevalence and occurrence of SFP that varied considerably among different regions and areas in China, this discrepancy may be explained by different local eating habits and food product usage. In addition, staphylococcal strain-dependent differences may also contribute to the variation.

Food commonly involved in staphylococcal intoxication include protein food (even salty) such as ham, raw or processed meat, puddings, pastries tuna, chicken, sandwich fillings, cream fillings, potato and meat salads, custards, raw bilk milk, milk products (especially unpasteurized milk), cheese products, and creamed potatoes [22, 23]. In China, raw meat, milk and dairy products, frozen products, and cooked food have been found as major food types contaminated by *S. aureus*, taking up 38%, 20%, 16%, and 14%, respectively [2, 8, 21]. In Europe, meat and related products have been the common food vehicles for SFP. In Japan, SFP frequently occurs in rice balls and Japanese-style desserts [24]. Food made by hand requiring no further cooking or handled frequently during preparation are major targets for *Staphylococcus* contamination and at highest risk for the production of bacterial toxins, which eventually cause SFP. Although *S. aureus* contamination can be readily avoided by heat treatment of food, its ability to grow in a wide range of temperatures (7–48.5°C), pH [4.2–9.3] and sodium chloride concentrations (up to 15% NaCl) facilitates the contamination and transmission of this organism to various types of food [16, 17, 25, 26].

2. Staphylococcal Enterotoxins (SEs)

Some *S. aureus* strains are capable of producing SEs, and several CoNS species have also been reported to produce SEs, including *Staphylococcus epidermis* and *Staphylococcus haemolyticus*, among others [27–29]. Classified into members of the pyrogenic toxin superantigen family [30], SEs are a series of extracellular single-chain proteins primarily produced in food or culture media and secreted by some *Staphylococcus* strains, and the causative agent of SFP, which after ingestion may cause intoxication exhibited by vomiting (emetic action on the abdominal viscera) and diarrhea (epithelium of the intestinal tract and inhibition of water absorption in intestine), even collapse in severe cases [15]. In most SFP cases, a single enterotoxigenic staphylococcal strain isolated from the contaminated food is suspected to be the responsible strain [31]. However, from a number of SFP outbreaks with symptoms similar to gastrointestinal syndromes mediated by SEs, only nonenterotoxigenic staphylococci have been isolated, which may be explained by the outgrown enterotoxigenic staphylococci by the nonenterotoxigenic ones. Containing low α-helix and high β-pleated sheet content, SEs have similar and flexible structure, with a low molecular weight between 24,000 and 30,000 Da [32]. One of the characteristic features of SEs is their heat stability (withstanding heating to 121°C for 10 min), as SEs are commonly produced ranging from 10°C to 50°C, with the optimum at 30°C to 40°C [24, 26, 33–35]. Consequently, when *Staphylococcus* strains grow in food under appropriate conditions and produce SEs undetectable by taste or smell, SEs remain active even when bacteria itself have been eliminated by heating. As concluded, effective ways include the prevention of food contamination from staphylococci, staphylococci organisms from growing, and SEs from being produced under either heating or freezing. SEs are also highly hydrophilic, with pH ranges from 4.8 to 9.0 (optimum between 5.3 and 7.0) and water activity (aw) ranges from 0.87 to 0.99 (optimum at 0.90) [24, 36, 37]. Being well studied and documented, SEs are found to possess a number of biological properties, including superantigenicity (induction of T-cell mitogenicity and human interferon), emetic activity, and pyrogenicity [38]. In addition to gastrointestinal symptoms in SFP as aforementioned, SEs have also been implicated in other diseases such as atopic eczema [39–41], rheumatoid arthritis [41–43], and urticaria [41, 44]. With binding of functional SEs (not presursor) to both the α-helical regions of the major histocompatibility complex (MHC) class II molecules outside the peptide binding groove of the antigen-presenting cells (APC) and the variable region (Vβ) on the T-cell receptor (TCR), a bridge between T cells and APC is formed, leading to nonspecific activation and proliferation of a large number of T cells [41, 45, 46] resulting in robust inflammatory cytokine release.

As classified according to the distinct immunological entities, until recently, a total of 25 types of SEs (A–V and X, with 3 subtypes for C) have been identified. The finding of SEs has been in accordance with the development of identification methodologies. In early studies, animal testing experiments had been commonly used for the observation of SE activity, requiring monkey, feline, or cavy models [47–49]. After feeding with contaminated food, animals exhibited abnormal behavior or gross morphology changes, and the SEs contained in food were determined by the number of vomiting events, the time until the first vomiting event, and behavioral changes [47]. These types of animal feeding experiments had been commonly used for the characterization of emetic activities of SEs, thus determining the roles for emetic activities of SEs during SFP. However, such methodologies were also significantly limited by low sensitivity and specificity, poor reproducibility, high expense and laboratory operation, diversity in animals, and incapability of quantification and accurate identification [47]. Therefore, the identification of individual single type of SEs was reported by the availability of serological assays. Based on the specific reaction between antigen and antibody, identification of SEs via specific antibody [50] was first reported in the 1930s. Ouchterlony double immunodiffusion, also known as agar gel immunodiffusion, had become applicable in 1948 [51], and the first serological study on distinct SEs was conducted in 1958 [52]. From 1960s to 1970s, a number of SEs with emetic activities were identified and reported by the employment of serological methodologies including Ouchterlony double immunodiffusion, radial immunodiffusion, and enzyme-linked immunosorbent assays (ELISA). With the development of molecular biotechnology such as PCR and even genome sequencing in the 1990s and 2000s, a large number of newly identified SEs (G–V and X) as well as variants (for instance, 17 distinct variants of *sel-x*) had been discovered on enterotoxin gene cluster (*egc*), staphylococcal pathogenicity islands (SaPIs), mobile genetic elements (MGEs), and even the bacterial genome (*sel-p* on genome of N315).

2.1. Classical SEs

Although investigation on the observation and detection of SEs from animal feeding experiments could be dated back from 1930s, the immunological characteristics were not clarified until a number of serological studies had been conducted in the 1950s and 1960s. These studies revealed 6 types (A, B, C_1, C_2, D, and E) of SEs that were characterized and further referred to as classic SEs (including C_3 reported in 1984). Antigenicity of SEs was validated when the antisera prepared from rabbits infected with SEs was demonstrated to protect cats [52], and later from a further serological study, the origins of two heat-resistant types of SEs had been verified [52, 53]. These extracellular immunologically distinct SEs responsible for the clinical manifestation of SFP were first referred to as type F (food poisoning) and type E (mostly produced by strains of "enteritis" origin) and later designated as A and B for a better sequential numbering system, with only type A commonly associated with food poisoning (A-1963). Sharing a similar basic three-dimensional structure, the 5 serological groups of SEs (A–E) exhibit nucleotide sequence identity of 50% to 85%, with types A, D, and E categorized into one group (52–83% amino acid identity) and types B and Cs falling into a separate, more closely related group (62–64% amino acid identity) [53, 54]. As origins and sources, types B and C are important causes of nonmenstrual-associated TSS, and types A and D are common causes of SFP with types B and C to a lesser extent [49].

2.1.1. *Staphylococcal Enterotoxin A (sea or SEA)*

SEA was first identified in 1959 from *S. aureus* strain FRI-196E [52, 55] and then named as type A [56], which has been considered to be the most commonly detected SE associated with food poisoning, with its minimal toxic dose for humans ranging from 20 to 100 ng [24]. Also, SEA has been verified to be responsible for a number of SFP outbreaks, including an extensive outbreak caused by ingestion of dairy products in Osaka, Japan, in 2000 [24, 56, 57]. After identification by serological methodology, the production of SEA in different media or condition had become a major concern. The maximal level of SEA production was found to occur during exponential phase. In semisolid BHI agar (pH 5.3], the production of SEA was acquired, and trace amounts of SEA and SEB were obtained with cellophane sac culture [58, 59]. As food samples were concerned, SEA production was detected in a number of meat samples (raw beef and pork, cooked beef and pork, and canned ham). Better growth of *S. aureus* and production of SEA were detected in cooked meat than raw meat despite no significant difference obtained, and such diversity may be explained by the bacterial competition between anaerobic and aerobic conditions (with the latter preferable for *S. aureus*) [50, 58, 59]. In milk, SEA production was also found to be associated with staphylococcal growth [60]. In fermented sausage, SEA production had been detected aerobically at pH 5.1 (with an inoculum of 4×10^7 cells/g sausage) but not anaerobically at pH 5.7 [61]. For co-growth food microorganisms, inhibition was observed to be more common than stimulation, including inhibition of both staphylococcal growth and SE formation with no apparent effect on growth [62]. However, such influence of other microorganisms on *Staphylococcus* was affected markedly by environmental conditions, and discrepancy on inhibition had been also noticed between plate test and meat slurries [58, 59]. Despite acquisition of appropriate staphylococcal growth in both pure culture and in the presence of other food microorganisms, production of SEA was only formed in pure culture [58, 59]. Although SEB was reported to be produced in much larger quantities and more diverse among strains than SEA, SEA had been implicated in a larger number of food poisoning cases [63]. Resembling a primary metabolite (with SEB as a secondary metabolite), SEA is secreted by the bacterium during the exponential phase of growth, with various of factors affecting its production, including salt concentration (NaCl, $NaNO_2$, and $NaNO_3$ showing no influence), surfactants (increase in SEA secretion), pH (optimal ranging from 6.5 to 7.0), and antimicrobial agents (inhibition by chloramphenicol or 2,4-dinitrophenol, with streptomycin or penicillin G exhibiting no influence), which may explain the higher frequency and incidence of SEA in food poisoning [63–66]. In addition, temperature and inoculum size play important roles in SEA production. As temperature was concerned, SEA production was detected under a broad range of temperatures from 10°C to 50°C (in BHI broth) but not at low temperatures (such as 8°C or 10°C as reported) [24]. SEA was detected in the exponential phase from 15°C to 37°C, and its production increased with the elevation of temperature. Also, SEA was detected in the stationary or death phase at 10°C despite acquisition of the lowest SEA concentration at this temperature. Similar to SEE, SEA contains 2 MHC-II binding sites (Zn^{2+} dependent) and thus possesses strong superantigenicity for T-cell activation [67, 68]. Carried by a polymorphic family of lysogenic phages [69, 70], the gene encoding SEA has a length of 771 bp, and its translational product is SEA precursor of 257 amino acids. With 24-residue N-terminal hydrophobic leader sequence further processed, the mature form of SEA was composed of 233 amino acids [38, 53, 68–74]. Unlike other classic SEs (*seb*, *sec*, and *sed*), expression of *sea* had been found to be independent of *agr* regulation [54, 75, 76].

2.1.2. Staphylococcal Enterotoxin B (seb and SEB)

Being the first identified (from *S. aureus* strain FRI-243) and the most studied SE, SEB was initially named as type E and subsequently designated as type B. As the most potent SE and requiring much lower quantities for toxic effect than synthetic chemicals, SEB are capable of causing multiorgan system failure and death at low concentration. As an exotoxin secreted by *S. aureus*, production of SEB had been reported from diverse clonal complexes, including CC8 (the most common), CC20, and CC59 [32, 77–79]. As a superantigen capable of cross-linking APCs and T cells to form a ternary complex between MHC-II and TCR at specific Vβ chain, SEB had been well studied as a causative agent for food poisoning, TSS, atopic dermatitis (common colonization of *S. aureus* and frequent occurrence of SEB-specific antibody from patients with AD), and respiratory diseases (asthma and nasal polyps) [32, 77, 80, 81]. As a well-characterized protein, SEB had been found to be extremely stable (retaining its activity even in acidic environment), water soluble, heat stable (among the most heat-stable proteins, with intact protein under 78°C to 80°C for 30 min), broadly pH tolerant [4–10], and resistant to proteolytic digestion (such as pepsin, trypsin, and papain) [47, 82, 83]. Nevertheless, SEB formation and production were influenced by a number of factors, including inhibition of SEB formation in BHI broth by medium filtrates (such as K_2HPO_4, KCl, $CoCl_2$, NaF, acriflavine, phenethyl alcohol, streptomycin sulfate, chloramphenicol, spermine phosphate, spermidine phosphate, and Tween-80) [82], decrease of SEB production by either temperature depletion (without affecting staphylococcal growth) or curing salt concentration elevation (more rapid reduction of SEB production than staphylococcal growth) [84], catabolite repression [85], and minerals (double SEB production was obtained when magnesium and potassium are under appropriate concentration) [34, 86, 87]. Generally, maximal SEB production occurs in postexponential growth. From early studies in the 1950s and 1960s, SEB was considered to be irrelevant to food poisoning [50, 88]. Located in either chromosome (strain FRI-243, FRI-277, or S6) or plasmid (strain DU-4916), *seb* is 705 bp in length, and the mature SEB consists of 239 amino acid residues sharing nucleotide and amino acid sequence homology with sec_1 and streptococcal pyrogenic exotoxin A [89–91]. Regulated by the staphylococcal two-component system, accessory gene regulatory (*agr*), the region between 59 and 93 nucleotides upstream of the transcription ignition site was found to be essential for transcription and expression [92–94]. *seb* had been commonly found in toxin-mediated foodborne and clinical *S. aureus* strains, and recently, *seb* (by PCR and Western blotting) was detected from 5% of 300 clinical *Staphylococcus* strains [95].

2.1.3. *Staphylococcal Enterotoxin C (sec and SEC)*

According to the new numbering system agreed on the American Society for Microbiology (ASM) meeting in 1962, the first verification of SEC was then reported in 1965, with its toxicity and specificity also confirmed [96]. In this study, the enterotoxins from *S. aureus* strains FRI-137 and FRI-361 were both discovered to react with a specific antibody; thus, strain FRI-137 (ATCC 19095) was selected as the prototype of SEC [96]. However, 2 years later, enterotoxins from strain FRI-137 and FRI-361 were purified as distinct enterotoxins [96, 98] and consequently labeled as SEC_1 (strain FRI-137) and SEC_2 (strain FRI-361). In 1984, the third enterotoxin C (SEC_3) was discovered from a *S. aureus* strain FRI-913 from prawn in England, which were serologically and chemically similar to SEC_1 and SEC_2 but identical by isoelectric focusing, radioimmunoassay (RIA), and N-terminal analysis [99, 100]. Despite cross-reactivity with same antibody, each of the 3 SEC had antibodies that reacted with minor determinants [99]. Located on chromosome (SaPIs), *sec* is composed of 801 bp and encodes a precursor protein of 267 amino acids, with the mature toxin of 239 amino acids [101, 102]. Aside from 3 types of classic SECs, additional *sec* variants (such as *sec*-bovine from SaPIbov) possessing >95% deduced amino acid homology among them had been also reported [103–107]. As SEC_1, SEC_2, and SEC_3 are all emetic enterotoxins with equal toxicity to that of SEA and SEB in both oral administration and intravenously [96, 99], SEC have been responsible for numerous SFP outbreaks (mostly caused by milk) [108]. Maximal production of SEC occurs during postexponential growth. SEC-positive strains of *S. aureus* are commonly associated with bovine, ovine, and caprine dairy products [109]. Yet interestingly, SEC expression has been noted to be reduced in cheeses [110]. From a recent study, milk environment was found to dramatically change the expression profiles of enterotoxin genes despite no influence on staphylococcal growth. In particular, SEC production was substantially reduced in milk compared to the laboratory medium on the protein level, which may be explained by the down-regulation of the *agr* system [111].

2.1.4. Staphylococcal Enterotoxin D (sed and SED)

In 1967, SED was first reported from *S. aureus* strain FRI-293 (which also produced SEC; thus, strain FRI-494 was selected as the prototype strain, also known as ATCC 23235) and its emetic activity in cats, and specific neutralization of biological activity by antisera had been verified [112]. The production of SED alone and in combination with SEA was considered to play a key role in food poisoning (ranking second in frequency after SEA) and recognized as one of the most commonly recovered enterotoxins in SFP outbreaks [112, 113]. Encoding a toxin of 228 amino acids, *sed* is located on a 27.6-kb penicillinase plasmid pIB485 [114]. SED was found to be partially activated by *agr* via RNA III-mediated reduction of Rot (repressor of toxin) during postexponential growth phase, as independent formation from *agr* was found under high concentration. As a consequence of *agr* regulation via quorum sensing, during growth in BHI broth, a modest postexponential induction ratio (<10-fold) was obtained as *sed* reached maximal production during transition from exponential to stationary phase of growth [113, 115]. With the existence of a consensus -10 sequence, a less conserved -35 sequence, and a TG dinucleotide motif, the presence of 52-bp sequence (from -34 to +18) and transcription from +1 to +18 were important for promoter function and *agr* regulation [116]. Aside from regulation by the *agr* system, NaCl stress was capable of decreasing *sed* expression, although no significant effect was further verified. However, regulation under NaCl stress may be highly strain specific variable [117]. As food samples were concerned, in cheese manufacturing (with starter culture including 10^3 CFU/ml of milk), *sed* expression was not induced even when inoculated at 10^6 CFU/ml (equal to 10^8 CFU/g of cheese), presenting a low level of expression and a prolonged pattern that was similar to SEA [113, 118]. In different ham products, when *S. aureus* was inoculated for optimal growth in cultivation broth for 7 days, continuous *sed* expression was observed throughout the entire incubation period for both boiled and smoked ham [115]. However, much less production of *sed* (9 times less) was detected in the latter. For Serrano ham, SED was only detected after 5 days of incubation (*sed* expression still too low to determine), similar to which a second increase had been obtained for boiled and smoked ham after the same time span of incubation [115].

2.1.5. Staphylococcal Enterotoxin E (see and SEE)

In 1971, SEE was reported from a food poisoning *S. aureus* strain FRI-326, which produced distinct SEE having no immunoreactivity with specific antibodies to other SEs [119]. Its toxicity (in rhesus monkeys), specificity, and neutralization with specific antibody were also validated [119]. Located on the phage, *see* is composed of 771 bp and encodes a precursor with a molecular weight of 29,358 Da, which was further processed to a mature extracellular form with a molecular weight of 26,425 Da [120]. Containing a single polypeptide chain, SEE consists of 259 amino acid residues (no free sulfhydryl groups found), with serine and threonine as the NH_2- and COH-terminal amino acids, respectively [121]. Under extreme acidic (pH 2) and basic (pH 12) conditions as well as heating, the toxicity (emetic activity) and antigenicity (serological activity) were found to be destroyed, which is likely due to conformational change [121].

2.2. Staphylococcal enterotoxin-like toxins

Before the 1990s, a total of 7 types of classic SEs (*sea, seb,* sec_1*,* sec_2*,* sec_3*, sed,* and *see*) had been known as causative agents of SFP in humans due to emetic activity. However, starting from the discovery of *seh* in 1994 (aside from discovery of *sef* on 1981), a large variety of novel SE or SE-related toxins (as well as variants) had been reported (G–V and X) based on genetic homology with classical SEs. In 2004, the International Nomenclature Committee for Staphylococcal Superantigens has proposed that only staphylococcal superantigens inducing emesis after oral administration in a primate model should be designated as SEs, whereas other related toxins lacking either emetic properties in a primate model or verification of emetic activity should be otherwise designated as staphylococcal enterotoxin-like toxin type X [122, 123].

2.2.1. *Staphylococcal Enterotoxin-Like Toxin Type F (sel-f and SEl-F)*

In 1981, Bergdoll et al. had noticed an enterotoxin-like protein recovered from 93.8% [61/65] *S. aureus* strains sampled from patients with TSS, representing the first evidence of *sel-f* [124]. With its purification and preparation of specific antibody, *sel-f* was also recovered from 11.5% [3/26] of laboratory *S. aureus* strains, compared with only 4.6% [4/87] from other sources, which suggested the association between SEl-F and TSS [124]. However, from an investigation on the spread of a TSS strain, a temporal association of antibodies to SEl-F with cessation of recurrences of TSS was found, indicating that its production may not either reach clinically significant levels during infection or is insufficient to cause TSS [125]. Generally, studies and reports on SEl-F have been rarely available.

2.2.2. *Staphylococcal Enterotoxin-Like Toxin Type G (sel-g and SEl-G)*

In 1998, SEl-G and SEl-I (from *S. aureus* strains FRI-572 and FRI-445, respectively) had been identified and characterized, including verification on emesis (eliciting emetic response in rhesus monkeys) and superantigenicity (proliferation of T cells) [126]. *sel-g* consists of 777 nucleotides and encodes a precursor protein of 258 amino acids, which has typical bacterial signal sequences and is then cleaved to form mature toxin with 233 amino acids and with a molecular weight of 27,043 Da [127, 128]. SEl-G showed higher homology to SpeA, SEB, SEC, and SSA (38–42% amino acid identity) and exhibited similar epitopes with SEC_1 [126].

2.2.3. *Staphylococcal Enterotoxin-Like Toxin Type H (sel-h and SEl-H)*

In 1994, the first discovery of *sel-h* from *S. aureus* strain D4508 was reported, with its nucleotide and amino acid sequences identified [129]. One year later, SEl-H was identified and purified from *S. aureus* strain FRI-569, which elicited an emetic response in monkeys and was found to be antigenically distinct from other existent SEs [51]. SEl-H shares about 35% amino acid identity with SEA, SED, and SEE [130]. As a superantigen homologous to SEA subfamily, SEl-H displays unique MHC-II binding properties. As a potent T-cell mitogen, SEl-H was capable of activating large amounts of T cells by cross-linking APC and T cells via Vα domain (Vα10, TRAV27) of TCR (with no TCR Vβ-specific expansion) by direct interaction between SEl-H and TCR Vα domain [131, 132]. With emetic activity, *sel-h* was commonly detected alone or together with *sea* [133] and responsible for a number of SFP outbreaks. In 1996, an outbreak was caused by cheese and *S. aureus* strains isolated from cheese were found to produce SEl-H [133]. From the SFP outbreak caused by reconstituted milk in Japan, SEl-H was also detected along with SEA. In a survey on 146 *S. aureus* strains isolated from humans, cows, and bovine in Japan, 7 and 4 strains were found to harbor $sea^{+}sel\text{-}h^{+}$ and *sel-h* alone, respectively [57, 133]. In December 2003, a suspected SFP outbreak involving 8 persons (3 adults and 5 children) with symptoms of vomiting, stomach cramps, and diarrhea shortly after lunch was caused by contaminated mashed potato, and *S. aureus* strains contained in raw bovine milk for preparation of mashed potato were found to produce sufficient SEl-H for food poisoning [134]. SEl-H production was influenced by a variety of factors, including aeration and pH conditions. Higher production level of SEl-H was acquired under aerobic incubation or pH controlled at 7.0, with decrease in SEl-H production under anaerobic condition or slight change of pH (such as 6.5 or 7.5) [135].

2.2.4. *Staphylococcal Enterotoxin-Like Toxin Type I (sel-i and SEl-I)*

As aforementioned, SEl-I were identified together with SEl-G in 1998 [126]. Unlike SEl-G, SEl-I was more similar to SEA, SED, and SEE (26–28% amino acid identity). *sel-i* consists of 729 nucleotides and encodes a precursor protein of 242 amino acids, which contains typical bacterial signal sequences and is further cleaved to form mature SEl-I of predicted 218 amino acids with a molecular weight of 24,928 Da [127, 128]. Although separated by DNA related to other SEs, linkage of *sel-g* and *sel-i* was discovered, and this enterotoxin gene cluster was designated as *egc*, with *sel-g* located 2002 bp downstream of *sel-i* [127]. In southern France, carriage of $sel\text{-}g^{+}sel\text{-}i^{+}$ and $sec^{+}sel\text{-}g^{+}sel\text{-}i^{+}$ was detected from 41.9% and 24.5% of 155 *S. aureus* strains isolated from various food samples [128]. In Taiwan, 14.5% [8/55] *S. aureus* strains of human origin and 9.4% [13/139] strains isolated from frozen food, Chinese sausage, and meal boxes were found to harbor *sel-g*, *sel-h*, and/or *sel-i*, suggesting a minor role that such SEs play in SFP outbreaks [136]. However, a discrepancy between the presence of *sel-g* and *sel-i* and the production of enough quantities of SEG and SEI was also noticed [128]. In 2004, 10.1% [11/109] wild *Staphylococcus* spp. stains were found to contain SEs and *egc*, and the *egc* from strain AB-8802 present variants of *sel-g* and *sel-i* (*sel-gv* and *sel-iv*) [137].

2.2.5. *Staphylococcal Enterotoxin-Like Toxin Type J (sel-j and SEl-J)*

In 1998, *sel-j* was first found to be located on the plasmid pIB485 encoding *sed*, which was separated from *sed* by 895 bp of intergenic region containing a perfect inverted repeat (with each arm of the repeat having a length of 21 bp) [138]. Most of *sel-j* was detected on *sed*-encoding plasmid, suggesting the coexistence of these 2 SEs and their relative contribution to the food poisoning symptomology [138]. With transcription in opposite directions, both *sel-j* and *sed* were capable of expression in *S. aureus* strains, with *sed* only under the transcriptional control of *agr* [138]. Containing 269 amino acid residues, sequence of SEl-J showed substantial similarity to the SE family of *sea, sed,* and *see.*

2.2.6. *Staphylococcal Enterotoxin-Like Toxin Type K (sel-k and SEl-K)*

Despite observation of the *sel-k* gene on SaPI1 (in 1998) and *egc* from *S. aureus* strains A900322 (in 2001), the first designation of *sel-k* from *S. aureus* TSS isolates MN NJ was reported in 2001, with its identification on SaPI3 together with *seb* [139, 140]. Possessing biochemical and biological properties similar to classic SEs, including superantigenicity (Vβ-specific T-cell activation), pyrogenicity, emesis, and lethality in primates, SEl-K was secreted by clinical *S. aureus* strains, with a molecular weight of 26,000 Da and a pI between 7.0 and 7.5 [48, 140]. An increase in the secretion of SEl-K was obtained when coexpressed with SEB (K-2014). However, regardless of the variation in SEl-K secretion amount *in vitro,* similar levels of SEl-K accumulation were found *in vivo* [141]. SEl-K was commonly detected in clinical isolates (more than half) and almost all USA300 strains. In addition, a genetic variation of *sel-k* was discovered, with 6 variants found among 20 clinical isolates [141].

2.2.7. *Staphylococcal Enterotoxin-Like Toxin Type L (sel-l and SEl-L)*

First noticed on *egc* from *S. aureus* strain A900322 [142], *sel-l* was identified on pathogenicity island SaPIbov (15,891 bp) from a bovine mastitis *S. aureus* isolate RF122 (*sel-l*) in 2001, with a molecular weight of 26,000 Da and an isoelectric point of 8.5 [105]. Lacking emetic activity, SEl-L was found to exhibit a number of biological properties similar to other SEs, including superantigenicity, pyrogenicity, enhancement of endotoxin shock, and lethality in rabbits when administered via subcutaneous mini-osmotic pumps, but the protein lacked emetic activity [105].

2.2.8. *Staphylococcal Enterotoxin-Like Toxin Type M (sel-m and SEl-M)*

In 2001, *sel-m* was reported to be located on the *egc* (enterotoxigenic gene cluster) together with *sel-g, sel-I, sel-k,* and *sel-l,* and SEl-M was found to exhibit superantigenicity activity with specific Vβ pattern [142]. However, the emetic activity has not been elucidated yet. Most clinical *S. aureus* strains harboring *egc* were found to carry such SEs regardless of the diseases, suggesting the potential derivation of SEs and the putative cluster of SE genes from *egc.*

2.2.9. *Staphylococcal Enterotoxin-Like Toxin Type N (sel-n and SEl-N)*

From the *egc* reported in 2001, *sel-n* was also found to be located between *sel-i* and *sel-g* [142]. A study was conducted on the cloning and expression of *sel-m* and *sel-n* from *S. aureus* strain FRI-1230, demonstrating that SEl-M and SEl-N were capable of stimulating T cells and inhibiting K562-ADM and B16 cells with an equivalent level to that of SEC_2 [143]. Although superantigenicity had been verified, the emetic activity of SEl-N is still unclear [144].

2.2.10. *Staphylococcal Enterotoxin-Like Toxin Type O (sel-o and SEl-O)*

In 2001, *sel-o* was identified from the *egc* cluster, on which other 4 SEs and 2 pseudogenes were also located, including *sel-i, sel-j, sel-m, sel-n, Ψent1,* and *Ψent2* [142]. However, the biological and biochemical properties of *sel-o* remains unclear despite validation of its superantigenicity [144].

2.2.11. *Staphylococcal Enterotoxin-Like Toxin Type P (sel-p and SEl-P)*

In 2001, *sel-p* (previously called *sep*) was first discovered from the genome of MRSA N315, and its biological properties were fully characterized in 2005 (with *sel-p* from *S. aureus* strain Sagal isolated from an SE unidentified food poisoning outbreak in Japan), including superantigenicity (induction of a substantial proliferative response and production of cytokines) and emetic activity (at relatively high dose as 50–150 μg/animal) [123, 145]. According to this study, SEl-P was detected in 60% of the 30 *sel-p*-positive *S. aureus* isolates, and all 10 strains harboring *seb* and *sel-p* produced SEB but not SEl-P, suggesting that inactivation of the *sel-p* locus associates with a particular SE genetic constitution [123]. Most recently, colonization with *sel-p*-positive MRSA increased the risk of bacteremia, which indicated *sel-p* as a predictive virulence factor for invasive disease [146].

2.2.12. *Staphylococcal Enterotoxin-Like Toxin Type Q (sel-q and SEl-Q)*

In 2002, a member of the new subfamily (group V), *sel-q* (from *S. aureus* strain MN NJ) was identified and located directly 5 of *sel-k*, with a molecular weight of 26,000 Da and isoelectric point between 7.5 and 8.0 [147]. Despite a lack of emetic activity (incapability in neither lethality in rabbits nor emetic activity in monkeys), *sel-q* had been found to possess superantigenicity, pyrogenicity, and ability to enhance endotoxin shock.

2.2.13. *Staphylococcal Enterotoxin-Like Toxin Type R (sel-r and SEl-R)*

In 2003, *sel-r* was recovered and identified from 4 *S. aureus* strains (Fukuoka 5, Fukuoka 6, Fukuoka 7, and Fukuoka 8) isolated from patients with nausea, vomiting, and diarrhea from a food poisoning outbreak occurred at a lunch-box shop in Fukuoka prefecture of Japan in September 1997 [148]. Located on 2 types of plasmid, pBI485 (and pBI485-like plasmids, encoding *sed* and *sel-j* as well) and pK0311 (pF5, pF6, and pF7), *sel-r* was found to most closely related to *sel-g* [148]. Investigation on the biological properties of SEl-R revealed its superantigenicity (T-cell stimulation activity via MHC-II) and emetic activity (induction of a reaction in animals within 5 h at 100 μg/kg) [148–150]. SEl-R production was also verified in seropositive *S. aureus* strains [148, 149]. A survey was conducted on the SEl-R production from 24 *sed*-positive *S. aureus* isolates, and *sel-r* expression was detectable from 22 isolates despite carriage of variant *sed* gene for seven strains lacking SED production [151].

2.2.14. *Staphylococcal Enterotoxin-Like Toxin Type S (sel-s and SEl-S)*

Two novel SE-like genes, *sel-s* and *sel-t,* had been reported on the plasmid pF5, where *sel-j* and *sel-r* were located. SEl-S (rSES) was characterized for biological properties, including superantigenicity (specific stimulation of human T cells via MHC-II APC) and emetic activity (induction of emetic reactions in monkeys) [150].

2.2.15. *Staphylococcal Enterotoxin-Like Toxin Type T (sel-t and SEl-T)*

As aforementioned, a first identification of *sel-t* was reported on plasmid pF5 harbored by *S. aureus* strain Fukuoka 5 from SFP. Similar to SEl-S, SEl-T was found to exhibit both superantigenicity and emetic activity (induction of a delayed reaction after 24 h or 5 days postadministration). Data from the emetic study on SEs involved in the SFP outbreak in Fukuoka in 1997 combined with emesis studies in house musk shrews (similar as in the monkeys) suggest that SEl-R and SEl-S were validated to be the causative toxins of vomiting [150].

2.2.16. *Staphylococcal Enterotoxin-Like Toxin Type U (sel-u and SEl-U)*

From sequencing of 24 *S. aureus* strains harboring *egc*, *sel-u* was identified on 4 of the tested strains [152]. SEl-U was found to result from sequence divergence in the *Ψent1* and *Ψent2* pseudogenes, as *sel-u* was located between *sel-iv* and *sel-n* in *egc* of strain 382F (AY158703) with replacement of the *Ψent1* and *Ψent2* between *sel-iv* and *sel-n* in *egc* of strain Mu50 (AP003363) [144, 152]. A variant *sel-u*, designated as *sel-u2*, was recovered from an atypical *egc* locus and generated by a limited deletion in the pseudogenes *Ψent1* and *Ψent2*, which contained superantigenicity for activation of T cell families Vβ 13.2 and Vβ 14 [144].

2.2.17. *Staphylococcal Enterotoxin-Like Toxin Type V (sel-v and SEl-V)*

In a broad surveillance on *egc* from 666 clinical *S. aureus* isolates, 63% [421/666] strains were positive for *egc* locus [144]. The archetypal *egc* harboring 5 SEs and 2 pseudogenes was found in 409 strains, and a novel SE-like toxin, designated as *sel-v*, was discovered from an atypical *egc* locus from *S. aureus* strain A900624 [144]. SEl-V was generated by recombination between *sel-m* and *sel-i*, and its superantigenicity for activation of T-cell families Vβ-6, Vβ-18 and Vβ-21 has also been validated.

2.2.18. *Staphylococcal Enterotoxin-Like Toxin Type X (sel-x and SEl-X)*

In 2011, *sel-x* was discovered from the core genome of 95% of phylogenetically diverse *S. aureus* strains with human and animal origins, including 17 distinct allelic variants (*sel-x*1 to *sel-x*14, *sel-xov*, *sel-xbov*1, and *sel-xbov*2). Acquisition of *sel-x* includes the horizontal transfer by a *S. aureus* progenitor, allelic diversification by point mutation, and assortative recombination, which explains the high genetic diversity of *sel-x*. With a unique predicted structure, SEl-X was well characterized by biological activities, including superantigenicity (activation of Vβ-specific T cells), pyrogenicity, and endotoxin enhancement. It is also noteworthy that SEl-X produced by strain USA300 (CA-MRSA) had been found to be responsible for the lethality in a rabbit model, which suggested a novel virulence determinant of CA-MRSA disease pathogenesis.

2.3. Pathogenicity Islands (PAIs) and *S. aureus* PIs (SaPIs)

2.3.1. PAIs

In 1986, before the first report of PAIs by Hacker et al. in Werner Goebel's group of Germany in 1994, two large segments had been found to be capable of deletion and thus enable the host bacterial to produce hemolysin and loss of P. fimbriae [139, 153, 154]. Considered to be foreign DNA segments integrated into the bacterial genome, such segments existed within pathogenic isolates (cause of virulence) but not on highly genetically similar nonpathogenic strains [153]. As a subclass of genomic islands, PAIs are defined as a group of gene clusters encoding bacterial virulence on a large DNA segment (ranging from 20 to 100 kb) located on the bacterial chromosome [139, 154–156]. PAIs are acquired by microorganisms via horizontal gene transfer via transduction, conjugation, and transformation. Acquisition of PAIs may rapidly and radically alter the genome of a bacterium, consequently strengthening or reducing its fitness within the host [154, 157]. Pathogens are capable of harboring one or more PAIs associated with one or more virulence genes. PAIs are capable of encoding genetic products, including secretory proteins (such as type III secretion system), cell surface proteins (such as erythrocytolysin, fimbriae, and heme binding factors), signal transduction systems, and regulation systems [139, 155–157].

As distinct DNA regions are present in the genome of pathogenic bacteria and absent in nonpathogenic strains (despite same or close species), typical PAIs are composed of mobility genes (such as integrases) commonly located at the beginning of the island and close to the tRNA locus or the respective attachment site. A number of virulence genes (V1–V4) are frequently interspersed with other mobility elements including insertion sequence (IS) elements (Isc, complete insertion element) or remnants of IS elements (ISd, defective insertion element) [153, 155, 156]. Commonly flanked by direct repeats (DR) and IS elements, PAIs are often genetically unstable and comprise some potential mobile components, such as IS elements, integrase, transposase, and plasmid replication initiation sites. As DNA sequences ranging from 16 to 20 bp (with maximum of 130 bp) with sequence repetition, DR plays a critical role in insertion and deletion (as recognition sites), such as integration of bacteriophages. Although combination of IS elements may be capable of mediating transfer of large DNA fragments, insertion mediated by IS also leads to inactivation of genes as well [153, 154, 156]. Consequently, PAIs are capable of deletion with distinct frequencies and loss of virulence traits encoded by PAIs are reported to occur at higher frequency than that encoded by mutation. PAIs are commonly inserted in the backbone genome of the host strain, typically located to specific sites such as tRNA loci or adjacent to tRNA genes, or sites associated with plasmid and phage integration, due to highly conserved genes encoding tRNAs among various bacterial species [153, 154].

PAIs differ from host chromosomes in GC content and codon usage, which may account for the discovery of novel PAIs and maintenance of the divergent nucleotide composition from the horizontally acquired DNA. Based on significant differences with respect to bacterial virulence, GC content, and codon usage, a hypothesis was proposed that such characteristics may be bestowed from DNA segments on the plasmid and phage; thus, acquisition of PAIs and the emergence of new pathogenic organisms can be correlated [153, 156].

2.3.2. *SaPIs*

As mobile pathogenicity islands with length ranging from 14 to 17 kb and carriage of genes for superantigen toxins and other virulence factors, SaPIs have been responsible for the TSS and other superantigen-related diseases, especially SE-like toxins. Located in specific loci of the chromosome and induced by bacteriophages, SaPIs are capable of incorporating small infective phage-like particles via a program of excision-replication-packaging. Containing most of the staphylococcal toxins and virulence factors, SaPIs facilitate the horizontal acquisition of MGEs and, thus, play an important role in the evolution of *Staphylococcus* [139, 154, 155].

2.3.2.1. *SaPI families*

Aside from SCC*mec* (introduction in detail as below), a large number of toxin genes have been found in SaPIs, including SE-like toxins and TSST. Several types of SaPIs have been identified. SaPI1 was found to be inserted in an *attC* site close to the *tyrB* gene and flanked by the region of *tst* gene, with a length of 15,233 bp [139]. The characteristic features of SaPI1 include mobility and instability, whereas SaPI2 was identified as a second locus. Transduction between SaPI1 and SaPI2 by via helper phage was demonstrated, and stable integration of these 2 SaPIs without phages had also been verified [139, 155, 156]. Inserted at the 3 end of the GMP synthase gene, SaPIbov was identified in a bovine isolate of *S. aureus* by PAI related to SaPI1, with a length of 15,891 bp and carriage of *sec*, *sel-l*, and *tst*. SaPI3 was identified to contain *sel-k* and *sel-q* [157]. SaPIs with similar structure between SaPI3 and SaPI1 had also been reported.

2.3.2.2. *νSa families*

Up to date, 7 conservative PAI types had been discovered in *S. aureus*, namely, νSa1 (including SaPI1 and SaPI3), νSa2 (SaPIbov), νSa3, νSa4 (including SaPI2), νSaα, νSaβ, and νSaγ [139, 155–157]. νSa1 to νSa4 were found to contain integrase genes as putative elements of genetic mobility. Derivation of *int* and *att* sites from phage genome was pointed out, as both were found in SaPIs. On the contrary, νSaα and νSaβ harbor transposase genes, which may be derived form transposons. Comprising SaPI1 and SaPI3, the νSa1 locus of CA-MRSA also carries a large number of genes encoding enterotoxins and TSST. Similarly, the loci of both νSa2 from CA-MRSA and SaPIbov were found to contain enterotoxins and TSST. Capable of high-frequency deletion and formation of an episomal circular DNA, νSa3 was identified in CA-MRSA MW2 and Mu50, and one type of νSa3 harbors novel allelic forms of *sec* and *sel-l*. With a lower frequency of excision than that of νSa3, the νSa4 family contains several allelic forms of a genomic island, and type I νSa4 carries *sec* and *sel-l* [157]. Despite the presence of both νSaα and νSaβ in all sequenced *S. aureus* genomes, the size and number of ORFs in νSaα, as well as the size and gene composition of νSaβ, were found to be highly variable, with neither SaPIs spontaneously excised from the chromosome [145]. The composition of νSaα includes 11 allelic forms of *set* genes (encoding exotoxins), *lukDE* genes (encoding leukotoxins), and lipoprotein gene clusters. However, all varieties of νSaβ contain a gene cluster for serine proteins and superantigen genes absent in CA-MRSA but present in HA-MRSA strains. Adjacent to short DRs, the locus *etd* PAI contains exfoliative toxins *etd* and *edin-B* (encoding exfoliative toxins), IS element, and restriction/modification system [158].

2.3.2.3. Development and evolution of pathogenicity

The mechanisms of horizontal gene transfer in prokaryotic cells include transduction, conjugation, and transformation, among which phage transduction has been the primary transmission drivers of genes among different species and thus plays an important role in the formation of PAI [159]. The formation of PAIs may include 5 stages [160] as follows: (i) acquisition of virulence gene via horizontal gene transfer regulated by an operon and derived from "gene pool" of varied environment; (ii) integration of foreign genes (commonly derived from various complex genes of different donors) into the bacterial chromosome or plasmid via site-specific recombination or other mechanisms, following similar rules to complete integration and shaping obvious structure of genetic island; (iii) evolvement of MGEs into PAI via restructure, gene elimination, and acquisition of other genetic materials, during which the gene components associated with mobility may be inactivation or lost, such as origin of plasmid replication, self-transmissible plasmid tra and phage int; (iv) induced expression of foreign genes under temperate environment; and (v) acquisition or elimination of genetic information constantly via a serious of recombination, insertion or elimination, by which PAI may retrieve MGE and obtain the ability of excision and transfer the whole PAI from chromosome to another recipient bacterium.

Evolution of the bacterial genome may significantly influence its pathogenicity, mainly including point mutations, recombination, and horizontal gene transfer. Despite slow evolution due to relatively low frequency of point mutation, the horizontal gene transfer of large genetic segments (such as PAIs and SaPIs) undoubtedly speeds up the exchange of bacterial genes (as "quantum leap" in short time), leading to the consequent appearance and spread of various novel mutations or variants [161, 162].

3. Novel perspectives of *Staphylococcus* associated with food safety

3.1. Antimicrobial resistance mediated by MGEs

Antibiotics, as compounds or substances that kill or inhibit the growth of microorganisms, have been regarded as one of the greatest contributions to medicine and humanity and used to treat a wide range of infectious diseases caused by bacteria for both animals and human beings [5, 165]. Abuse of existing antibiotics contributes to the spread of antibiotic resistance and poses a predicament for the treatment of several bacterial infections, including therapy for individuals with food poisoning. In China, as one of the currently worst areas for antibiotics abuse, the annual prescription of antibiotics, including both clinical and veterinary treatment, is approaching 140 grams per person and has been roughly estimated to be 10 times higher per capita than that in United Kingdom [6, 7, 163, 164]. From a retrospective study conducted on 1,739 *Staphylococcus* isolates from a hospital in Guangzhou, China, from 2001 to 2010, antimicrobial resistance of tested drugs (exclusively for teicoplanin and vancomycin) was commonly observed among the isolates examined, with high resistance rates for β-lactamases (94.0% and 73.7% for penicillin and oxacillin, respectively) and resistance percentages for cefoxitin, chloramphenicol, ciprofloxacin, clindamycin, erythromycin, gentamicin, trimethoprim-sulfamethoxazole, and tetracycline ranging from 83.9% to 19.4% [165]. As a consequence, antibiotic resistance in microorganisms still remains one of the core concerns in global public health, with methicillin-resistant staphylococci (MRS) strains representing one important group, commonly considered as "super bugs" [7]. Since their first discovery in 1961, MRS (including MRSA and MRCNS) have become among the most prevalent pathogens causing nosocomial infections throughout the world [13, 16, 17, 19, 166]. With the first report of an MRSA-mediated gastrointestinal illness outbreak [174], MRS strains have been considered a major contributor to both health-care-associated and foodborne illnesses. MRS had been identified from contamination of various food samples, such as milk, pork, chicken, veal, beef, turkey, and lamb meat [168–170] as well as in food production animals, such as cattle, chickens, pigs, and cows and are closely connected with the newly discovered MRSA designated as livestock-associated MRSA (LA-MRSA) [171–173]. Carriage of MRS strains in a wide variety of food and food production livestock may not be limited to only food hazard but also poses a significant occupational risk for the industrial staff, such as handlers, asymptomatic carriers, and uncolonized individuals. MRS strains show resistance to nearly all β-lactam antibiotics and commonly multiple other drugs due to the *mecA* and other resistance genes carried by an MGE designated as staphylococcal cassette chromosome *mec* (SCC*mec*). Additionally, the role of integrons as a mobile genetic mechanism in horizontal transfer of antibiotic resistance has also been well established [174–177].

3.1.1. SCCmec

As one of the major foodborne infectious pathogens, *S. aureus* (in particular, MRSA) has been considered to be a potential "super bug," posing a challenge to hospital infection control and a threat to global food safety. Due to the carriage of the *mecA* gene encoding a novel specific penicillin-binding protein (PBP2a), which exhibits a decreased binding affinity to antibiotics, MRSA presents resistance to virtually all β-lactam antibiotics [178]. Evolution from methicillin-susceptible *S. aureus* (MSSA) to MRSA occurred with the acquisition of a genomic island, the staphylococcal cassette chromosome (SCC*mec*). Harboring *mec*A and a large number of functional and regulatory genes, SCC*mec* is an MGE present in *Staphylococcus* species. With accurate excision and integration mediated by site-specific recombinase genes *ccrA* and *ccrB*, SCC*mec* is capable of integration into the bacterial chromosome, leading to the rapid spread of antimicrobial resistance (to β-lactam or other antibiotics) among staphylococcal strains.

MRSA was found shortly after the common use of methicillin, which was first licensed to treat penicillin-resistant *S. aureus* infections in Britain in 1959 [179], resulting in outbreaks of MRSA occurring worldwide. In the 1980s, an extraordinary large chromosomal DNA segment greater than 30 kb carrying *mec* was found to contain no allelic equivalence in MSSA strains by direct chromosome analysis of MRSA strains; this region was designated as *mec* DNA [180–183]. In 1987, the sequence of *mecA* gene cloned from a Japanese MRSA strain was determined [184,

185]. In consideration of the threat caused by this pathogen, the characteristics of MRSA were studied and SCC*mec* was found to be an MGE in MRSA. Additionally, the types of SCC*mec* were found to be genetically diverse. In 1999, the cloning and determination of the structure of the entire *mec* DNA sequence from a Japanese *S. aureus* strain N315 (first isolated in 1982) was reported [186]. Based on the structure, it was reported that *mec* DNA was a novel genomic element designated as staphylococcal cassette chromosome *mec* (SCC*mec*) driven by two site-specific recombinase genes referred to as cassette chromosome recombinases A (*ccrA*) and cassette chromosome recombinases B (*ccrB*) [178]. This was the first time that SCC*mec* was defined, and *ccrA* and *ccrB* were proposed as a novel set of recombinases, defining a new family of staphylococcal genomic elements. After the discovery of SCC*mec*, various types of SCC*mec* were continuously found by scientists around the world. In 2001, the identification of two additional types of SCC*mec* was isolated in other countries of the world and designated as type II SCC*mec* (found in N315) [187]. The two additional types of SCC*mec* were type I found in NCTC10442, which is the first MRSA isolate in England in 1961, and type III found in 85/2082 isolated in New Zealand in 1985. In 2002, a novel type of SCC*mec* designated as type IV was identified from CA-MRSA strains [188]. According to the foundation of type IV SCC*mec*, a new type of SCC*mec* designated as VI, which was originally mistaken as type IV SCC*mec*, was explored [189, 190]. In 2004, type V SCC*mec* was found in the chromosome of a CA-MRSA strain (WIS [WBG8318]) isolated in Australia [191]. Since 2008, novel types of SCC*mec* were found globally. Types VII and VIII SCC*mec* were identified in MRSA strains JCSC6082 (a Swedish isolate) [192] and C10682 (a Canadian isolate) [193], respectively. Novel types of SCC*mec* designated as types IX and X were identified in MRSA strains JCSC6943 and JCSC6945, respectively [194]. The latest type of SCC*mec* to be classified was designated as type XI; it was discovered in the MRSA strain LGA251 genome with a divergent *mec*A homologue (*mec*ALGA251) [195].

3.1.1.1. Structure and types

As a major member of the SCC family and a carrier for gene exchange in staphylococci strains, SCC*mec* is located near the replication origin of the *Staphylococcus* chromosome and inserted at *attB* site located at the 3 end of a novel ORF with unclear function (*orfX*). Typical SCC*mec* comprise 3 basic genetic elements: (i) *Ccr* complex, composed of two site-specific recombinase genes (*ccrA* and *ccrB*), and surrounding ORFs. Via site-specific recombination mediated by *ccrA* and *ccrB*, multiple antibiotic resistance and heavy metal resistance genes are capable of insertion into SCC*mec*. SCC*mec* is further integrated to the staphylococcal chromosome by accurate excision and integration, leading to adaption of the bacterial host to different environments and pressure of antibiotic selection. According to the different types of *ccrA* and *ccrB*, *Ccr* complex was classified into 8 allotypes: type 1 for *ccrA1* and *ccrB1*, type 2 for *ccrA2* and *ccrB2*, type 3 for *ccrA3* and *ccrB3*, type 4 for *ccrA4* and *ccrB4*, type 5 for *ccrC1*, type 6 for *ccrA5* and *ccrB3*, type 7 for *ccrA1* and *ccrB6*, and type 8 for *ccrA1* and *ccrB3*. (ii) *Mec* complex harboring *mecA* and related regulatory genes. Based on the regulatory genes located upstream and downstream of *mecA* and the difference of insertion sequences, *mec* complex was classified to five classes: class A carried the integrated *mecI-mecR1-mecA-IS431* structure, class B carried devoid the *IS1272-ΔmecR1-mecA-IS431* structure that contains integration insertion sequence, class C carried the *IS431-mecA-ΔmecR1-IS431* structure that contains two copies of insertion sequence *IS431*, class D carried the *IS431-mecA-ΔmecR* structure, and class E carried the *blaZ-mecALGA251-mecR1LGA251-mecILGA251* structure. Class C *Mec* complex was divided into two different class designations: class C1 (two *IS431*s were arranged in the same direction) and class C2 (two *IS431*s were arranged in the opposite direction) by the inserted direction of *IS431*. (iii) A junkyard region (J region) is located between *Ccr* complex and *Mec* complex. According to its location in SCC*mec*, J region was classified into J1 (also known as L-C region), J2 (also known as C-M region), and J3 (also known as M-R region) region, located at the upstream of *ccr* gene complex and downstream of *Ccr* complex and the upstream of *Mec* complex and downstream of *Mec* complex, respectively.

Up to date, SCC*mec* elements are classified into types I to XI based on the nature of *Ccr* and *Mec* complex and are further classified into different subtypes in accordance with diverse J region. Types and subtypes of SCC*mec* are described in detail as follows: (i) Type I SCC*mec* was first discovered and had a long history dating back to the 2000s. The characteristic MRSA strain carrying type I SCC*mec* was identified, designated as NCTC10442, which was recovered in the United Kingdom in 1961, representing the first MRSA strain [187]. Type I SCC*mec*, carrying class B *Mec* complex and type 1 *Ccr* complex, carried a *pls* regulator in J1 region. A subtype within type I SCC*mec* was designated as IA, containing a plasmid pUB110 located in J3 region [202]. According to the nomenclature proposed in 2006 [271], type I SCC*mec* was designated as 1B.1.1 and its subtype IA was designated as 1B.1.2. (ii) The characteristic MRSA strain carrying type II SCC*mec* had been identified and designated as N315, which was first isolated in 1982 and discovered in 1999 [178, 186, 196, 197]. Type II SCC*mec* harbored class A *Mec* complex and type 2 *Ccr* complex. In J3 region, an integrated copy of staphylococcal plasmid pUB110 was found and a *kdp* regulator was found in J1 region. A number of subtypes were designated as IIA, IIB, IIC, IID, IIE, and IIb and a variant in type II SCC*mec* in consideration with the difference of J1 and J3 regions. According to the nomenclature reported in 2006 [196], type II SCC*mec* was named 2A.1.1 and type IIb was designated as 2A.2. IIA, IIB, IIC, IID, and IIE were designated as 2A.3.1, 2A.3.2, 2A.3.3, 2A.3.4, and 2A.3.5, respectively. The variant of type II was designated as 2A.1.2. (iii) A Zelanian isolate designated as 85/2082 first isolated in 1985 was found carrying type III SCC*mec*, which was first discovered in 2001 together with type I SCC*mec* [196] and was known as the representative MRSA strain of type III SCC*mec* until now. Type III SCC*mec* carries class A *Mec* complex, type 3 *Ccr* complex, and an integrated copy of plasmid pT181 encoding tetracycline and mercury resistance in J3 region. Regarding the difference within the J3 region, there were several subtypes in type III SCC*mec* designated as IIIA and IIIB and two variants designated as IIIC and IIID. According to the 2006 nomenclature [196], type III SCC*mec* was designated as 3A.1.1, IIIA was designated as 3A.1.2, and IIIB was designated as 3A.1.3. (iv) The two commonly characteristic MRSA strains carrying type IV SCC*mec*, which was first discovered in 2002, were designated as CA05 (JCSC1986) and 8/6-3p (JCSC1978) [188]. Type IV SCC*mec* was found to have a unique combination of class B *Mec* complex and type 2 *Ccr* complex, and transposon Tn4001 was found in J3 region of type IV SCC*mec*. Diversity in subtypes of type IV SCC*mec* was obtained, including IVa, IVb, IVc, IVd, IVE, IVF, IVA, IVg, IVh, IVi, IVj, and IV1. Based on the 2006 nomenclature [196], IVa, IVb, IVc, and IVd were designated as 2B.1.1, 2B.2.1, 2B.3.1, and 2B.4, respectively. IVE was named 2B.

3.3, IVF was designated as 2B.2.2, and IVA was designated as 2B.N.2. IVg, IVh, IVi, IVj, and IVk were designated as 2B.5.1, 2B.6.1, 2B.7.1, and 2B.8.1, and IV1 was designated as 2B.new.1. (v) The CA-MRSA strain WIS (JCSC3624) isolated in Australia was the characteristic MRSA strain carrying type V SCC*mec* [191], which carried class C2 *Mec* complex and type 5 *Ccr* complex. No subtype had been found so far within the group of type V SCC*mec*. According to the nomenclature proposed in 2006 [196], type V SCC*mec* was designated as 5C.1. (vi) In 2001, type VI SCC*mec* was first identified from a pediatric MRSA clone named HDE288, which was first reported in 1992 [189, 190]. Type VI SCC*mec* carries a class B *Mec* complex and type 4 *Ccr* complex. Until now, no subtype of type VI SCC*mec* has been found, which was designated as 4B according to the 2006 nomenclature [196]. (vii) CA-MRSA strain designated as JCSC6082 (p5747/2002) was isolated in 2002 [198] and identified to carry a type VII SCC*mec* in 2008 [192]. Type VII SCC*mec* carries a class C1 *Mec* complex that was different from class C2 *Mec* complex carried by type V SCC*mec* and type 5 *Ccr* complex. There was no subtype of type VII SCC*mec* reported. According to the nomenclature put forward in 2006 [196], type VII SCC*mec* was designated as 5C1. (viii) Type VIII SCC*mec* was first identified from a Canadian MRSA strain designated as C10682 isolated in 2003 [193]. It harbored a novel combination of class A *Mec* complex and type 4 *Ccr* complex. No subtype of type VIII SCC*mec* has been found currently. According to the nomenclature proposed in 2006 [196], type VIII SCC*mec* was designated as 4A. (ix) Type IX SCC*mec* was first reported to be identified in a MRSA strain designated as JCSC6943 isolated from a Thailand participant [194]. It was found carrying class C2 *Mec* complex and type 1 *Ccr* complex. No subtype of type IX SCC*mec* was found so far. According to the nomenclature put forward in 2006 [196], type IX SCC*mec* was designated as 1C2. (x) Together with type IX SCC*mec*, type X SCC*mec* carried class C1 *Mec* complex and novel type 7 *Ccr* complex was identified in a Canadian MRSA strain designated as JCSC6945 [194]. So far, no subtype of type X SCC*mec* has been found. According to the 2006 nomenclature [196], type X SCC*mec* was designated as 7C1. (xi) Type XI SCC*mec* isolated from MRSA strain LGA254 in southwest England of 2007 was a novel type different from other SCC*mec* with carriage of distinct class E *Mec* complex and type 8 *Ccr* complex. No subtype was found in type XI SCC*mec*, and according to the 2006 nomenclature, it was designated as 8E [195, 196, 199].

A thorough understanding of the molecular epidemiology and evolution of MRSA may aid in the further identification, control, prevention, and therapy of *Staphylococcus*-mediated human diseases, necessitating SCC*mec* typing as an essential tool for discrimination of different types and subtypes. Currently, there are several SCC*mec* typing methods available for the global evolutionary study of MRSA, with multiplex PCR as the major and widely used methodology. Multiplex PCR was first developed in 1988 and put in use to distinguish different types and subtypes of SCC*mec* in 2002 [198, 200]. In various multiplex PCR strategies, specificity of primer design has been the major concern determining the application of SCC*mec* typing. The multiplex PCR assay described in 2006 [201] was applicable for unique and specific typing of types and subtypes I, II, III, IVa, IVb, IVc, IVd, and V, respectively. After years of validation, this multiplex PCR strategy had been demonstrated to be a rapid, simple, and feasible method for SCC*mec* typing and serves as a useful tool for further prevention and control of *Staphylococcus*-mediated infections by clinicians and epidemiologists. However, with emergence of novel SCC*mec* (11 types and various subtypes to date), inclusive and novel SCC*mec* typing methodologies are desperately required.

3.1.1.2. *Prevalence and occurrence*

As different types and subtypes of SCC*mec* have been verified to influence the multidrug resistance and the antimicrobial MIC of β-lactam, a thorough understanding of the prevalence of SCC*mec* may aid in the further identification, control, prevention, and therapy of *Staphylococcus*-mediated human diseases. Consequently, surveillance of SCC*mec* has been performed globally in past decades. As the first identified type, type I SCC*mec* was nonpredominant in the 1970s, which was reported in a limited number of areas, including Brazil, Iran, Japan, Philippines, Spain, Switzerland, and the United States [202–210]. Type II SCC*mec* had been commonly found in Japan, Korea (occasionally in China), and the United States [206, 207, 211–215] and occasionally detected in Algeria, Brazil, China, Iran, Turkey, and Thailand [205, 216–221]. Type III SCC*mec* has been most frequently found among HA-MRSA and remains the predominant type in many countries or areas including Asia (China, Hong Kong, Iran, Malaysia, Singapore, Taiwan, and Thailand), Europe (Poland, Portugal, and Turkey), and South America (Brazil) [205, 208–214, 217, 220, 222–231]. Types IV and V have been implicated as CA-MRSA-associated SCC*mec*. A large number of variants (subtypes) have been reported within type IV, which is also the predominant type in Algeria, Brazil, Denmark, Korea, New Zealand, Portugal, Philippines, Sweden, Switzerland, Spain, and the United States [202–204, 209, 219, 221, 231–238]. Other types of SCC*mec* are rarely detected and reported [166]. According to our preliminary studies, from 2001 to 2006 in Guangzhou, analysis of the distribution of SCC*mec* type in 262 *Staphylococcus* strains demonstrated that the classic nosocomial SCC*mec* type (I–III) dominated among the tested strains, and none of the tested strain carried type IV or V. For MRSA strains, 3 and 198 strains belonged to SCC*mec* types II and III, respectively, with 8 strains untypeable. For MRCNS strains, 9, 24, and 12 strains were classified as SCC*mec* types I, II, and III respectively, with 8 strains untypeable. From a retrospective study conducted on 1,739 *Staphylococcus* isolates from a local hospital in Guangzhou from 2001 to 2010, SCC*mec* typing was performed on 263 randomly selected MRSA strains. Type III SCC*mec* was most frequently observed with an identification rate of 94.7% [249/263], with type II detected in 4 isolates (one individual isolate in 2001, 2002, 2005, and 2008, respectively) and 10 untypeable MRSA strains were recorded [165]. However, diversity in SCC*mec* types had been obtained from SCC*mec* surveillance of MRSA from another medical setting in Guangzhou from 2009 to 2012, as types I, II, III, IIIA, IV, V, and VI SCC*mec* carriage were found to be 17.6%, 56.8%, 6.2%, 10.7%, 4.1%, and 2.1%, respectively.

3.1.2. *Other resistance determinants in Staphylococcus*

Aside from SCC*mec*, the role of integrons as a mobile genetic mechanism in the horizontal transfer of antimicrobial genes or determinants among microorganisms has been recently well characterized, established, and documented, which may contribute to the broad distribution and spread of antibiotic resistance and ultimate emergence and unleashing of "super bugs" [174–177]. A complete and functional integron platform comprises three elements: (i) the integrase gene (*intI*) encoding an integrase, (ii) a proximal primary recombination site *attI*, and (iii) a promoter gene (Pc) functionally demonstrated for all integrons [240]. Several classes of integrons have been identified and distinguished by differences and divergence in the *intI* sequences, and integron classes 1 to 3 are so-called multiresistant integron (RIs) with a capability of acquiring identical gene cassettes [173]. Class 4 integron is considered to be a distinct type of integron and termed super integron (SI), which was found on the small chromosome of *Vibrio cholerae* and known to be an integral component of various γ-proteobacterial genomes [17, 241, 242]. As a direct result of the linkage to Tn402-like transposons and associated with Tn3 transposon family (Tn21 or Tn1696), the class 1 integron platform has been the most ubiquitous among microbes and remains the focus of numerous studies, with a large variety of clinical Gram-negative organisms and a few Gram-positive bacteria reported to harbor this integron class [243–245]. The first observation of class 1 integron within *Staphylococcus* spp. was reported in 2004, with species including *Staphylococcus lentus, Staphylococcus nepalensis*, and *Staphylococcus xylosus* [246]. In Guangzhou, class 1 integrons were commonly found in MRSA strains (31.6%, 83/263) during 2001 to 2010, with decreasing identification rates observed [6, 7, 165, 166, 247]. From 2001 to 2004 in Guangzhou, the detection rate of class 1 integron for MRSA and MRCNS was 51.7% [46/89] and 56.6% [30/53], respectively [6, 13, 165, 247, 248]. From 2007 to 2010, class 1 integron was found in MRS isolates based on a series of studies of systematic integron investigation in hundreds of staphylococci strains from 2001 to 2006 [165, 247, 248]. Nevertheless, only 38.3% [46/120] of MRSA isolates carried class 1 integron. Undoubtedly, the commonly detected integron-based antimicrobial resistance mechanisms have contributed to the evolution of the resistance of MRSA and may further lead to dissemination of new waves of "super bugs." Class 2 integron has an organization similar to that of class 1 but is associated with the Tn7 transposon family [174, 249]. Class 3 integron contains a comparable structure to that of class 2 integron and up to date has only been found in a limited number of microorganisms, including *Pseudomonas, Alcaligenes, Serratia marcescens*, and *Klebsiella pneumoniae* [249–252]. Class 4 integron harbors hundreds of gene cassettes encoding adaptations that extend beyond antibiotic resistance and pathogenicity [253]. The remaining classes of integrons may also contain antibiotic resistance gene cassettes, but knowledge of their worldwide prevalence remains limited [240, 254]. As a genetic element existing in 9% of bacteria and representatives from a broad range of phyla and environments, integrons play a core role in antibiotic resistance among clinical organisms and contribute to the evolution and adaption of bacteria.

3.1.3. Mobility and evolution of MGEs in staphylococci

As a commonly found MGE with an antibiotic resistance gene (*mecA*) and site-specific recombinase genes (*ccrA* and *ccrB*), SCC*mec* has been classified into 11 types, various subtypes, and variants and plays a core role in antibiotic resistance, molecular epidemiology, and evolution of staphylococci. Through recognition of recombination sites (*attB, attSCC, attI1, attC*, secondary sites, etc.) and via this site-specific recombination event, MGEs are capable of capturing foreign genes. The mobility of MGEs is defined as being associated with mobile DNA elements (transposons or plasmids) and antibiotic resistance genes in addition to having a small array size and substantial heterogeneity in recombination sites [187, 229] From

Southern hybridization analysis in preliminary studies, 58 staphylococci isolates were found to harbor one copy of class 1 integron on the chromosomal instead of plasmid DNA compared with their frequent location on plasmids for facilitation of conjugative-mediated transfer [13]. As natural capture systems and assembly platforms, MGEs in *Staphylococcus* (SCC*mec* or integrons system) allow bacteria to incorporate foreign genes and convert them to functional proteins by ensuring the correct expression. Despite affinity for self-transposition, integron systems are commonly associated with the transposons and conjugative plasmids serving as vehicles for the intra- and interspecies transmission of genetic material as well as gene cassettes capable of mobilizing to other integrons or to secondary sites in the bacterial genome [255]. This event has been regarded as a key mechanism in the dissemination and spread of resistance genes responsible for the swift spread of resistance genes and the rapid evolution of resistance to a wide range of unrelated antibiotics among diverse bacteria [251, 256]. Any ORF existing in the environmental "gene pool" is conceivably capable of being structured into the bacterial genome through the recombination platforms, and MGEs consequently have the potentially limitless capacity to exchange and stockpile functional genes, which enables rapid adaptation to selective pressure and may ultimately endow additional fitness and advantage to the bacterial host. In addition, a vast number of MGEs (such as conjugative plasmids, transposons, insertion sequences, and even entire chromosome) and the captured genes comprise the vast reservoirs of integrons and lead to the longstanding concept of a single massive "gene pool" that is available and temporally shared among bacteria [73]. The common observation of MGEs in microorganisms from the general environment and its enormous sequence diversity detected from such microbes, as well as various products unrelated to antibiotic resistance, strongly suggests that MGEs are ancient genomic structural elements and have played a general role in evolution and adaptation for a considerable period of time [43].

As a genomic island (G island) and MGE demarcated by a pair of DRs and inverted repeats, SCC*mec* has a set of site-specific recombinase genes (*ccrA* and *ccrB*) required for its movement and is inserted at the 3 end of *orfX* and located adjacent to the replication origin [220]. In the chromosome of staphylococci, SCC*mec* may have evolved from a primordial mobile element SCC, into which the *mec* complex was inserted. However, the function of the putative SCC*mec* may not be limited as the conveyer of antimicrobial resistance (mediated by *mec* complex) alone, and this MGE may serve as a vehicle for the exchange of useful genes for the better survival for staphylococci in various environments. In addition, SCC*mec* is a general genetic information exchange system of staphylococci with *ccrA* and *ccrB* involved in the recombination events (integration and excision), which plays a significant role in the evolution of *Staphylococcus*. MGEs serve as the reservoir for various genes and possess the function of interspecies genetic exchange. It is interesting to speculate whether multiple MGEs carried by staphylococci would speed up the rate of gene exchange or genome evolution, although these hypotheses require further investigation. From previous surveys, the influence of carriage of multiple MGEs on antimicrobial resistance had been investigated in MRSA. The presence of multiple MGEs was found to be strongly correlated with antimicrobial resistance, including erythromycin, gentamicin, tetracycline, and trimethoprim-sulfamethoxazole, which further limits the therapeutic options for deep-seated *Staphylococcus* infection and diseases. For treatment of complicated *Staphylococcus* infections, gentamicin is commonly prescribed by many clinicians in combination with vancomycin due to enhanced efficacy based on synergistic antibacterial activity [257]. For penicillin-allergic patients, erythromycin has been frequently used. As the first choice for suspected CA-MRSA cutaneous infections, trimethoprim-sulfamethoxazole has also been commonly used in combination with rifampin for MRSA in carriers despite the high recurrence (up to 50%) and frequent emergent resistance of this organism.

Up to date, the most known functional genes carried by MGES are found to encode resistance to the oldest groups of antibiotics (such as tetracycline, streptomycin, and spectinomycin) that have been discontinued in clinical settings for decades but still available in veterinary practice. Although the indiscriminate use of these older antibiotics is no longer occurring in the clinical setting, their use in veterinary medicine may contribute to a novel and significant concern in food safety. Abuse of antibiotics leads to the emergence of antibiotic resistance and poses a predicament for the future treatment of bacterial infection, with MGEs undoubtedly facilitating the rapid spread and dissemination of a vast number of resistance genes among microorganisms.

3.1.4. Livestock-Associated MRSA (LA-MRSA)

As a common pathogen for both clinical medicine and food safety, MRSA was first reported as hospital associated before the 1990s and thus designated as HA-MRSA. Since the 1990s, CA-MRSA strains have increasingly been reported among groups of patients with no apparent connection to hospitals. It is noteworthy that a large number of such CA-MRSA-infected patients or carriers were pediatric associated. Aside from HA-MRSA and CA-MRSA, LA-MRSA has been recently documented and is known to be more persistent in food products from swine and cattle [258], which is also responsible for pneumonia, endocarditis, and necrotizing fasciitis by LA-MRSA carriers [259]. Nowadays, LA-MRSA acts as an increasing risk for public health and a challenge to livestock farming and related food products. LA-MRSA was mostly found among animals (particularly pigs) and humans with frequent contact to livestock farming or livestock food products [260–262]. After the first isolation of MRSA from livestock (cows with mastitis) [263], a extremely limited number of reports were focused on LA-MRSA. However, after an initial LA-MRSA case occurred in humans, described in 2005 [262], LA-MRSA have been the focus of numerous recent studies. In 2007, a transmission of MRSA (ST1, spa-type t127) between cows and humans was reported, verifying the transmission between animals and humans [264]. Afterwards, different types of LA-MRSA have been continuously discovered globally, and the prevalence and occurrence of LA-MRSA vary significantly in different areas. In Europe and America, the majority of LA-MRSA strains belong to sequence type (ST) 398, whereas ST9 is frequently discovered in Asia [265–268]. Both livestock and humans are potential carriers of LA-MRSA, and individuals working in animal clinics and livestock production environments with direct contact or exposed to MRSA-positive animals or ingestion of the MRSA-positive livestock food products have an increased risk of becoming MRSA carriers [269, 270]. A high risk of animal to human transmission of ST398 was found to result from direct association between animal and/or human MRSA carriages in the farm setting [261, 269, 271–273] despite much lower occurrence of transmission between humans by LA-MRSA and that of HA-MRSA [274–276].

As a clone of typical LA-MRSA, ST398-LA-MRSA has been responsible for serious infections and outbreaks worldwide [277, 278]. Containing various spa-types, ST398-LA-MRSA strains are mostly found to carry type IV or V SCC*mec*, which are nontypeable by standard PFGE using *SmaI* digestion due to protection from digestion by the presence of a restriction/methylation system [271, 279, 280]. According to the virulent properties of ST398 strains, most animal-associated ST398-LA-MRSA strains lack the major virulence factors in staphylococci, such as Panton-Valentine leukocidin (PVL), TSS toxin 1, and exfoliative toxins [281]. However, various resistance genes commonly present in staphylococci of human and animal origins are also recovered in ST398-LA-MRSA strains, including the β-lactamase gene cluster *blaZ-blaI-blaR*, the tetracycline resistance genes *tetM* and *tetK*, the macrolide-lincosamide-streptogramin B (MLSB) resistance genes *ermA*, *ermB*, and *ermC*, the lincosamide resistance gene *lnuA*, and arrays *aacA-aphD* or *aadD* for resistance to gentamicin-tobramycin-kanamycin or kanamycin-neomycin, respectively [281–285]. In addition, novel resistance genes were also discovered in ST398-LA-MRSA strains, such as *dfrK* (trimethoprim resistance), ABC transporter genes *vgaC* and *vgaE* (pleuromutilin-lincosamide-eptogramin A resistance), and *apmA* (apramycin resistance) [286].

Emergence, spread, and dissemination of ST398-LA-MRSA from animals, as well as its transmission between humans and animals, strongly suggest that the antimicrobial resistance caused by veterinary antibiotic abuse poses a hazard to both humans and animals regarding food safety challenges associated with animal origins.

3.2. Viable Putative but Nonculturable (VPNC)

In nature, bacteria exist in various states such as normal growth state, dead state, dormant state, and VPNC state, which was first reported in 1986 [287]. Differing significantly from the "starvation survival" state, VPNC state [previously known as viable but nonculturable (VBNC)] is a specific state under which bacteria remain alive but fail to form a colony on routine bacteriological media that normally support their growth. Consequently, routine bacteriological detection methodology fails to detect the VPNC bacteria. However, given the right conditions, bacteria in the VPNC state remain active and can "resuscitate" to the normal state. Hence, the VPNC food spoilage or pathogenic bacteria are considered to be a stealth source of contamination, posing a significant concern for traditional surveillance and control methodologies of foodborne pathogens.

3.2.1. Induction and resuscitation

Entering into the VPNC state is considered to be a survival mechanism for nonsporulation bacteria under a number of harsh environmental conditions, which is described in detail as follows: [1] Nutrient starvation [288]. Without essential nutrients, bacterial growth and metabolism may be terminated and thus enter the death-like status. Nutrient starvation, such as the absence of carbon source or nitrogen source, which is an extreme condition for the growth of bacteria, can induce the VPNC state. [2] Extreme temperature [289]. The appropriate temperature for typical bacterial growth ranges from 20°C to 37°C, and termination of growth usually occurs under extremely high or low temperature. The temperature of 4°C or -20°C, at which bacteria stop growth and metabolism, is frequently used for induction of VPNC state. The combination of nutrient starvation and low temperature has also been widely applied as an induction condition. [3] pH value [290]. Most microorganisms grow in neutral and slightly acidic or alkaline pH conditions. Strong acidity or alkalinity may lead to bacterial death-like states, which has been occasionally used to induce the VPNC state. [4] Salinity [291]. As an extreme condition for bacterial growth, high salinity has been found to enable the entering of VPNC state. [5] Osmotic stress [291]. Extremely high osmotic stress was reported to be applied for VPNC state induction. [6] Oxygen availability [292, 293]. In an aerobic environment, anaerobic bacteria would enter into the VPNC state and the absence of oxygen would induce the VPNC state of aerobic bacteria and vice versa. [7] Existence of heavy metals [294, 295]. [8] Common food preservatives (cryopreservation, vacuum preservation, etc.). Currently, numerous bacteria are reported to have the ability to enter into VPNC state, such as *Salmonella* spp. [287, 296, 297], *Enterococcus* spp. [298–300], *Vibrio* spp. [301–309], *Campylobacter* spp. [310], *Pseudomonas* spp. [307, 311–313], *Shigella* spp. [314, 315], *Lactobacillus* spp. [316, 318], *Escherichia coli* [313, 316], and *Staphylococcus* spp. [318–324]. Furthermore, it has been well established and documented that bacteria in VPNC state can resuscitate and regain culturability when provided with appropriate conditions [308, 325, 326]. A variety of processes, including elevation of temperature gradually or directly [308], heat shock treating [325], adding nutrients [326], and adding organic matter (Tween-20, Tween-80, catalase, sodium pyruvate, etc.) were found to be applicable for resuscitation from the VPNC to normal state. The resuscitated bacteria are comparatively similar to their exponential-phase bacterial counterparts.

Currently, only 2 species of *Staphylococcus, S. aureus,* and *S. epidermidis,* were capable of entry into VPNC state [318–324]. In 2009, formation of VPNC *S. aureus* by radiation was reported for the first time, representing the first evidence of *Staphylococcus* cells entering the VPNC state [318]. One year later, induction of VPNC state by starvation of the *Staphylococcus* cells at low temperature (4°C) was also obtained [319]. Resuscitation of *S. aureus* strain under VPNC state was induced by temperature upshift (from 4°C to 22°C) or rich medium supplemented with sodium pyruvate [319, 320]. The prevention of resuscitation was observed by deficiencies in catalase or superoxide dismutase, indicating the relation of VPNC formation of *S. aureus* to oxidative stress [319, 320], constituting the initial studies on the mechanism of the formation and resuscitation of *S. aureus* in VPNC state. In addition, *S. aureus* cells in biofilm were found to enter into a VPNC state under antibiotic pressure (vancomycin or quinupristin/dalfopristin) [320, 321], suggesting that central venous catheter (CVC) or medical implant-associated biofilms may be potential reservoirs for *S. aureus* and *S. epidermidis* in the VPNC state [323]. Thus, both biofilm formation and VPNC induction may augment clinical challenges associated with antibacterial treatment options. *S. epidermidis* biofilms were reported to enter into the VPNC state when grown in excess glucose presumably due to accumulation of acidic compounds as the degradation products of glucose metabolism. This process was counteracted by high extracellular levels of calcium and magnesium added to the culture medium allowing modulation of the proportions of VPNC bacteria within *S. epidermidis* biofilms [324]. Although the induction and resuscitation of *Staphylococcus* cells in VPNC state has been verified, relatively little is known with respect to inducing and resuscitating condition, necessitating further investigation into this fascinating bacterial survival strategy.

3.2.2. Characteristics and mechanisms

Remaining metabolically or physiologically active, bacteria in VPNC state maintain cell integrity but exhibit dwarfing, which contribute to protect against a wide variety of stressors. The maintenance of metabolic activity and continuous gene expression under VPNC state [327, 328] indicates that potentially ingested bacteria may still be capable of causing foodborne illnesses. Such microorganisms also possess the capacity to regain culturability *in vivo* [329], exhibiting high ATP level, membrane potential [298], and retained plasmids, presenting higher autolytic capability than exponentially growing cells. The outer membrane protein profile also alters with entry into VPNC state [330]. Due to the diversity of VPNC bacteria, various characteristics among different species of microorganisms are being discovered worldwide. Regarding the mechanism of the VPNC state, the up- or down-regulation of genes and proteins associated with VPNC status compared to the exponential phase and the resuscitated status is considered to be potential factors for entering and exiting of VPNC state. However, it is currently unclear as to which genes are essential for these processes.

As for the pathogenicity of *Staphylococcus* cells under VPNC state, the viable cell numbers and gene expression had been found to remain constant in VPNC state by examination of epi-fluorescence microscopy, flow cytometry, and reverse transcription-PCR (RT-PCR) [320, 321]. This finding implied that *S. aureus* cells are likely still pathogenic in VPNC state and thus pose a significant concern on its threat to food safety.

3.2.3. Detection and identification

VPNC pathogenic bacteria are considered to be a threat to public health and food safety due to incapability of detection by the "gold standard" methodology for identification of food-associated microorganisms. Hence, the development, evaluation, validation, and further application of rapid and accurate detection methodology for VPNC bacteria are considered to be the leading concerns for the surveillance of bacterial cells in VPNC state as well as further understanding of the mechanisms on their survival and persistence in the extreme environment. The conventional detection method for VPNC bacteria was the combination of acridine orange direct count (AODC) (for total bacterial cell number counting), bright-field microscopy with nalidixic acid (for metabolically active cell number counting), and plate counting (for determination of culturability). The occurrence of entry of bacterial cells into VPNC state was validated and confirmed when colony counts were totally depleted on culture plates (with no observed colonies), which was designated as nonculturable, whereas the total bacterial and metabolically active cells still remained countable. Despite the limited application of nalidixic acid on Gram-negative microorganisms, the novel LIVE/DEAD Bacterial Viability Kit with requirement on differential fluorescence was employed for the detection of both Gram-positive and Gram-negative bacteria [331]. In consideration of the carcinogenesis and expense of fluorescence substances, the development of molecular assays, such as random amplified polymorphism DNA and RT-PCR [328], was recently applied to identify bacterial cells in the VPNC state. As *Staphylococcus* species were concerned, an immunosensing system using impedance spectroscopy measurements was recently developed and applied for rapid verification and quantification of *S. aureus* cells in the VPNC state [322], with high sensitivity and specificity obtained.

In conclusion, foodborne pathogens, especially *S. aureus* strains, which contain various virulence genes, are capable of forming VPNC state and resuscitating into active and pathogenic state under specific conditions, posing a significant threat to food safety. The "farm to table" process includes food ingredients, processing, transportation, and storage, which involves a large variety of conditions. A number of such conditions (such as low temperature during refrigeration) may be sufficient for entry into the VPNC state, complicating the use of routine diagnostics by resulting in high "false-negative" rates of pathogen detection. However, once resuscitation occurs under proper conditions, foodborne pathogens remain active and virulent, which thus are highly likely to cause food poisoning outbreaks.

4. Concluding remarks

Ingestion of food is the major (although, not only) way for human beings to obtain nutrient substances for basic living; therefore, the quality and safety of food have recently become a major concern. Considered to be an expanding global problem and leading topic in public health, food safety is no longer limited to foodborne illnesses but has been extended to all safety issues associated with "farm to table" food approaches. In the past decade, a large number of worldwide cases or reports have been available regarding food containing unhealthy, harmful, or toxic substances (other than food poisoning outbreaks). Foodborne microorganisms, previously limited to pathogenic bacteria and toxic substances produced in food, have played a critical role in food safety. However, now due to diversity in the genus and species of microbes, variety of mechanisms on the regulation of growth and survival, and complexity of ecosystem involving polymicrobial interaction and environmental factors, a number of novel microbial issues associated with food safety have been recently acknowledged. Microorganisms may very well be capable of surviving the journey from farm to table via various evasion mechanisms at various food processing stages, including source (antimicrobial resistance caused by the use of drugs in veterinary medicine or livestock feed), processing (formation of biofilm and further survival of bacterial elimination), storage (formation of VPNC state and "false-negative" detection), and even after cooking (production of heat-stable toxins that remain active despite elimination of host bacteria). The contributions of aforementioned and novel evasion mechanisms with respect to food safety undoubtedly require further investigation in vitro and in vivo for improved diagnostic and decontamination procedures.

Author details

Zhenbo Xu[1,2*], Brian M. Peters[3], Bing Li[1], Lin Li[1] and Mark E. Shirtliff[2]

*Address all correspondence to: zhenbo.xu@hotmail.com

1 College of Light Industry and Food Sciences, South China University of Technology, Guangzhou, China

2 Department of Microbial Pathogenesis, School of Dentistry, University of Maryland, Baltimore, MD, USA

3 Department of Clinical Pharmacy, College of Pharmacy, University of Tennessee Health Sciences Center, Memphis, TN, USA

References

[1] Centers for Disease Control and Prevention, Food Safety [Internet]. 2015. Available from: http://www.cdc.gov/foodsafety [Accessed 2015-08-31]

[2] Xu Z, Li L, Chu J. Development and application of loop-mediated isothermal amplification assays on rapid detection of various types of staphylococci strains. Food Res Int. 2012;47:166–173. DOI: 10.1016/j.foodres.2011.04.042

[3] Scallan E, Griffin PM, Angulo FJ. Foodborne illness acquired in the United States—unspecified agents. Emerg Infect Dis. 2011;17:16–22. DOI: 10.3201/eid1701.091101p2

[4] Summary of Report of the French Sanitary Agencies [Internet]. 2003. Available from: http://www.invs.sante.fr/publications/2004/inf_origine_alimentaire/grilleLecture.pdf

[5] Martyn DK, Ian M, Gill VH. Food borne illness in Australia: the OzFoodNet experience. Clin Infect Dis. 2008;47:392–400. DOI: 10.1086/589861

[6] Xu Z, Li L, Shi L. Class 1 integron in staphylococci. Mol Biol Rep. 2011;38:5261–5279. DOI: 10.1007/s11033-011-0676-7

[7] Xu Z, Li L, Shirtliff ME. Resistance class 1 integron in clinical methicillin-resistant *Staphylococcus aureus* strains in southern China, 2001–2006. Clin Microbiol Infect. 2011;17:714–718. DOI: 10.1111/j.1469-0691.2010.03379.x

[8] Xu Z, Liu X, Li L, Li B. Development of *Staphylococcus aureus* enterotoxin in food borne bacteria. Mod Food Sci. 2013;29:2317–2324

[9] Xu Z, Li L, Zhao X. Development and application of a novel multiplex polymerase chain reaction (PCR) assay for rapid detection of various types of staphylococci strains. Afr J Microbiol Res. 2011;5:1869–1873

[10] You R, Gui Z, Xu Z. Methicillin-resistance *Staphylococcus aureus* detection by an improved rapid PCR assay. Afr J Microbiol Res. 2012;6:7131–7133

[11] Zhao X, Li Y, Wang L. Development and application of a loop-mediated isothermal amplification method on rapid detection *Escherichia coli* O157 strains from food samples. Mol Biol Rep. 2010;37:2183–2188. DOI: 10.1007/s11033-009-9700-6

[12] Ben-Ami R, Navon-Venezia S, Schwartz D. Infection of a ventriculoatrial shunt with phenotypically variable *Staphylococcus epidermidis* masquerading as polymicrobial bacteremia due to various coagulase-negative staphylococci and Kocuria varians. J Clin Microbiol. 2003;41:2444–2447. DOI: 10.1128/JCM.41.6.2444-2447.2003

[13] Xu Z, Shi L, Alam MJ. Integron-bearing methicillin-resistant coagulase-negative staphylococci in South China, 2001–2004. FEMS Microbiol Lett. 2008;278:223–230. DOI: 10.1111/j.1574-6968.2007.00994.x

[14] Nouwen JL, van Belkum A, de Marie S. Clonal expansion of *Staphylococcus epidermidis* strains causing Hickman catheter-related infections in a hemato-oncologic department. J Clin Microbiol. 1998;36:2696–2702

[15] Sospedra I, Soler C, Manes J. Analysis of staphylococcal enterotoxin A in milk by matrix-assisted laser desorption/ionization-time of flight mass spectrometry. Anal Bioanal Chem. 2011;400:1525–1531. DOI: 10.1007/s00216-011-4906-6

[16] Alarcon B, Vicedo B, Aznar R. PCR-based procedures for detection and quantification of *Staphylococcus aureus* and their application in food. J Appl Microbiol. 2006;100:352–364. DOI: 10.1111/j.1365-2672.2005.02768.x

[17] Shimizu A, Fujita M, Igarashi H. Characterization of *Staphylococcus aureus* coagulase type VII isolates from staphylococcal food poisoning outbreaks (1980–1995) in Tokyo, Japan, by pulsed-field gel electrophoresis. J Clin Microbiol. 2000;38:3746–3749

[18] Scallan E, Hoekstra RM, Angulo FJ. Foodborne illness acquired in the United States—major pathogens. Emerg Infect Dis. 2011;17:7–15. DOI: 10.3201/eid1701.P11101

[19] Smyth DS, Kennedy J, Twohig J. *Staphylococcus aureus* isolates from Irish domestic refrigerators possess novel enterotoxin and enterotoxin-like genes and are clonal in nature. J Food Protect. 2006;69:508–515

[20] Asao T, Kumeda Y, Kawai T. An extensive outbreak of staphylococcal food poisoning due to low-fat milk in Japan: estimation of enterotoxin A in the incriminated milk and powdered skim milk. Epidemiol Infect. 2003;130:33–40. DOI: 10.1017/S0950268802007951

[21] Deng Y, Liu C, Li B. Review of methicillin-resistant *Staphylococcus aureus* and its detection in food safety. Mod Food Sci. 2015;31:259–266

[22] Bad Bug Book (second edition)—Foodborne Pathogenic Microorganisms and Natural Toxins Handbook [Internet]. 2012. Available from: http://www.fda.gov/food/foodborneillnesscontaminants/causesofillnessbadbugbook/default.htm

[23] U.S. Department of Agriculture Food Safety and Inspection Service, Basics for Handling Food Safely [Internet]. 2011. Available from: http://www.fsis.usda.gov/PDF/Basics_for_Safe_Food_Handling.pdf

[24] Tsutsuura S, Shimamura Y, Murata M. Temperature dependence of the production of staphylococcal enterotoxin A by *Staphylococcus aureus*. Biosci Biotechnol Biochem. 2013;77:30–37. DOI: 10.1271/bbb.120391

[25] Bergdoll MS. Enterotoxins. In: Adlam C, Easmon CFS, editors. Staphylococci and staphylococcal infections. London: Academic Press; 1983. pp. 559–598.

[26] Schmitt M, Schuler-Schmid U, Schmidt-Lorenz W. Temperature limits of growth, TNase and enterotoxin production of *Staphylococcus aureus* strains isolated from foods. Int J Food Microbiol. 1990;11:1–19. DOI: 10.1016/0168-1605(90)90036-5

[27] Bautista L, Gaya P, Medina M. A quantitative study of enterotoxin production by sheep milk staphylococci. Appl Environ Microbiol. 1988;54:566–569

[28] Becker K, Keller B, von Eiff C. Enterotoxigenic potential of *Staphylococcus intermedius*. Appl Environ Microbiol. 2001;67:5551–5557. DOI: 10.1128/AEM.67.12.5551-5557.2001

[29] Jay JM. Microbiological food safety. Crit Rev Food Sci Nutr. 1992;31:177–190. DOI: 10.1080/10408399209527567

[30] Pimenta-Martins MG, Furtado RF, Heneine LG. Development of an amperometric immunosensor for detection of staphylococcal enterotoxin type A in cheese. J Microbiol Methods. 2012;91:138–143. DOI: 10.1016/j.mimet.2012.05.016

[31] Noleto AL, Bergdoll MS. Staphylococcal enterotoxin production in the presence of non-enterotoxigenic staphylococci. Appl Environ Microbiol. 1980;39:1167–1171

[32] Fries BC, Varshney AK. Bacterial toxins—staphylococcal enterotoxin B. Microbiol Spectr. 2013;1. DOI: 10.1128/microbiolspec.AID-0002-2012

[33] Pexara A, Burriel A, Govaris A. *Staphylococcus aureus* and staphylococcal enterotoxins in foodborne diseases. J Hellenic Vet Med Soc. 2010;61:316–322

[34] McLean RA, Lilly HD, Alford JA. Effects of meat-curing salts and temperature on production of staphylococcal enterotoxin B. J Bacteriol. 1968;95:1207–1211

[35] Vandenbosch LL, Fung DY, Widomski M. Optimum temperature for enterotoxin production by *Staphylococcus aureus* S-6 and 137 in liquid medium. Appl Microbiol. 1973;25:498–500

[36] Smith JL, Buchanan RL, Palumbo SA. Effect of food environment on staphylococcal enterotoxin synthesis: a review. J Food Protect 1983;46:545–555

[37] Notermans S, Heuvelman CJ. Combined effect of water activity, pH and suboptimal temperature on growth and enterotoxin production of *Staphylococcus aureus*. J Food Sci. 1983;48:1832–1835. DOI: 10.1111/j.1365-2621.1983.tb05096.x

[38] Betley MJ, Mekalanos JJ. Nucleotide sequence of the type A staphylococcal enterotoxin gene. J Bacteriol. 1988;170:34–41

[39] Bunikowski R, Mielke M, Skarabis H. Prevalence and role of serum IgE antibodies to the *Staphylococcus aureus*-derived superantigens SEA and SEB in children with atopic dermatitis. J Allergy Clin Immunol. 1999;103:119–124. DOI: 10.1016/S0091-6749(99)70535-X

[40] Herz U, Bunikowski R, Mielke M. Contribution of bacterial superantigens to atopic dermatitis. Int Arch Allergy Immunol. 1999;118:240–241. DOI: 10.1159/000024085

[41] Rasooly R, Do PM. *In vitro* cell-based assay for activity analysis of staphylococcal enterotoxin A in food. FEMS Immunol Med Microbiol. 2009;56:172–178. DOI: 10.1111/j.1574-695X.2009.00561.x

[42] Bunning VK, Lindsay JA, Archer DL. Chronic health effects of microbial foodborne disease. World Health Stat Q. 1997;50:51–56

[43] Howell MD, Diveley JP, Lundeen KA. Limited T-cell receptor beta-chain heterogeneity among interleukin 2 receptor-positive synovial T cells suggests a role for superantigen in rheumatoid arthritis. Proc Natl Acad Sci U S A. 1991;88:10921–10925. DOI: 10.1073/pnas.88.23.10921

[44] Ye YM, Hur GY, Park HJ. Association of specific IgE to staphylococcal superantigens with the phenotype of chronic urticaria. J Kor Med Sci. 2008;23:845–851. DOI: 10.3346/jkms.2008.23.5.845

[45] Hu DL, Omoe K, Sashinami H. Immunization with a nontoxic mutant of staphylococcal enterotoxin A, SEAD227A, protects against enterotoxin-induced emesis in house musk shrews. J Infect Dis. 2009;199:302–310. DOI: 10.1086/596065

[46] Jozefczyk Z, Robbins RN, Spitz JM. Antibodies to staphylococcal enterotoxin in laboratory personnel. J Clin Microbiol. 1980;11:438–439

[47] Crawley GJ, Black JN, Gray I. Clinical chemistry of staphylococcal enterotoxin poisoning in monkeys. Appl Microbiol. 1966;14:445–450

[48] Omoe K, Hu DL, Ono HK. Emetic potentials of newly identified staphylococcal enterotoxin-like toxins. Infect Immun. 2013;81:3627–3631. DOI: 10.1128/IAI.00550-13

[49] Hu DL, Nakane A. Mechanisms of staphylococcal enterotoxin-induced emesis. Eur J Pharmacol. 2014;722:95–107. DOI: 10.1016/j.ejphar.2013.08.050

[50] Casman EP, Bennett RW. Detection of staphylococcal enterotoxin in food. Appl Microbiol. 1965;13:181–189

[51] Su YC, Wong AC. Identification and purification of a new staphylococcal enterotoxin, H. Appl Environ Microbiol. 1995;61:1438–1443

[52] Casman EP. Serologic studies of staphylococcal enterotoxin. Public Health Rep. 1958;73:599–609. DOI: 10.2307/4590200

[53] Borst DW, Betley MJ. Promoter analysis of the staphylococcal enterotoxin A gene. J Biol Chem. 1994;269:1883–1888

[54] Borst DW, Betley MJ. Phage-associated differences in staphylococcal enterotoxin A gene (sea) expression correlate with sea allele class. Infect Immun. 1994;62:113–118

[55] Casman EP. Further serological studies of staphylococcal enterotoxin. J Bacteriol. 1960;79:849–856

[56] Casman EP, Bergdoll MS, Robinson J. Designation of staphylococcal enterotoxins. J Bacteriol. 1963;85:715–716

[57] Tetsuya I, Naoto T, Keiji Y. Mass outbreak of food poisoning disease caused by small amounts of staphylococcal enterotoxins A and H. Appl Environ Microb. 2005;71:2793–2795. DOI: 10.1128/AEM.71.5.2793-2795.2005

[58] Casman EP, Bennett RW. Culture medium for the production of staphylococcal enterotoxin A. J Bacteriol. 1963;86:18–23

[59] Casman EP, Mccoy DW, Brandly PJ. Staphylococcal growth and enterotoxin production in meat. Appl Microbiol. 1963;11:498–500

[60] Silverman SJ, Knott AR, Howard M. Rapid, sensitive assay for staphylococcal enterotoxin and a comparison of serological methods. Appl Microbiol. 1968;16:1019–1023

[61] Barber LE, Deibel RH. Effect of pH and oxygen tension on staphylococcal growth and enterotoxin formation in fermented sausage. Appl Microbiol. 1972;24:891–898

[62] McCoy DW. Influence of food microorganisms on staphylococcal growth and enterotoxin production in meat. Appl Microbiol. 1966;14:372–377

[63] Markus ZH, Silverman GJ. Factors affecting the secretion of staphylococcal enterotoxin A. Appl Microbiol. 1970;20:492–496

[64] Reiser R, Conaway D, Bergdoll MS. Detection of staphylococcal enterotoxin in foods. Appl Microbiol. 1974;27:83–85

[65] Niskanen A, Lindroth S. Comparison of different purification procedure for extraction of staphylococcal enterotoxin A from foods. Appl Environ Microbiol. 1976;32:455–464

[66] Warren JR. Comparative kinetic stabilities of staphylococcal enterotoxin types A, B, and C_1. J Biol Chem. 1977;252:6831–6834

[67] Cavallin A, Arozenius H, Kristensson K. The spectral and thermodynamic properties of staphylococcal enterotoxin A, E, and variants suggest that structural modifications

are important to control their function. J Biol Chem. 2000;275:1665–1672. DOI: 10.1074/jbc.275.3.1665

[68] Noleto AL, Malburg JL, Bergdoll MS. Production of staphylococcal enterotoxin in mixed cultures. Appl Environ Microbiol. 1987;53:2271–2274

[69] Betley MJ, Lofdahl S, Kreiswirth BN. Staphylococcal enterotoxin A gene is associated with a variable genetic element. Proc Natl Acad Sci U S A. 1984;81:5179–5183. DOI: 10.1073/pnas.81.16.5179

[70] Betley MJ, Mekalanos JJ. Staphylococcal enterotoxin A is encoded by phage. Science. 1985;229:185–187. DOI: 10.1126/science.3160112

[71] Maina EK, Hu DL, Tsuji T. Staphylococcal enterotoxin A has potent superantigenic and emetic activities but not diarrheagenic activity. Int J Med Microbiol. 2012;302:88–95. DOI: 10.1016/j.ijmm.2012.01.003

[72] Shafer WM, Iandolo JJ. Staphylococcal enterotoxin A: a chromosomal gene product. Appl Environ Microbiol. 1978;36:389–391

[73] Christianson KK, Tweten RK, Iandolo JJ. Transport and processing of staphylococcal enterotoxin A. Appl Environ Microbiol. 1985;50:696–697

[74] Huang IY, Hughes JL, Bergdoll MS. Complete amino acid sequence of staphylococcal enterotoxin A. J Biol Chem. 1987;262:7006–7013

[75] Akhtar M, Park CE, Rayman K. Effect of urea treatment on recovery of staphylococcal enterotoxin A from heat-processed foods. Appl Environ Microbiol. 1996;62:3274–3276

[76] Yamashita K, Kanazawa Y, Ueno M. Significance of the detection of staphylococcal enterotoxin A gene in low fat milk which caused a serious outbreak of food poisoning. Shokuhin Eiseigaku Zasshi. 2003;44:186–190. DOI: 10.3358/shokueishi.44.186

[77] Krakauer T, Stiles BG. The staphylococcal enterotoxin (SE) family: SEB and siblings. Virulence. 2013;4:759–773. DOI: 10.4161/viru.23905

[78] Kashiwada T, Kikuchi K, Abe S. Staphylococcal enterotoxin B toxic shock syndrome induced by community-acquired methicillin-resistant *Staphylococcus aureus* (CA-MRSA). Intern Med. 2012;51:3085–3088. DOI: 10.2169/internalmedicine.51.7295

[79] Wang CC, Lo WT, Hsu CF. Enterotoxin B is the predominant toxin involved in staphylococcal scarlet fever in Taiwan. Clin Infect Dis. 2004;38:1498–1502. DOI: 10.1086/392501

[80] Li J, Yang J, Lu YW. Possible role of staphylococcal enterotoxin B in the pathogenesis of autoimmune diseases. Viral Immunol. DOI: 10.1089/vim.2015.0017

[81] Yang M, Kostov Y, Bruck HA. Gold nanoparticle-based enhanced chemiluminescence immunosensor for detection of staphylococcal enterotoxin B (SEB) in food. Int J Food Microbiol. 2009;133:265–271. DOI: 10.1016/j.ijfoodmicro.2009.05.029

[82] Friedman ME. Inhibition of staphylococcal enterotoxin B formation in broth cultures. J Bacteriol. 1966;92:277–278

[83] Jamlang EM, Bartlett ML, Snyder HE. Effect of pH, protein concentration, and ionic strength on heat inactivation of staphylococcal enterotoxin B 1. Appl Microbiol. 1971;22:1034–1040

[84] Keller GM, Hanson RS, Bergdoll MS. Effect of minerals on staphylococcal enterotoxin B production. Infect Immun. 1978;20:158–160

[85] Morse SA, Mah RA, Dobrogosz WJ. Regulation of staphylococcal enterotoxin B1. J Bacteriol. 1969;98:4–9

[86] Schumacher-Perdreau F, Akatova A, Pulverer G. Detection of staphylococcal enterotoxin B and toxic shock syndrome toxin: PCR versus conventional methods. Zentralbl Bakteriol. 1995;282:367–371. DOI: 10.1016/S0934-8840(11)80706-1

[87] De Boer ML, Chow AW. Toxic shock syndrome toxin 1-producing *Staphylococcus aureus* isolates contain the staphylococcal enterotoxin B genetic element but do not express staphylococcal enterotoxin B. J Infect Dis. 1994;170:818–827. DOI: 10.1093/infdis/170.4.818

[88] Tweten RK, Iandolo JJ. Transport and processing of staphylococcal enterotoxin B. J Bacteriol. 1983;153:297–303

[89] Huang IY, Bergdoll MS. The primary structure of staphylococcal enterotoxin B. 3. The cyanogen bromide peptides of reduced and aminoethylated enterotoxin B, and the complete amino acid sequence. J Biol Chem. 1970;245:3518–3525

[90] Shafer WM, Iandolo JJ. Chromosomal locus for staphylococcal enterotoxin B. Infect Immun. 1978;20:273–278

[91] Shafer WM, Iandolo JJ. Genetics of staphylococcal enterotoxin B in methicillin-resistant isolates of *Staphylococcus aureus*. Infect Immun. 1979;25:902–911

[92] Ranelli DM, Jones CL, Johns MB. Molecular cloning of staphylococcal enterotoxin B gene in *Escherichia coli* and *Staphylococcus aureus*. Proc Natl Acad Sci U S A. 1985;82:5850–5854. DOI: 10.1073/pnas.82.17.5850

[93] Johns MJ, Khan SA. Staphylococcal enterotoxin B gene is associated with a discrete genetic element. J Bacteriol. 1988;170:4033–4039

[94] Mahmood R, Khan SA. Role of upstream sequences in the expression of the staphylococcal enterotoxin B gene. J Biol Chem. 1990;265:4652–4656

[95] Ataee RA, Karami A, Izadi M. Molecular screening of staphylococcal enterotoxin B gene in clinical isolates. Cell J. 2011;13:187–192

[96] Bergdoll MS, Borja CR, Avena RM. Identification of a new enterotoxin as enterotoxin C. J Bacteriol. 1965;90:1481–1485

[97] Borja CR, Bergdoll MS. Purification and partial characterization of enterotoxin C produced by *Staphylococcus aureus* strain 137. Biochemistry. 1967;6:1467–1473

[98] Avena RM, Bergdoll MS. Purification and some physicochemical properties of enterotoxin C, *Staphylococcus aureus* strain 361. Biochemistry. 1967;6:1474–1480. DOI: 10.1021/bi00857a033

[99] Reiser RF, Robbins RN, Noleto AL. Identification, purification, and some physicochemical properties of staphylococcal enterotoxin C3. Infect Immun. 1984;45:625–630

[100] Bohach GA, Schlievert PM. Expression of staphylococcal enterotoxin C1 in *Escherichia coli*. Infect Immun. 1987;55:428–432

[101] Hovde CJ, Hackett SP, Bohach GA. Nucleotide sequence of the staphylococcal enterotoxin C3 gene: sequence comparison of all three type C staphylococcal enterotoxins. Mol Gen Genet. 1990;220:329–333. DOI: 10.1007/BF00260504

[102] Schmidt JJ, Spero L. The complete amino acid sequence of staphylococcal enterotoxin C1. J Biol Chem. 1983;258:6300–6306

[103] Marr JC, Lyon JD, Roberson JR. Characterization of novel type C staphylococcal enterotoxins: biological and evolutionary implications. Infect Immun. 1993;61:4254–4262

[104] Spero L, Morlock BA. Biological activities of the peptides of staphylococcal enterotoxin C formed by limited tryptic hydrolysis. J Biol Chem. 1978;253:8787–8791

[105] Orwin PM, Fitzgerald JR, Leung DY. Characterization of *Staphylococcus aureus* enterotoxin L. Infect Immun. 2003;71:2916–2919. DOI: 10.1128/IAI.71.5.2916-2919.2003

[106] Bohach GA, Handley JP, Schlievert PM. Biological and immunological properties of the carboxyl terminus of staphylococcal enterotoxin C1. Infect Immun. 1989;57:23–28

[107] Mantynen V, Niemela S, Kaijalainen S. MPN-PCR-quantification method for staphylococcal enterotoxin c1 gene from fresh cheese. Int J Food Microbiol. 1997;36:135–143

[108] Valihrach L, Alibayov B, Demnerova K. Production of staphylococcal enterotoxin C in milk. Int Dairy J. 2013;30:103–107. DOI: 10.1016/j.idairyj.2013.01.003

[109] Wilson IG, Cooper JE, Gilmour A. Detection of enterotoxigenic *Staphylococcus aureus* in dried skimmed milk: use of the polymerase chain reaction for amplification and detection of staphylococcal enterotoxin genes entB and entC1 and the thermonuclease gene nuc. Appl Environ Microbiol. 1991;57:1793–1798

[110] Cretenet M, Nouaille S, Thouin J. *Staphylococcus aureus* virulence and metabolism are dramatically affected by *Lactococcus lactis* in cheese matrix. Environ Microbiol Rep. 2011;3:340–351. DOI: 10.1111/j.1758-2229.2010.00230.x

[111] Valihrach L, Alibayov B, Zdenkova K. Expression and production of staphylococcal enterotoxin C is substantially reduced in milk. Food Microbiol. 2014;44:54–59. DOI: 10.1016/j.fm.2014.05.020.

[112] Casman EP, Bennett RW, Dorsey AE. Identification of a fourth staphylococcal enterotoxin, enterotoxin D. J Bacteriol. 1967;94:1875–1882

[113] Duquenne M, Fleurot I, Aigle M. Tool for quantification of staphylococcal enterotoxin gene expression in cheese. Appl Environ Microbiol. 2010;76:1367–1374. DOI: 10.1128/AEM.01736-09

[114] Bayles KW, Iandolo JJ. Genetic and molecular analyses of the gene encoding staphylococcal enterotoxin D. J Bacteriol. 1989;171:4799–4806

[115] Marta D, Wallin-Carlquist N, Schelin J. Extended staphylococcal enterotoxin D expression in ham products. Food Microbiol. 2011;28:617–620. DOI: 10.1016/j.fm.2010.11.013

[116] Zhang S, Stewart GC. Characterization of the promoter elements for the staphylococcal enterotoxin D gene. J Bacteriol. 2000;182:2321–2325. DOI: 10.1128/JB.182.8.2321-2325.2000

[117] Sihto HM, Tasara T, Stephan R. Temporal expression of the staphylococcal enterotoxin D gene under NaCl stress conditions. FEMS Microbiol Lett. DOI: 10.1093/femsle/fnv024

[118] Pereira JL, Salzberg SP, Bergdoll MS. Production of staphylococcal enterotoxin D in foods by low-enterotoxin-producing staphylococci. Int J Food Microbiol. 1991;14:19–25. DOI: 10.1016/0168-1605(91)90033-L

[119] Bergdoll MS, Borja CR, Robbins RN. Identification of enterotoxin E. Infect Immun. 1971;4:593-595

[120] Couch JL, Soltis MT, Betley MJ. Cloning and nucleotide sequence of the type E staphylococcal enterotoxin gene. J Bacteriol. 1988;170:2954–2960

[121] Borja CR, Fanning E, Huang IY. Purification and some physicochemical properties of staphylococcal enterotoxin E. J Biol Chem. 1972;247:2456–2463

[122] Lina G, Bohach GA, Nair SP. Standard nomenclature for the superantigens expressed by *Staphylococcus*. J Infect Dis. 2004;189:2334–2336. DOI: 10.1086/420852

[123] Omoe K, Imanishi K, Hu DL. Characterization of novel staphylococcal enterotoxin-like toxin type P. Infect Immun. 2005;73:5540–5546. DOI: 10.1128/IAI.73.9.5540-5546.2005

[124] Bergdoll MS, Crass BA, Reiser RF. A new staphylococcal enterotoxin, enterotoxin F, associated with toxic-shock-syndrome *Staphylococcus aureus* isolates. Lancet. 1981;1:1017–1021. DOI: 10.1016/S0140-6736(81)92186-3

[125] Arnow PM, Chou T, Weil D. Spread of a toxic-shock syndrome-associated strain of *Staphylococcus aureus* and measurement of antibodies to staphylococcal enterotoxin F. J Infect Dis. 1984;149:103–107. DOI: 10.1093/infdis/149.1.103

[126] Munson SH, Tremaine MT, Betley MJ. Identification and characterization of staphylococcal enterotoxin types G and I from *Staphylococcus aureus*. Infect Immun. 1998;66:3337–3348

[127] Jarraud S, Cozon G, Vandenesch F. Involvement of enterotoxins G and I in staphylococcal toxic shock syndrome and staphylococcal scarlet fever. J Clin Microbiol. 1999;37:2446–2449

[128] Omoe K, Ishikawa M, Shimoda Y. Detection of seg, seh, and sei genes in *Staphylococcus aureus* isolates and determination of the enterotoxin productivities of *S. aureus* isolates Harboring seg, seh, or sei genes. J Clin Microbiol. 2002;40:857–862. DOI: 10.1128/JCM.40.3.857-862.2002

[129] Ren K, Bannan JD, Pancholi V. Characterization and biological properties of a new staphylococcal exotoxin. J Exp Med. 1994;180:1675–1683. DOI: 10.1084/jem.180.5.1675

[130] Jorgensen HJ, Mathisen T, Lovseth A. An outbreak of staphylococcal food poisoning caused by enterotoxin H in mashed potato made with raw milk. FEMS Microbiol Lett. 2005;252:267–272. DOI: 10.1016/j.femsle.2005.09.005

[131] Nilsson H, Bjork P, Dohlsten M. Staphylococcal enterotoxin H displays unique MHC class II-binding properties. J Immunol. 1999;163:6686–6693

[132] Petersson K, Pettersson H, Skartved NJ. Staphylococcal enterotoxin H induces V alpha-specific expansion of T cells. J Immunol. 2003;170:4148–4154

[133] Rosec JP, Gigaud O. Staphylococcal enterotoxin genes of classical and new types detected by PCR in France. Int J Food Microbiol. 2002;77:61–70. DOI: 10.1016/S0168-1605(02)00044-2

[134] Pereira ML, Do CLS, Santos EJD. Enterotoxin H in staphylococcal food poisoning. J Food Protect. 1996;3:448–561

[135] Su YC, Wong AC. Production of staphylococcal enterotoxin H under controlled pH and aeration. Int J Food Microbiol. 1998;39:87–91. DOI: 10.1016/S0168-1605(97)00118-9

[136] Chen TR, Chiou CS, Tsen HY. Use of novel PCR primers specific to the genes of staphylococcal enterotoxin G, H, I for the survey of *Staphylococcus aureus* strains isolated from food-poisoning cases and food samples in Taiwan. Int J Food Microbiol. 2004;92:189–197. DOI: 10.1016/j.ijfoodmicro.2003.10.002

[137] Blaiotta G, Ercolini D, Pennacchia C. PCR detection of staphylococcal enterotoxin genes in *Staphylococcus* spp. strains isolated from meat and dairy products. Evidence for new variants of seG and seI in S. aureus AB-8802. J Appl Microbiol. 2004;97:719–730. DOI: 10.1111/j.1365-2672.2004.02349.x

[138] Zhang S, Iandolo JJ, Stewart GC. The enterotoxin D plasmid of *Staphylococcus aureus* encodes a second enterotoxin determinant (sej). FEMS Microbiol Lett. 1998;168:227–233

[139] Lindsay JA, Ruzin A, Ross HF. The gene for toxic shock toxin is carried by a family of mobile pathogenicity islands in *Staphylococcus aureus*. Mol Microbiol. 1998;29:527–543. DOI: 10.1046/j.1365-2958.1998.00947.x

[140] Orwin PM, Leung DY, Donahue HL. Biochemical and biological properties of staphylococcal enterotoxin K. Infect Immun. 2001;69:360–366. DOI: 10.1128/IAI.69.1.360-366.2001

[141] Aguilar JL, Varshney AK, Wang X. Detection and measurement of staphylococcal enterotoxin-like K (SEl-K) secretion by *Staphylococcus aureus* clinical isolates. J Clin Microbiol. 2014;52:2536–2543. DOI: 10.1128/JCM.00387-14

[142] Jarraud S, Peyrat MA, Lim A. egc, a highly prevalent operon of enterotoxin gene, forms a putative nursery of superantigens in *Staphylococcus aureus*. J Immunol. 2001;166:669–677. DOI: 10.4049/jimmunol.166.1.669

[143] Pan YQ, Ding D, Li DX. Expression and bioactivity analysis of staphylococcal enterotoxin M and N. Protein Expr Purif. 2007;56:286–292. DOI: 10.1016/j.pep.2007.08.005

[144] Thomas DY, Jarraud S, Lemercier B. Staphylococcal enterotoxin-like toxins U2 and V, two new staphylococcal superantigens arising from recombination within the enterotoxin gene cluster. Infect Immun. 2006;74:4724–4734. DOI: 10.1128/IAI.00132-06

[145] Kuroda M, Ohta T, Uchiyama I. Whole genome sequencing of methicillin-resistant *Staphylococcus aureus*. Lancet. 2001;357:1225–1240. DOI: 10.1016/S0140-6736(00)04403-2

[146] Calderwood MS, Desjardins CA, Sakoulas G. Staphylococcal enterotoxin P predicts bacteremia in hospitalized patients colonized with methicillin-resistant *Staphylococcus aureus*. J Infect Dis. 2014;209:571–577. DOI: 10.1093/infdis/jit501

[147] Orwin PM, Leung DY, Tripp TJ. Characterization of a novel staphylococcal enterotoxin-like superantigen, a member of the group V subfamily of pyrogenic toxins. Biochemistry. 2002;41:14033–14040. DOI: 10.1021/bi025977q

[148] Omoe K, Hu DL, Takahashi-Omoe H. Identification and characterization of a new staphylococcal enterotoxin-related putative toxin encoded by two kinds of plasmids. Infect Immun. 2003;71:6088–6094. DOI: 10.1128/IAI.71.10.6088-6094.2003

[149] Omoe K, Imanishi K, Hu DL. Biological properties of staphylococcal enterotoxin-like toxin type R. Infect Immun. 2004;72:3664–3667. DOI: 10.1128/IAI.72.6.3664-3667.2004

[150] Ono HK, Omoe K, Imanishi K. Identification and characterization of two novel staphylococcal enterotoxins, types S and T. Infect Immun. 2008;76:4999–5005. DOI: 10.1128/IAI.00045-08

[151] Lis E, Podkowik M, Schubert J. Production of staphylococcal enterotoxin R by *Staphylococcus aureus* strains. Foodborne Pathogens Dis. 2012;9:762–766. DOI: 10.1089/fpd.2012.1185

[152] Letertre C, Perelle S, Dilasser F. Identification of a new putative enterotoxin SEU encoded by the egc cluster of *Staphylococcus aureus*. J Appl Microbiol. 2003;95:38–43. DOI: 10.1046/j.1365-2672.2003.01957.x

[153] Schmidt H, Hensel M. Pathogenicity islands in bacterial pathogenesis. Clin Microbiol Rev. 2004,17:14–56. DOI: 10.1128/CMR.17.1.14-56.2004

[154] Ubeda C, Barry P, Penades JR. A pathogenicity island replicon in *Staphylococcus aureus* replicates as an unstable plasmid. Proc Natl Acad Sci U S A. 2007;104:14182–14188. DOI: 10.1073/pnas.0705994104

[155] Ruzin A, Lindsay J, Novick RP. Molecular genetics of SaPI1—a mobile pathogenicity island in *Staphylococcus aureus*. Mol Microbiol. 2001;41:365–377. DOI: 10.1046/j.1365-2958.2001.02488.x

[156] Yarwood JM, McCormick JK, Paustian ML. Characterization and expression analysis of *Staphylococcus aureus* pathogenicity island 3. Implications for the evolution of staphylococcal pathogenicity islands. J Biol Chem. 2002;277:13138–13147. DOI: 10.1074/jbc.M111661200

[157] Baba T, Takeuchi F, Kuroda M. Genome and virulence determinants of high virulence community-acquired MRSA. Lancet. 2002;359:1819–1827. DOI: 10.1016/S0140-6736(02)08713-5

[158] Yamaguchi T, Nishifuji K, Sasaki M. Identification of the *Staphylococcus aureus* etd pathogenicity island which encodes a novel exfoliative toxin, ETD, and EDIN-B. Infect Immun. 2002;70:5835–5845. DOI: 10.1128/IAI.70.10.5835-5845.2002

[159] Boyd EF, Brussow H. Common themes among bacteriophage-encoded virulence factors and diversity among the bacteriophages involved. Trends Microbiol. 2002;10:521–529. DOI: 10.1016/S0966-842X(02)02459-9

[160] Hacker J, Carniel E. Ecological fitness, genomic islands and bacterial pathogenicity. A Darwinian view of the evolution of microbes. EMBO Rep. 2001;2:376–381. DOI: 10.1093/embo-reports/kve097

[161] Groisman EA, Ochman H. Pathogenicity islands: bacterial evolution in quantum leaps. Cell. 1996;87:791–794. DOI: 10.1016/S0092-8674(00)81985-6

[162] Hentschel U, Hacker J. Pathogenicity islands: the tip of the iceberg. Microbes Infect. 2001;3:545–548. DOI: 10.1016/S1286-4579(01)01410-1

[163] Hussein AI, Ahmed AM, Sato M. Characterization of integrons and antimicrobial resistance genes in clinical isolates of Gram-negative bacteria from Palestinian hospitals. Microbiol Immunol. 2009;53:595–602. DOI: 10.1111/j.1348-0421.2009.00168.x

[164] Zhong N, Gui Z, Xu L. Solvent-free enzymatic synthesis of 1,3-diacylglycerols by direct esterification of glycerol with saturated fatty acids. Lipids Health Dis. 2013;12:65. DOI: 10.1186/1476-511X-12-65

[165] Deng Y, Liu J, Peters BM. Antimicrobial resistance investigation on *Staphylococcus* strains in a local hospital in Guangzhou, China, 2001–2010. Microb Drug Resist. 2015;21:102–104. DOI: 10.1089/mdr.2014.0117

[166] Xu Z, Li L, Alam MJ. First confirmation of integron-bearing methicillin-resistant *Staphylococcus aureus*. Curr Microbiol. 2008;57:264–268. DOI: 10.1007/s00284-008-9187-8

[167] Jones TF, Kellum ME, Porter SS. An outbreak of community-acquired foodborne illness caused by methicillin-resistant *Staphylococcus aureus*. Emerg Infect Dis. 2002;8:82–84. DOI: 10.3201/eid0801.010174

[168] Andreoletti O, Budka H, Buncic S. Foodborne antimicrobial resistance as a biological hazard. EFSA J. 2008;765:2–87.

[169] de Boer E, Zwartkruis-Nahuis JT, Wit B. Prevalence of methicillin-resistant *Staphylococcus aureus* in meat. Int J Food Microbiol. 2009;134:52–56. DOI: 10.1016/j.ijfoodmicro.2008.12.007

[170] Kwon NH, Park KT, Moon JS. Staphylococcal cassette chromosome mec (SCCmec) characterization and molecular analysis for methicillin-resistant *Staphylococcus aureus* and novel SCCmec subtype IVg isolated from bovine milk in Korea. J Antimicrob Chemother. 2005;56:624–632. DOI: 10.1093/jac/dki306

[171] Khanna T, Friendship R, Dewey C. Methicillin resistant *Staphylococcus aureus* colonization in pigs and pig farmers. Vet Microbiol. 2008;128:298–303. DOI: 10.1016/j.vetmic.2007.10.006

[172] Lee JH. Occurrence of methicillin-resistant *Staphylococcus aureus* strains from cattle and chicken, and analyses of their mecA, mecR1 and mecI genes. Vet Microbiol. 2006;114:155–159. DOI: 10.1016/j.vetmic.2005.10.024

[173] Vanderhaeghen W, Cerpentier T, Adriaensen C. Methicillin-resistant *Staphylococcus aureus* (MRSA) ST398 associated with clinical and subclinical mastitis in Belgian cows. Vet Microbiol. 2010;144:166–171. DOI: 10.1016/j.vetmic.2009.12.044

[174] Hall RM, Stokes HW. Integrons: novel DNA elements which capture genes by site-specific recombination. Genetica. 1993;90:115–132. DOI: 10.1007/BF01435034

[175] Hall RM, Brown HJ, Brookes DE. Integrons found in different locations have identical 5 ends but variable 3 ends. J Bacteriol. 1994;176:6286–6294.

[176] Hall RM, Collis CM. Mobile gene cassettes and integrons: capture and spread of genes by site-specific recombination. Mol Microbiol. 1995;15:593–600. DOI: 10.1111/j.1365-2958.1995.tb02368.x

[177] Stokes HW, Hall RM. A novel family of potentially mobile DNA elements encoding site-specific gene-integration functions: integrons. Mol Microbiol. 1989;3:1669–1683. DOI: 10.1111/j.1365-2958.1989.tb00153.x

[178] Katayama Y, Ito T, Hiramatsu K. A new class of genetic element, staphylococcus cassette chromosome mec, encodes methicillin resistance in *Staphylococcus aureus*. Antimicrob Agents Chemother. 2000;44:1549–1555. DOI: 10.1128/AAC.44.6.1549-1555.2000

[179] Graham D. Intellectual property rights and the life science industries: past, present and future. World Sci. 2009;63:0013–0117

[180] Beck WD, Berger-Bachi B, Kayser FH. Additional DNA in methicillin-resistant *Staphylococcus aureus* and molecular cloning of mec-specific DNA. J Bacteriol. 1986;165:373–378

[181] Chikramane SG, Matthews PR, Noble WC. Tn554 inserts in methicillin-resistant *Staphylococcus aureus* from Australia and England: comparison with an American methicillin-resistant group. J Gen Microbiol. 1991;137:1303–1311. DOI: 10.1099/00221287-137-6-1303

[182] Dubin DT, Matthews PR, Chikramane SG. Physical mapping of the mec region of an American methicillin-resistant *Staphylococcus aureus* strain. Antimicrob Agents Chemother. 1991;35:1661–1665. DOI: 10.1128/AAC.35.8.1661

[183] Skinner S, Inglis B, Matthews PR. Mercury and tetracycline resistance genes and flanking repeats associated with methicillin resistance on the chromosome of *Staphylococcus aureus*. Mol Microbiol. 1988;2:289–292. DOI: 10.1111/j.1365-2958.1988.tb00030.x

[184] Matsuhashi M, Song MD, Ishino F. Molecular cloning of the gene of a penicillin-binding protein supposed to cause high resistance to beta-lactam antibiotics in *Staphylococcus aureus*. J Bacteriol. 1986;167:975–980

[185] Song MD, Wachi M, Doi M. Evolution of an inducible penicillin-target protein in methicillin-resistant *Staphylococcus aureus* by gene fusion. FEBS Lett. 1987;221:167–171. DOI: 10.1016/0014-5793(87)80373-3

[186] Ito T, Katayama Y, Hiramatsu K. Cloning and nucleotide sequence determination of the entire mec DNA of pre-methicillin-resistant *Staphylococcus aureus* N315. Antimicrob Agents Chemother. 1999;43:1449–1458

[187] Ito T, Katayama Y, Asada K. Structural comparison of three types of staphylococcal cassette chromosome mec integrated in the chromosome in methicillin-resistant *Staphylococcus aureus*. Antimicrob Agents Chemother. 2001;45:1323–1336. DOI: 10.1128/AAC.45.5.1323-1336.2001

[188] Ma XX, Ito T, Tiensasitorn C. Novel type of staphylococcal cassette chromosome mec identified in community-acquired methicillin-resistant *Staphylococcus aureus* strains. Antimicrob Agents Chemother. 2002;46:1147–1152. DOI: 10.1128/AAC.46.4.1147-1152.2002

[189] Oliveira DC, Milheirico C, de Lencastre H. Redefining a structural variant of staphylococcal cassette chromosome mec, SCCmec type VI. Antimicrob Agents Chemother. 2006;50:3457–3459. DOI: 10.1128/AAC.00629-06

[190] Oliveira DC, Tomasz A, de Lencastre H. The evolution of pandemic clones of methicillin-resistant *Staphylococcus aureus*: identification of two ancestral genetic backgrounds and the associated mec elements. Microb Drug Resist. 2001;7:349–361. DOI: 10.1089/10766290152773365

[191] Ito T, Ma XX, Takeuchi F. Novel type V staphylococcal cassette chromosome mec driven by a novel cassette chromosome recombinase, ccrC. Antimicrob Agents Chemother. 2004;48:2637–2651. DOI: 10.1128/AAC.48.7.2637-2651.2004

[192] Berglund C, Ito T, Ikeda M. Novel type of staphylococcal cassette chromosome mec in a methicillin-resistant *Staphylococcus aureus* strain isolated in Sweden. Antimicrob Agents Chemother. 2008;52:3512–3516. DOI: 10.1128/AAC.00087-08

[193] Zhang K, McClure JA, Elsayed S. Novel staphylococcal cassette chromosome mec type, tentatively designated type VIII, harboring class A mec and type 4 ccr gene complexes in a Canadian epidemic strain of methicillin-resistant *Staphylococcus aureus*. Antimicrob Agents Chemother. 2009;53:531–540. DOI: 10.1128/AAC.01118-08

[194] Li S, Skov RL, Han X. Novel types of staphylococcal cassette chromosome mec elements identified in clonal complex 398 methicillin-resistant *Staphylococcus aureus* strains. Antimicrob Agents Chemother. 2011;55:3046–3050. DOI: 10.1128/AAC.01475-10

[195] Garcia-Alvarez L, Holden MT, Lindsay H. Methicillin-resistant *Staphylococcus aureus* with a novel mecA homologue in human and bovine populations in the UK and Denmark: a descriptive study. Lancet Infect Dis. 2011;11:595–603. DOI: 10.1016/S1473-3099(11)70126-8

[196] Chongtrakool P, Ito T, Ma XX. Staphylococcal cassette chromosome mec (SCCmec) typing of methicillin-resistant *Staphylococcus aureus* strains isolated in 11 Asian countries: a proposal for a new nomenclature for SCCmec elements. Antimicrob Agents Chemother. 2006;50:1001–1012. DOI: 10.1128/AAC.50.3.1001-1012.2006

[197] Savolainen K, Korhonen TK, and Kuusela P. PLS, a large surface protein encoded by mec DNA of methicillin resistant *Staphylococcus aureus*, prevents bacterial adhesion *in*

vitro. Ninth International Symposium on Staphylococci and Staphylococcal Infections Kolding, Denmark; 2000.

[198] Oliveira DC, de Lencastre H. Multiplex PCR strategy for rapid identification of structural types and variants of the mec element in methicillin-resistant *Staphylococcus aureus*. Antimicrob Agents Chemother. 2002;46:2155–2161. DOI: 10.1128/AAC.46.7.2155-2161.2002

[199] Katayama Y, Ito T, Hiramatsu K. Genetic organization of the chromosome region surrounding mecA in clinical staphylococcal strains: role of IS431-mediated mecI deletion in expression of resistance in mecA-carrying, low-level methicillin-resistant *Staphylococcus haemolyticus*. Antimicrob Agents Chemother. 2001;45:1955–1963. DOI: 10.1128/AAC.45.7.1955-1963.2001

[200] Garcia-Alvarez L. Assessment of the role of cattle movements and other risk contacts on the spread of *Staphylococcus aureus* strain types between UK dairy farms [thesis]. Cambridge: University of Cambridge; 2009.

[201] Kim JS, Song W, Kim HS. Association between the methicillin resistance of clinical isolates of *Staphylococcus aureus*, their staphylococcal cassette chromosome mec (SCCmec) subtype classification, and their toxin gene profiles. Diagn Microbiol Infect Dis. 2006;56:289–295. DOI: 10.1016/j.diagmicrobio.2006.05.003

[202] Cabrera EC, Ramirez-Argamosa DT, Rodriguez RDM. Prevalence of community-acquired methicillin-resistant *Staphylococcus aureus* from inmates of the Manila City Jail, characterization for SCCmec type and occurrence of Panton-Valentine leukocidin gene. Philipp Sci Lett. 2010;3:4-12.

[203] Laplana LM, Cepero MA, Ruiz J. Molecular typing of *Staphylococcus aureus* clinical isolates by pulsed-field gel electrophoresis, staphylococcal cassette chromosome mec type determination and dissemination of antibiotic resistance genes. Int J Antimicrob Agents. 2007;30:505–513. DOI: 10.1016/j.ijantimicag.2007.06.020

[204] Stranden AM, Frei R, Adler H. Emergence of SCCmec type IV as the most common type of methicillin-resistant *Staphylococcus aureus* in a university hospital. Infection. 2009;37:44–48. DOI: 10.1007/s15010-008-7430-7

[205] Namvar AE, Afshar M, Asghari B. Characterisation of SCCmec elements in methicillin-resistant *Staphylococcus aureus* isolated from burn patients. Burns. 2014;40:708–712. DOI: 10.1016/j.burns.2013.09.010

[206] Barada K, Hanaki H, Yamaguchi Y. Trends of beta-lactam antibiotic susceptibility in blood-borne methicillin-resistant *Staphylococcus aureus* (MRSA) and their linkage to the staphylococcal cassette chromosome mec (SCCmec) type. J Infect Chemother. 2007;13:213–218. DOI: 10.1007/s10156-007-0523-x

[207] Hu DL, Maina EK, Omoe K. Superantigenic toxin genes coexist with specific staphylococcal cassette chromosome mec genes in methicillin-resistant *Staphylococcus aureus*. Tohoku J Exp Med. 2011;225:161–169. DOI: 10.1620/tjem.225.161

[208] Reiter KC, Machado AB, Freitas AL. High prevalence of methicillin-resistant *Staphylococcus aureus* with SCCmec type III in cystic fibrosis patients in southern, Brazil. Rev Soc Bras Med Trop. 2010;43:377–381

[209] Lima DF, Brazao NB, Folescu TW. Panton-Valentine leukocidin (PVL) gene carriage among *Staphylococcus aureus* strains with SCCmec types I, III, IV, and V recovered from cystic fibrosis pediatric patients in Brazil. Diagn Microbiol Infect Dis. 2014;78:59–62. DOI: 10.1016/j.diagmicrobio.2013.10.004

[210] Gomez E, Chiang T, Hogan PA. Methicillin-resistant *Staphylococcus aureus* SCCmec type and its association with clinical presentation, severity, and length of stay among patients with complicated skin and skin structure infections. Adv Infect Dis. 2014;4:111–115.

[211] Krzyszton-Russjan J, Empel J, Leski T. Clonal structure of the methicillin-resistant *Staphylococcus aureus* (MRSA) population in Poland: revision and update. Microb Drug Resist. 2005;11:127–136. DOI: 10.1089/mdr.2005.11.127

[212] Moon SY, Lee HJ, Lee MS. Molecular characteristics of methicillin-resistant *Staphylococcus aureus* blood isolates: clonal spread of staphylococcal cassette chromosome mec type IVA between the community and the hospital. Microb Drug Resist. 2010;16:217–222. DOI: 10.1089/mdr.2010.0010

[213] Kim J, Jeong JH, Cha HY. Detection of diverse SCCmec variants in methicillin-resistant *Staphylococcus aureus* and comparison of SCCmec typing method. Clin Microbiol Infect. 2007;13:1128–1130. DOI: 10.1111/j.1469-0691.2007.01806.x

[214] Ho CM, Ho MW, Lee CY. Clonal spreading of methicillin-resistant SCCmec *Staphylococcus aureus* with specific spa and dru types in central Taiwan. Eur J Clin Microbiol Infect Dis. 2012;31:499–504. DOI: 10.1007/s10096-011-1338-3

[215] Heltshe SL, Saiman L, Popowitch EB. Outcomes and treatment of chronic methicillin-resistant *Staphylococcus aureus* differs by staphylococcal cassette chromosome mec (sccmec) type in children with cystic fibrosis. Pediatr Infect Dis J. 2014

[216] Noto MJ, Archer GL. A subset of *Staphylococcus aureus* strains harboring staphylococcal cassette chromosome mec (SCCmec) type IV is deficient in CcrAB-mediated SCCmec excision. Antimicrob Agents Chemother. 2006;50:2782–2788. DOI: 10.1128/AAC.00032-06

[217] Lulitanond A, Chanawong A, Sribenjalux P. Preliminary report of SCCmec-types and antimicrobial susceptibilities of methicillin-resistant *Staphylococcus aureus* isolates from a university hospital in Thailand. Southeast Asian J Trop Med Public Health. 2010;41:920–927.

[218] Chu H, Zhao L, Zhang Z. Antibiotic resistance and molecular epidemiology of methicillin-resistant *Staphylococcus aureus* from lower respiratory tract: multi-resistance and high prevalence of SCCmec III type. Cell Biochem Biophys. 2013;67:795–801. DOI: 10.1007/s12013-013-9542-7

[219] Matos PD, Schuenck RP, Cavalcante FS. Accuracy of phenotypic methicillin susceptibility methods in the detection of *Staphylococcus aureus* isolates carrying different SCCmec types. Mem Inst Oswaldo Cruz. 2010;105:931–934. DOI: 10.1590/S0074-02762010000700017

[220] Teodoro CR, Mattos CS, Cavalcante FS. Characterization of MLS(b) resistance among *Staphylococcus aureus* and *Staphylococcus epidermidis* isolates carrying different SCCmec types. Microbiol Immunol. 2012;56:647–650. DOI: 10.1111/j.1348-0421.2012.00481.x

[221] Djoudi F, Bonura C, Benallaoua S. Panton-Valentine leukocidin positive sequence type 80 methicillin-resistant *Staphylococcus aureus* carrying a staphylococcal cassette chromosome mec type IVc is dominant in neonates and children in an Algiers hospital. N Microbiol. 2013;36:49–55

[222] Szczepanik A, Koziol-Montewka M, Al-Doori Z. Spread of a single multiresistant methicillin-resistant *Staphylococcus aureus* clone carrying a variant of staphylococcal cassette chromosome mec type III isolated in a university hospital. Eur J Clin Microbiol Infect Dis. 2007;26:29–35. DOI: 10.1007/s10096-006-0237-5

[223] Japoni A, Jamalidoust M, Farshad S. Characterization of SCCmec types and antibacterial susceptibility patterns of methicillin-resistant *Staphylococcus aureus* in southern Iran. Jpn J Infect Dis. 2011;64:28–33

[224] Fatholahzadeh B, Emaneini M, Gilbert G. Staphylococcal cassette chromosome mec (SCCmec) analysis and antimicrobial susceptibility patterns of methicillin-resistant *Staphylococcus aureus* (MRSA) isolates in Tehran, Iran. Microb Drug Resist. 2008;14:217–220. DOI: 10.1089/mdr.2008.0822

[225] Kilic A, Guclu AU, Senses Z. Staphylococcal cassette chromosome mec (SCCmec) characterization and Panton-Valentine leukocidin gene occurrence for methicillin-resistant *Staphylococcus aureus* in Turkey, from 2003 to 2006. Antoine Van Leeuwenhoek. 2008;94:607–614. DOI: 10.1007/s10482-008-9278-3

[226] Ahmad N, Ruzan IN, Abd GM. Characteristics of community- and hospital-acquired methicillin-resistant *Staphylococcus aureus* strains carrying SCCmec type IV isolated in Malaysia. J Med Microbiol. 2009;58:1213–1218. DOI: 10.1099/jmm.0.011353-0

[227] Peng Q, Hou B, Zhou S. Staphylococcal cassette chromosome mec (SCCmec) analysis and antimicrobial susceptibility profiles of methicillin-resistant *Staphylococcus aureus* (MRSA) isolates in a teaching hospital, Shantou, China. Afr J Microbiol Res. 2010;4:844–848.

[228] Chen Y, Liu Z, Duo L. Characterization of *Staphylococcus aureus* from distinct geographic locations in China: an increasing prevalence of spa-t030 and SCCmec type III. PLoS One. 2014;9:e96255. DOI: 10.1371/journal.pone.0096255

[229] Pan SC, Wang JT, Lauderdale TL. Epidemiology and staphylococcal cassette chromosome mec typing of methicillin-resistant *Staphylococcus aureus* isolates in Taiwan: a multicenter study. J Formosan Med Assoc. 2014;113:409–416. DOI: 10.1016/j.jfma.2012.05.012

[230] Neela V, Ghasemzadeh MH, van Belkum A. First report on methicillin-resistant *Staphylococcus aureus* of Spa type T037, sequence type 239, SCCmec type III/IIIA in Malaysia. Eur J Clin Microbiol Infect Dis. 2010;29:115–117. DOI: 10.1007/s10096-009-0813-6

[231] Park C, Lee DG, Kim SW. Predominance of community-associated methicillin-resistant *Staphylococcus aureus* strains carrying staphylococcal chromosome cassette mec type IVA in South Korea. J Clin Microbiol. 2007;45:4021–4026. DOI: 10.1128/JCM.01147-07

[232] Berglund C, Sjoberg L, Soderquist B. Predominance of staphylococcal cassette chromosome mec (SCCmec) type IV among methicillin-resistant *Staphylococcus aureus* (MRSA) in a Swedish county and presence of unknown SCCmec types with Panton-Valentine leukocidin genes. Clin Microbiol Infect. 2005;11:447–456. DOI: 10.1111/j.1469-0691.2005.01150.x

[233] de A TP, Pacheco RL, Costa SF. Prevalence of SCCmec type IV in nosocomial bloodstream isolates of methicillin-resistant *Staphylococcus aureus*. J Clin Microbiol. 2005;43:3435–3437. DOI: 10.1128/JCM.43.7.3435-3437.2005

[234] Conceicao T, Tavares A, Miragaia M. Prevalence and clonality of methicillin-resistant *Staphylococcus aureus* (MRSA) in the Atlantic Azores islands: predominance of SCCmec types IV, V and VI. Eur J Clin Microbiol Infect Dis. 2010;29:543–550. DOI: 10.1007/s10096-010-0892-4

[235] Bartels MD, Boye K, Rohde SM. A common variant of staphylococcal cassette chromosome mec type IVa in isolates from Copenhagen, Denmark, is not detected by the BD GeneOhm methicillin-resistant *Staphylococcus aureus* assay. J Clin Microbiol. 2009;47:1524–1527. DOI: 10.1128/JCM.02153-08

[236] Valsesia G, Rossi M, Bertschy S. Emergence of SCCmec type IV and SCCmec type V methicillin-resistant *Staphylococcus aureus* containing the Panton-Valentine leukocidin genes in a large academic teaching hospital in central Switzerland: external invaders or persisting circulators? J Clin Microbiol. 2010;48:720–727. DOI: 10.1128/JCM.01890-09

[237] Ammons DR, Puttagunta R, Granados JC. An exploratory study of methicillin-resistant *Staphylococcus aureus* and SCCmec elements obtained from a community setting

along the Texas border with Mexico. Curr Microbiol. 2010;60:321–326. DOI: 10.1007/s00284-009-9544-2

[238] Williamson DA, Roberts SA, Ritchie SR. Clinical and molecular epidemiology of methicillin-resistant *Staphylococcus aureus* in New Zealand: rapid emergence of sequence type 5 (ST5)-SCCmec-IV as the dominant community-associated MRSA clone. PLoS One. 2013;8:e62020. DOI: 10.1371/journal.pone.0062020

[239] Sobhy N, Aly F, Abd EKO. Community-acquired methicillin-resistant *Staphylococcus aureus* from skin and soft tissue infections (in a sample of Egyptian population): analysis of mec gene and staphylococcal cassette chromosome. Braz J Infect Dis. 2012;16:426–431. DOI: 10.1016/j.bjid.2012.08.004

[240] O'Halloran F, Lucey B, Cryan B. Molecular characterization of class 1 integrons from Irish thermophilic *Campylobacter* spp. J Antimicrob Chemother. 2004;53:952–957. DOI: 10.1093/jac/dkh193

[241] Mazel D, Dychinco B, Webb VA. A distinctive class of integron in the *Vibrio cholerae* genome. Science. 1998;280:605–608. DOI: 10.1126/science.280.5363.605

[242] Rowe-Magnus DA, Guerout AM, Ploncard P. The evolutionary history of chromosomal super-integrons provides an ancestry for multiresistant integrons. Proc Natl Acad Sci U S A. 2001;98:652–657. DOI: 10.1073/pnas.98.2.652

[243] Xu Z, Li L, Shirtliff ME. First report of class 2 integron in clinical Enterococcus faecalis and class 1 integron in *Enterococcus faecium* in South China. Diagn Microbiol Infect Dis. 2010;68:315–317. DOI: 10.1016/j.diagmicrobio.2010.05.014

[244] Labbate M, Case RJ, Stokes HW. The integron/gene cassette system: an active player in bacterial adaptation. Methods Mol Biol. 2009;532:103–125. DOI: 10.1007/978-1-60327-853-9_6

[245] Barlow RS, Pemberton JM, Desmarchelier PM. Isolation and characterization of integron-containing bacteria without antibiotic selection. Antimicrob Agents Chemother. 2004;48:838–842. DOI: 10.1128/AAC.48.3.838-842.2004

[246] Nandi S, Maurer JJ, Hofacre C. Gram-positive bacteria are a major reservoir of class 1 antibiotic resistance integrons in poultry litter. Proc Natl Acad Sci U S A. 2004;101:7118–7122. DOI: 10.1073/pnas.0306466101

[247] Xu Z, Shi L, Zhang C. Nosocomial infection caused by class 1 integron-carrying *Staphylococcus aureus* in a hospital in South China. Clin Microbiol Infect. 2007;13:980–984. DOI: 10.1111/j.1469-0691.2007.01782.x

[248] Xu Z, Li L, Shirtliff ME. Resistance class 1 integron in clinical methicillin-resistant *Staphylococcus aureus* strains in southern China, 2001–2006. Clin Microbiol Infect. 2011;17:714–718. DOI: 10.1111/j.1469-0691.2010.03379.x

[249] Senda K, Arakawa Y, Ichiyama S. PCR detection of metallo-beta-lactamase gene (blaIMP) in Gram-negative rods resistant to broad-spectrum beta-lactams. J Clin Microbiol. 1996;34:2909–2913

[250] Nemergut DR, Robeson MS, Kysela RF. Insights and inferences about integron evolution from genomic data. BMC Genomics. 2008;9:261. DOI: 10.1186/1471-2164-9-261

[251] Correia M, Boavida F, Grosso F. Molecular characterization of a new class 3 integron in *Klebsiella pneumoniae*. Antimicrob Agents Chemother. 2003;47:2838–2843. DOI: 10.1128/AAC.47.9.2838-2843.2003

[252] Arakawa Y, Murakami M, Suzuki K. A novel integron-like element carrying the metallo-beta-lactamase gene blaIMP. Antimicrob Agents Chemother. 1995;39:1612–1615. DOI: 10.1128/AAC.39.7.1612

[253] Rowe-Magnus DA, Guerout AM, Mazel D. Super-integrons. Res Microbiol. 1999;150:641–651. DOI: 10.1016/S0923-2508(99)00127-8

[254] Nield BS, Holmes AJ, Gillings MR. Recovery of new integron classes from environmental DNA. FEMS Microbiol Lett. 2001;195:59–65. DOI: 10.1111/j.1574-6968.2001.tb10498.x

[255] Rowe-Magnus DA, Mazel D. Integrons: natural tools for bacterial genome evolution. Curr Opin Microbiol. 2001;4:565–569. DOI: 10.1016/S1369-5274(00)00252-6

[256] Mazel D. Integrons: agents of bacterial evolution. Nat Rev Microbiol. 2006;4:608–620. DOI: 10.1038/nrmicro1462

[257] Mulazimoglu L, Drenning SD, Muder RR. Vancomycin-gentamicin synergism revisited: effect of gentamicin susceptibility of methicillin-resistant *Staphylococcus aureus*. Antimicrob Agents Chemother. 1996;40:1534–1535.

[258] Stefani S, Chung DR, Lindsay JA. Methicillin-resistant *Staphylococcus aureus* (MRSA): global epidemiology and harmonisation of typing methods. Int J Antimicrob Agents. 2012;39:273–282. DOI: 10.1016/j.ijantimicag.2011.09.030

[259] Graveland H, Duim B, van Duijkeren E. Livestock-associated methicillin-resistant *Staphylococcus aureus* in animals and humans. Int J Med Microbiol. 2011;301:630–634. DOI: 10.1016/j.ijmm.2011.09.004

[260] Frana TS, Beahm AR, Hanson BM. Isolation and characterization of methicillin-resistant *Staphylococcus aureus* from pork farms and visiting veterinary students. PLoS One. 2013;8:e53738. DOI: 10.1371/journal.pone.0053738

[261] Smith TC, Male MJ, Harper AL. Methicillin-resistant *Staphylococcus aureus* (MRSA) strain ST398 is present in Midwestern U.S. swine and swine workers. PLoS One. 2009;4:e4258. DOI: 10.1371/journal.pone.0004258

[262] Voss A, Loeffen F, Bakker J. Methicillin-resistant *Staphylococcus aureus* in pig farming. Emerg Infect Dis. 2005;11:1965–1966. DOI: 10.3201/eid1112.050428

[263] Devriese LA, Hommez J. Epidemiology of methicillin-resistant *Staphylococcus aureus* in dairy herds. Res Vet Sci. 1975;19:23–27

[264] Juhasz-Kaszanyitzky E, Janosi S, Somogyi P. MRSA transmission between cows and humans. Emerg Infect Dis. 2007;13:630–632. DOI: 10.3201/eid1304.060833

[265] Cui S, Li J, Hu C. Isolation and characterization of methicillin-resistant *Staphylococcus aureus* from swine and workers in China. J Antimicrob Chemother. 2009;64:680–683. DOI: 10.1093/jac/dkp275

[266] de Neeling AJ, van den Broek MJ, Spalburg EC. High prevalence of methicillin resistant *Staphylococcus aureus* in pigs. Vet Microbiol. 2007;122:366–372. DOI: 10.1016/j.vetmic.2007.01.027

[267] Neela V, Zafrul AM, Mariana NS. Prevalence of ST9 methicillin-resistant *Staphylococcus aureus* among pigs and pig handlers in Malaysia. J Clin Microbiol. 2009;47:4138–4140. DOI: 10.1128/JCM.01363-09

[268] Sergio DM, Koh TH, Hsu LY. Investigation of methicillin-resistant *Staphylococcus aureus* in pigs used for research. J Med Microbiol. 2007;56:1107–1109. DOI: 10.1099/jmm.0.47283-0

[269] Lewis HC, Molbak K, Reese C. Pigs as source of methicillin-resistant *Staphylococcus aureus* CC398 infections in humans, Denmark. Emerg Infect Dis. 2008;14:1383–1389. DOI: 10.3201/eid1409.071576

[270] Morgan M. Methicillin-resistant *Staphylococcus aureus* and animals: zoonosis or humanosis? J Antimicrob Chemother. 2008;62:1181–1187. DOI: 10.1093/jac/dkn405

[271] Smith TC, Pearson N. The emergence of *Staphylococcus aureus* ST398. Vector Borne Zoonotic Dis. 2011;11:327–339. DOI: 10.1089/vbz.2010.0072

[272] van den Broek IV, van Cleef BA, Haenen A. Methicillin-resistant *Staphylococcus aureus* in people living and working in pig farms. Epidemiol Infect. 2009;137:700–708. DOI: 10.1017/S0950268808001507

[273] Graveland H, Wagenaar JA, Heesterbeek H. Methicillin resistant *Staphylococcus aureus* ST398 in veal calf farming: human MRSA carriage related with animal antimicrobial usage and farm hygiene. PLoS One. 2010;5:e10990. DOI: 10.1371/journal.pone.0010990

[274] Wassenberg MW, Bootsma MC, Troelstra A. Transmissibility of livestock-associated methicillin-resistant *Staphylococcus aureus* (ST398) in Dutch hospitals. Clin Microbiol Infect. 2011;17:316–319. DOI: 10.1111/j.1469-0691.2010.03260.x

[275] Bootsma MC, Wassenberg MW, Trapman P. The nosocomial transmission rate of animal-associated ST398 methicillin-resistant *Staphylococcus aureus*. J R Soc Interf. 2011;8:578–584. DOI: 10.1098/rsif.2010.0349

[276] Graveland H, Wagenaar JA, Bergs K. Persistence of livestock associated MRSA CC398 in humans is dependent on intensity of animal contact. PLoS One. 2011;6:e16830. DOI: 10.1371/journal.pone.0016830

[277] Cuny C, Friedrich A, Kozytska S. Emergence of methicillin-resistant *Staphylococcus aureus* (MRSA) in different animal species. Int J Med Microbiol. 2010;300:109–117. DOI: 10.1016/j.ijmm.2009.11.002

[278] Kluytmans JA. Methicillin-resistant *Staphylococcus aureus* in food products: cause for concern or case for complacency? Clin Microbiol Infect. 2010;16:11–15. DOI: 10.1111/j. 1469-0691.2009.03110.x

[279] Bens CC, Voss A, Klaassen CH. Presence of a novel DNA methylation enzyme in methicillin-resistant *Staphylococcus aureus* isolates associated with pig farming leads to uninterpretable results in standard pulsed-field gel electrophoresis analysis. J Clin Microbiol. 2006;44:1875–1876. DOI: 10.1128/JCM.44.5.1875-1876.2006

[280] Vanderhaeghen W, Hermans K, Haesebrouck F. Methicillinresistant *Staphylococcus aureus* (MRSA) in food production animals. Epidemiol Infect. 2010;138:606–625

[281] Monecke S, Coombs G, Shore AC. A field guide to pandemic, epidemic and sporadic clones of methicillin-resistant *Staphylococcus aureus*. PLoS One. 2011;6:e17936. DOI: 10.1371/journal.pone.0017936

[282] Fessler A, Scott C, Kadlec K. Characterization of methicillin-resistant *Staphylococcus aureus* ST398 from cases of bovine mastitis. J Antimicrob Chemother. 2010;65:619–625. DOI: 10.1093/jac/dkq021

[283] Fessler A, Kadlec K, Hassel M. Characterization of methicillin-resistant *Staphylococcus aureus* isolates from food and food products of poultry origin in Germany. Appl Environ Microbiol. 2011;77:7151–7157

[284] Kadlec K, Ehricht R, Monecke S. Diversity of antimicrobial resistance pheno- and genotypes of methicillin-resistant *Staphylococcus aureus* ST398 from diseased swine. J Antimicrob Chemother. 2009;64:1156–1164. DOI: 10.1093/jac/dkp350

[285] Argudin MA, Fetsch A, Tenhagen BA. High heterogeneity within methicillin-resistant *Staphylococcus aureus* ST398 isolates, defined by Cfr9I macrorestriction-pulsed-field gel electrophoresis profiles and spa and SCCmec types. Appl Environ Microbiol. 2010;76:652–658. DOI: 10.1128/AEM.01721-09

[286] Kadlec K, Fessler A. Novel and uncommon antimicrobial resistance genes in livestock-associated methicillin-resistant *Staphylococcus aureus*. Clin Microbiol Infect. 2012;18:745–755

[287] Roszak DB, Grimes DJ, Colwell RR. Viable but non-culturable forms of Salmonella enteritidis in aquatic systems. Can J Microbiol. 1984;30:334–338.

[288] Cook KL, Bolster CH. Survival of *Campylobacter jejuni* and *Escherichia coli* in groundwater during prolonged starvation at low temperatures. Appl Microbiol. 2007;103:573–583

[289] Besnard V, Federighi M, Declerq E. Environmental and physico-chemical factors induce VBNC state in *Listeria monocytogenes*. Vet Res. 2002;33:359–370

[290] Cunningham E, O'Byrne C, Oliver JD. Effect of weak acids on *Listeria monocytogenes* survival: evidence for a viable but nonculturable state in response to low pH. Food Control. 2009;20:1141–1144. DOI: 10.1016/j.foodcont.2009.03.005

[291] Asakura H, Kawamoto K, Haishima Y. Differential expression of the outer membrane protein W (OmpW) stress response in enterohemorrhagic *Escherichia coli* O157:H7 corresponds to the viable but non-culturable state. Res Microbiol. 2008;159:709–717. DOI: 10.1016/j.resmic.2008.08.005

[292] Kana BD, Gordhan BG, Downing KJ. The resuscitation-promoting factors of *Mycobacterium tuberculosis* are required for virulence and resuscitation from dormancy but are collectively dispensable for growth *in vitro*. Mol Microbiol. 2008;67:672–684. DOI: 10.1111/j.1365-2958.2007.06078.x

[293] Mascher F, Hase C, Moenne-Loccoz Y. The viable-but-nonculturable state induced by abiotic stress in the biocontrol agent *Pseudomonas fluorescens* CHA0 does not promote strain persistence in soil. Appl Environ Microbiol. 2000;66:1662–1667. DOI: 10.1128/AEM.66.4.1662-1667.2000

[294] Ghezzi JI, Steck TR. Induction of the viable but nonculturable condition in *Xanthomonas campestris* pv. *campestris* in liquid microcosms and sterile soil. FEMS Microbiol Ecol. 1999;30:203–208.

[295] Del CR, Russi P, Mara P. *Xanthomonas axonopodis* pv. *citri* enters the VBNC state after copper treatment and retains its virulence. FEMS Microbiol Lett. 2009;298:143–148. DOI: 10.1111/j.1574-6968.2009.01709.x

[296] Reissbrodt R, Rienaecker I, Romanova JM. Resuscitation of *Salmonella enterica* serovar *typhimurium* and enterohemorrhagic *Escherichia coli* from the viable but nonculturable state by heat-stable enterobacterial autoinducer. Appl Environ Microbiol. 2002;68:4788–4794. DOI: 10.1128/AEM.68.10.4788-4794.2002

[297] Cho JC, Kim SJ. Green fluorescent protein-based direct viable count to verify a viable but non-culturable state of *Salmonella typhi* in environmental samples. J Microbiol Methods. 1999;36:227–235. DOI: 10.1016/S0167-7012(99)00038-X

[298] Signoretto C, Lleo MM, Tafi MC. Cell wall chemical composition of Enterococcus faecalis in the viable but nonculturable state. Appl Environ Microbiol. 2000;66:1953–1959. DOI: 10.1128/AEM.66.5.1953-1959.2000

[299] Lleo MM, Bonato B, Tafi MC. Resuscitation rate in different enterococcal species in the viable but non-culturable state. J Appl Microbiol. 2001;91:1095–1102. DOI: 10.1046/j.1365-2672.2001.01476.x

[300] Wery N, Pourcher AM, Stan V. Survival of *Listeria monocytogenes* and *Enterococcus faecium* in sludge evaluated by real-time PCR and culture methods. Lett Appl Microbiol. 2006;43:131–136. DOI: 10.1111/j.1472-765X.2006.01946.x

[301] Du M, Chen J, Zhang X. Characterization and resuscitation of viable but nonculturable *Vibrio alginolyticus* VIB283. Arch Microbiol. 2007;188:283–288. DOI: 10.1007/s00203-007-0246-5

[302] Eguchi M, Fujiwara E, Miyamoto N. Survival of *Vibrio anguillarum* in freshwater environments: adaptation or debilitation? J Infect Chemother. 2000;6:126–129. DOI: 10.1007/s101560000027

[303] McDougald D, Rice SA, Weichart D. Nonculturability: adaptation or debilitation. FEMS Microbiol Ecol. 1998;25:1–9. DOI: 10.1111/j.1574-6941.1998.tb00455.x

[304] Senoh M, Ghosh-Banerjee J, Ramamurthy T. Conversion of viable but nonculturable *Vibrio cholerae* to the culturable state by co-culture with eukaryotic cells. Microbiol Immunol. 2010;54:502–507

[305] Ramaiah N, Ravel J, Straube WL. Entry of *Vibrio harveyi* and *Vibrio fischeri* into the viable but nonculturable state. J Appl Microbiol. 2002;93:108–116. DOI: 10.1046/j.1365-2672.2002.01666.x

[306] Jiang X, Chai TJ. Survival of *Vibrio parahaemolyticus* at low temperature under starvation conditions and subsequent resuscitation of viable, nonculturable cells. Appl Environ Microbiol. 1996;62:1300-1305

[307] Oliver JD. Recent findings on the viable but nonculturable state in pathogenic bacteria. FEMS Microbiol Rev. 2010;34:415–425. DOI: 10.1111/j.1574-6976.2009.00200.x

[308] Vattakaven T, Bond P, Bradley G. Differential effects of temperature and starvation on induction of the viable-but-nonculturable state in the coral pathogens *Vibrio shiloi* and *Vibrio tasmaniensis*. Appl Environ Microbiol. 2006;72:6508–6513. DOI: 10.1128/AEM.00798-06

[309] Whitesides MD, Oliver JD. Resuscitation of *Vibrio vulnificus* from the viable but nonculturable state. Appl Environ Microbiol. 1997;63:1002–1005

[310] Churruca E, Girbau C, Martinez I. Detection of *Campylobacter jejuni* and *Campylobacter coli* in chicken meat samples by real-time nucleic acid sequence-based amplification with molecular beacons. Int J Food Microbiol. 2007;117:85–90. DOI: 10.1016/j.ijfoodmicro.2007.02.007

[311] Khan NH, Ahsan M, Taylor WD. Culturability and survival of marine, freshwater and clinical *Pseudomonas aeruginosa*. Microbes Environ. 2010;25:266–274. DOI: 10.1264/jsme2.ME09178

[312] Lowder M, Unge A, Maraha N. Effect of starvation and the viable-but-nonculturable state on green fluorescent protein (GFP) fluorescence in GFP-tagged *Pseudomonas fluorescens* A506. Appl Environ Microbiol. 2000;66:3160–3165. DOI: 10.1128/AEM.66.8.3160-3165.2000

[313] Lowder M, Oliver JD. The use of modified GFP as a reporter for metabolic activity in *Pseudomonas putida*. Microb Ecol. 2001;41:310–313. DOI: 10.1007/s002480000094

[314] Rahman I, Shahamat M, Chowdhury MA. Potential virulence of viable but nonculturable *Shigella dysenteriae* type 1. Appl Environ Microbiol. 1996;62:115–120

[315] Nicolo MS, Gioffre A, Carnazza S. Viable but nonculturable state of foodborne pathogens in grapefruit juice: a study of laboratory. Foodborne Pathogens Dis. 2011;8:11–17. DOI: 10.1089/fpd.2009.0491

[316] Nybom SM, Collado MC, Surono IS. Effect of glucose in removal of microcystin-LR by viable commercial probiotic strains and strains isolated from Dadih fermented milk. J Agric Food Chem. 2008;56:3714–3720. DOI: 10.1021/jf071835x

[317] Suzuki K, Iijima K, Asano S. Induction of viable but nonculturable state in beer spoilage lactic acid bacteria. J Inst Brew. 2006;112:295–301. DOI: 10.1002/j.2050-0416.2006.tb00734.x

[318] Caillet S, Ursachi L, Shareck F. Effect of gamma radiation and oregano essential oil on murein and ATP concentration of *Staphylococcus aureus*. J Food Sci. 2009;74:M499–M508. DOI: 10.1111/j.1750-3841.2009.01368.x

[319] Masmoudi S, Denis M, Maalej S. Inactivation of the gene katA or sodA affects the transient entry into the viable but non-culturable response of *Staphylococcus aureus* in natural seawater at low temperature. Mar Pollut Bull. 2010;60:2209–2214. DOI: 10.1016/j.marpolbul.2010.08.017

[320] Pasquaroli S, Zandri G, Vignaroli C. Antibiotic pressure can induce the viable but non-culturable state in *Staphylococcus aureus* growing in biofilms. J Antimicrob Chemother. 2013;68:1812–1817. DOI: 10.1093/jac/dkt086

[321] Pasquaroli S, Citterio B, Cesare AD. Role of daptomycin in the induction and persistence of the viable but non-culturable state of *Staphylococcus aureus* biofilms. Pathogens. 2014;3:759–768. DOI: 10.3390/pathogens3030759

[322] Bekir K, Barhoumi H, Braiek M. Electrochemical impedance immunosensor for rapid detection of stressed pathogenic *Staphylococcus aureus* bacteria. Environ Sci Pollut Res Int. DOI: 10.1007/s11356-015-4761-7

[323] Zandri G, Pasquaroli S, Vignaroli C. Detection of viable but non-culturable staphylococci in biofilms from central venous catheters negative on standard microbiological

assays. Clin Microbiol Infect. 2012;18:E259–E261. DOI: 10.1111/j. 1469-0691.2012.03893.x

[324] Cerca F, Andrade F, Franca A. *Staphylococcus epidermidis* biofilms with higher proportions of dormant bacteria induce a lower activation of murine macrophages. J Med Microbiol. 2011;60:1717–1724. DOI: 10.1099/jmm.0.031922-0

[325] Gupte AR, De Rezende CL, Joseph SW. Induction and resuscitation of viable but nonculturable *Salmonella enterica* serovar *typhimurium* DT104. Appl Environ Microbiol. 2003;69:6669–6675. DOI: 10.1128/AEM.69.11.6669-6675.2003

[326] Aagot N, Nybroe O, Nielsen P. An altered *Pseudomonas* diversity is recovered from soil by using nutrient-poor *Pseudomonas*-selective soil extract media. Appl Environ Microbiol. 2001;67:5233–5239. DOI: 10.1128/AEM.67.11.5233-5239.2001

[327] Oliver JD. The viable but nonculturable state in bacteria. J Microbiol. 2005;43:93–100

[328] Maalej S, Denis M, Dukan S. Temperature and growth phase effects on *Aeromonas hydrophila* survival in natural seawater microcosms: role of protein synthesis and nucleic acid content on viable but temporarily nonculturable response. Microbiology. 2004;150:181–187

[329] Anuchin AM, Mulyukin AL, Suzina NE. Dormant forms of *Mycobacterium smegmatis* with distinct morphology. Microbiology. 2009;155:1071–1079. DOI: 10.1099/mic. 0.023028-0

[330] Muela A, Seco C, Camafeita E. Changes in Escherichia coli outer membrane subproteome under environmental conditions inducing the viable but nonculturable state. FEMS Microbiol Ecol. 2008;64:28–36. DOI: 10.1111/j.1574-6941.2008.00453.x

[331] Bunthof CJ, Bloemen K, Breeuwer P. Flow cytometric assessment of viability of lactic acid bacteria. Appl Environ Microbiol. 2001;67:2326–2335. DOI: 10.1128/AEM. 67.5.2326-2335.2001

2

Spread and Control of Prion Diseases in the Food and Feed Chains

Qinchun Rao and Yun-Hwa Peggy Hsieh

Abstract

Prion diseases, or transmissible spongiform encephalopathies (TSEs), are a group of chronic, progressive, and fatal neurodegenerative disorders that affect a variety of mammalian species. This chapter discusses the issues raised by two foodborne prion diseases, namely bovine spongiform encephalopathy (BSE) and variant Creutzfeldt–Jakob disease (vCJD), particularly those related to their spread in cattle and humans, the contamination of specified risk material (SRM) in meat, the relevant regulations, and appropriate detection methods for surveillance.

Keywords: Bovine spongiform encephalopathy, variant Creutzfeldt–Jakob disease, specified risk material, central nervous system tissues, regulations, detection methods

1. Introduction

Prion diseases or transmissible spongiform encephalopathies (TSEs) are a group of chronic, progressive, and fatal neurodegenerative disorders that affect a variety of mammalian species. They include scrapie in sheep and goats, chronic wasting disease in deer and elk, bovine spongiform encephalopathy (BSE or so-called "mad cow disease") in cattle, and variant Creutzfeldt–Jakob disease (vCJD) in humans.[1] Since the term "prion" was coined by Dr. Stanley Prusiner in 1982,[2] the extensive research on prion diseases has investigated their molecular biology,[3] causes,[4] pathogenesis,[5] genetics,[6] types,[7] biochemical mechanisms,[8] and therapies,[9] among other aspects. A prion is a proteinaceous infectious particle that lacks nucleic acid.[7] Abnormal prions (PrP^{Sc}) are the pathologic isoforms of prions and are expressed in specified risk material (SRM) such as bovine central nervous system (CNS) tissue.[10] PrP^{Sc} has a high β-sheet content, is extremely resistant to heat and proteases, and is insoluble in non-denaturing detergents.[11, 12] Infection of prion diseases occurs naturally via the oral route as well as by blood transfusion and maternal routes.[13] This chapter reviews the research into two foodborne prion diseases, BSE and vCJD, related to their spread in cattle and humans, the contamination of SRM in meat, the relevant regulations, and the detection methods available for surveillance.

2. Spread of BSE and vCJD

Research into clinical BSE cases has reported that 84.3% of the BSE infectivity is associated with bovine CNS tissue.[14] Another study found that the CNS tissue carries 89.7% of the infectivity.[15] With regard to the pathogenesis of BSE, animal by-products such as meat and bone meal (MBM), which have been used as nutritional supplements in livestock feedstuffs for more than 100 years, are major carriers of the infectious agent PrP^{Sc}.[16, 17] Significant changes in the manufacturing processes of feedstuffs that took place in the 1970s, including the introduction of mechanical systems that permitted continuous flow production and solvent extraction of fats, enabled PrP^{Sc} to enter the livestock feed chain.[18, 19] The recycling of animal by-products from BSE-infected animals caused a wide spread infection of BSE in cattle,[19] subsequently affecting the food chain through the consumption of the PrP^{Sc}-contaminated meat products.[20] To date, more than 190,000 BSE cases have been reported in 28 countries since it was first identified in the UK in around 1986,[21] with approximately 97% of the cases being reported in the UK (Figure 1).[22] It has been estimated that the total number of BSE-infected cattle is around one million, of which around 75% entered the human food chain during the 1980s and early to mid-1990s.[23, 24] BSE has had a huge impact on the beef industry worldwide.[25-30] In the USA, estimates of the total loss of beef and offal exports ranged from \$3.2 billion to \$4.7 billion in 2004 after the nation's first BSE case was confirmed at the end of 2003.[31] Although the link between BSE and vCJD through eating prion-contaminated meat has not been conclusively proven, the consumption of food of bovine origin contaminated with PrP^{Sc} has been strongly linked to the occurrence of vCJD in humans.[20] Both vCJD and BSE share the same infectious agent, PrP^{Sc}.[32, 33] A total of 229 vCJD patients have been identified in 12 countries since it was first discovered in the UK in 1996 (Figure 1),[34, 35] of whom about 77% were in the UK (Figure 1).[36] It should be noted that following the successful containment of the BSE epidemic through the imposition of strict feed controls for ruminant animals, the numbers of cases of both BSE and vCJD in the UK have declined significantly since 1992 and 2000, respectively (Figure 1).

3. Contamination of SRM in meat products

During the animal slaughter process, two major SRM, brain and spinal cord material, can easily contaminate meat products.[37-40] In general, there are four main pathways for the contamination of CNS-based SRM in meat products, which can occur at any stage during these processes: animal stunning, animal pithing, carcass dressing, and advanced meat recovery. The risks associated with each process have been studied and are discussed in turn below in

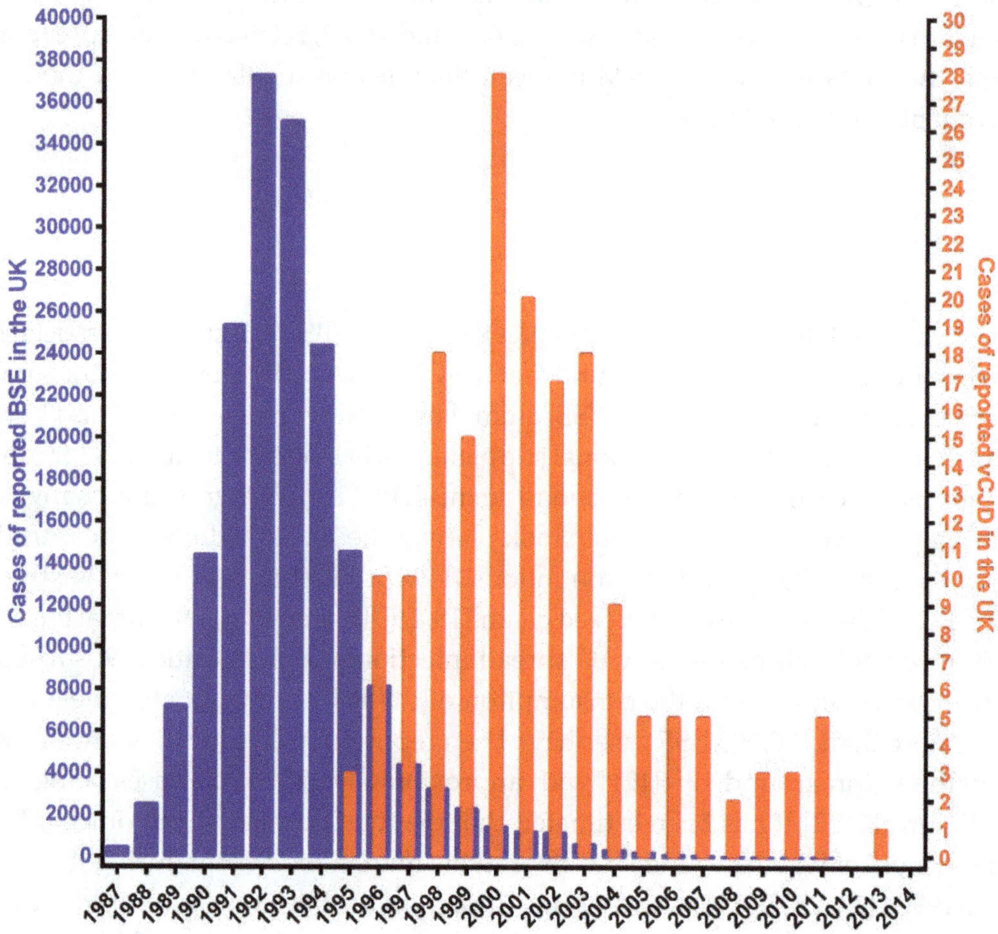

Figure 1. Number of reported BSE-cases (blue) and number of reported vCJD-cases (red) in the UK (1987-present)[22, 36]

the following sections. The major results of the relevant studies for each of these four pathways are summarized in Table 1.

3.1. Animal stunning

The stunning method has been widely used in many countries and areas of the world for some time as it slaughters the animals humanely. Stunning renders the livestock unconscious before slaughter, but because the heart of the stunned animal continues pumping for several minutes after stunning, during this time any CNS tissue that enters the jugular venous blood could still be spread throughout the body, contaminating both muscle and bone marrow via the blood circulation.[41] Several CNS markers that have a molecular weight similar to the PrP^{Sc} (MW: 30 kDa),[3] including syntaxin 1B and glial fibrillary acidic protein (GFAP),[42, 43] have been detected in non-CNS tissue after stunning. It thus seems likely that PrP^{Sc} could also be present in the edible carcass after stunning. Two types of captive bolt stunner, penetrative and non-penetrative, are widely used to stun domestic animals prior to bleeding (Figure 2). Most abattoirs prefer to use penetrative captive bolt stunners, either with or without air injection, to render cattle quickly and painlessly unconscious before slaughter.

Routes by which meat becomes contaminated by SRM			Location of SRM	SRM-contaminated samples (%)	Species		Detection methods						References
					Cattle	Sheep	Microscopy	Macroscopy	Immunoassay				
									ELISA[a]	IB[b]	IHC[c]	ICH[d]	
Animal stunning	Penetrative stunning with air injection		Left and right branches of the main pulmonary artery	2.5–5	√			√					[46]
			Right ventricle	33 (*n*[h]=1050)	√			√					[51]
			Jugular vein	26.7 (*n*=15)	√		√		Syntaxin 1-B Annexin V			S100β	[48]
			Jugular vein	13.3 (*n*=15)		√	√		Syntaxin 1-B			S100β	[49]
			Aorta	18.2 (*n*=11)		√						NF[e] S100	[47]
			Heart, lung, liver and kidney					√					[52]
	Penetrative stunning without air injection	Pneumatic-powered stunning	Heart	12 (*n*=450)	√			√					[51]
		Cartridge-operated stunning	Heart	1 (*n*=480)	√			√					[51]
			Jugular vein	13.3 (*n*=15)		√						S100β	[49]
			Pulmonary artery and right ventricle	8.5 (*n*=412)				√					[56]
			Jugular vein	4 (*n*=100)	√		√		GFAP[f]			S100	[55]
			Jugular vein	1 (*n*=360)	√				GFAP				[178]
			Carcass surface	100 (*n*=30)	√				GFAP				[43]
	Non-penetrative stunning		Jugular vein	2 (*n*=100)	√		√		GFAP			S100	[55]
			Carcass surface	83.3 (*n*=30)	√				GFAP				[43]
Animal pithing			Jugular vein	6.3 (*n*=16)	√		√		Syntaxin 1-B Annexin V			S100β	[48]
			Pulmonary artery and right ventricle	4.1 (*n*=314)				√					[56]
Carcass dressing			Hand-held screen, tray, apron, and internal surfaces of the carcass along the cut vertebral surface		√				S100β				[69]
			Sawing water, areas of thoracal and cervical regions of the carcass		√					NSE[g]			[56]

Routes by which meat becomes contaminated by SRM	Location of SRM	SRM-contaminated samples (%)	Species		Detection methods						References
			Cattle	Sheep	Microscopy	Macroscopy	Immunoassay				
							ELISA[a]	IB[b]	IHC[c]	ICH[d]	
	Captive bolt, floor of the receiving platform, aprons, hands, knives, captive bolt aperture, head wash water, spinal cord channel, saw, table striploin, conveyor belt, and beef hide		√				Syntaxin 1B GFAP				[42]
	Captive bolt gun/hole, apron, hands, saw, spinal cord channel, and meat		√				Syntaxin 1B GFAP				[67]
	Carcasses, loin, forerib, minced meat, worktable surface, saw, and air samples		√				S100β GFAP				[179]
	Split vertebral face and saw		√				GFAP				[69]
AMRS[i]	Meat	35	√								[77]
	Meat	41.2 (n=17)	√						NF Synapto physin GFAP		[75]
	Meat	17.2 (n=279)	√						GFAP		[76]

[a]ELISA, enzyme-linked immunosorbent assay; [b]IB, immunoblotting; [c]IHC, immunohistochemistry; [d]ICH, immunocytochemistry; [e]NF, neurofilament; [f]GFAP, glial fibrillary acidic protein; [g]NSE, neuron-specific enolase; [h]n, sample size; [i]AMRS, advanced meat recovery system.

Note: NF is heteropolymer intermediate filaments that are composed of three neurofilament subunits (NF-L, NF-M and NF-H), which are expressed mainly in neurons.[180]

S100 protein (MW: 10.5 kDa) is a calcium-binding protein found predominantly in the vertebrate nervous system.[181] Syntaxin 1B (MW: 33 kDa) which is an integral membrane protein exclusively and abundantly expressed in nervous tissues.[182]

Table 1. Summary of the results of SRM contamination through the four major routes during animal slaughtering and meat processing

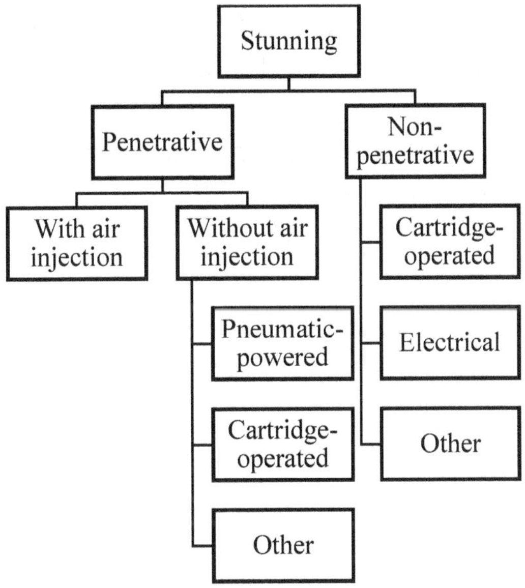

Figure 2. Classification of animal stunning techniques

3.1.1. Penetrative stunning with air injection

This stunning technique involves the injection of compressed air into the cranium of cattle to effectively disrupt the brain structure. In 2001, it was estimated that 15% of cattle were stunned using an air-injection stunner in the USA.[44] However, penetrative stunning with air injection poses a major risk of spreading the PrP^{Sc} from the BSE-infected CNS tissue to the edible carcass. Similarly, it is well known that severe penetrating injuries to the human brain, such as a gunshot wound, can cause acute brain damage and produce cerebral emboli that end up in the lung.[45]

For each BSE-infected animal stunned with this type of stunner, there is a 31.2, 16.3, 3.3, and 0.7% probability that a portion of the PrP^{Sc} would be transferred to the blood, heart, lung, and liver, respectively,[44] which is significant because it has been estimated that 60% of the liver, 50% of the heart, 25% of the kidney, and 5% of the blood are potentially available for human consumption.[44] Studies have also found that an air-injection stunner forced visible pieces of CNS tissue into the circulatory system of stunned cattle and sheep.[46, 47] This device has been shown to cause the formation of grossly visible brain tissue in the left and right branches of the main pulmonary artery in 2.5–5% of cattle,[46] in the jugular venous blood of cattle and sheep,[48-50] and in the aortic blood from sheep.[47] Particles of brain tissue as large as 20 μm in diameter can pass through the ovine pulmonary capillary network and enter the systemic arterial circulation during stunning, suggesting that this technique could easily disperse PrP^{Sc} throughout the edible carcass.[47] It has also been reported that CNS tissue has been observed in 33% of cattle hearts,[51] and in the kidneys of cattle stunned with air-injection stunners.[52]

Another factor to consider is that on occasion when operation of air-injection stunners fails, leading to an increase of CNS emboli formation and deposition, the probability of PrP^{Sc} transfer could be nearly 50% higher in blood, twice as high for the heart, and 10 times higher for the lung and liver.[44] Furthermore, a slaughterer who uses this type of stunner for too long or uses it successively to immobilize the same cattle could significantly increase the CNS tissue dispersion in blood vessels of the carcass.[51] It is also common for this type of stunner to press brain material out of the brain cavity through the bolt hole in the animal skull and splash onto the operator's face.[41, 53] Based on these results, the use of this type of stunner has been prohibited in many BSE-reporting countries and areas.

3.1.2. Penetrative stunning without air injection

This type of captive bolt stunner has a sharp-rimmed steel bolt and is powered by either compressed air (pneumatic-powered) or a blank cartridge that produces sufficient penetration force to initiate trauma to the cortex. It is the preferred tool for stunning cattle in the EU; 79 and 96% of the slaughterhouses in Europe use this method for cattle and calves, respectively.[54] However, the use of pneumatic-powered stunners and cartridge-fired stunners produced visible CNS clots in 12 and 1% of the cattle heart samples, respectively.[51] In addition, after using this type of stunner, CNS tissue was detected in the jugular venous blood from sheep,[49, 50] in venous blood from cattle,[55] in bovine pulmonary arterial system,[56] and on the surface of beef carcasses.[43]

In a study designed to trace the spread of CNS tissue in the cattle body after stunning with a cartridge-fired stunner, a marker microorganism, *Pseudomonas fluorescens*, was inoculated into the bovine brain via the stunning procedure.[57] The researchers reported finding this bacterium in the animal's blood, spleen, liver, kidney, lymph node, lung, and spinal cord. Using a similar approach, two microorganisms, *Escherichia coli* and *P. fluorescens*, have been detected in the blood, liver, lung, spleen, lymph node, longissimus muscle, and on the carcass surface of stunned sheep.[58] These studies demonstrate that the use of penetrative stunning without air injection could still spread CNS tissue throughout other portions of the carcass.

As with air-injection stunning, multiple use of this type of stunner also poses a high risk of spreading PrP^{Sc} from the infected brain to the edible carcass. In 2004, as few as 71% of slaughterhouses in the USA using penetrating captive bolt stunning successfully stunned 99–100% of the cattle with one shot and only 50% of abattoirs in Canada achieved the same target.[59] In 2005, this percentage decreased to 55% in US beef plants.[60] Since then, the percentage has increased gradually in both countries, reaching 91% in the USA by 2011.[61] In the EU, about 4–6.6% of captive bolt stunning in cattle requires a second stun.[54] Such practices could increase the risk of CNS tissue contamination in meat products. Although this type of stunner exerts less impulsive force on the stunned animal compared with air-injection stunners, it remains a threat to meat safety because this type of stunner is still in use worldwide.

3.1.3. Non-penetrative stunning

Non-penetrative stunning devices include the non-penetrating cartridge-operated stunner and the electrical stunner. Several studies have reported that no CNS tissue was detected in either aortic or venous blood samples collected from cattle or sheep after using these two devices.[48-50] However, other studies have detected brain tissue in venous blood samples from stunned cattle,[55] and on the surface of beef carcasses[43] after non-penetrating stunning. There also is an increased risk to operators as some animals may be inadequately stunned with non-penetrating stunning and recover their consciousness during slaughtering. Electrical stunning is the main alternative method for stunning cattle, but it is expensive to install and meat quality defects such as petechial hemorrhages have been linked to electrical stunning in sheep.[62]

3.2. Animal pithing

Pithing is the insertion of an elongated rod-shaped instrument into the cranial cavity of a stunned animal to further lacerate the CNS tissue.[63, 64] This operation prevents the animal stunned by a penetrating captive bolt from recovering permanently.[52] Pithing is considered by the industry to be a more effective technique for maintaining the safety of the operator and was used in 70% of UK abattoirs as recently as 1997.[50] However, in 2001, this technique, which has never been used in the USA,[52] was banned in the EU.[65] Brain tissue has been detected in venous blood[48, 50] and arterial blood samples[56] from cattle after pithing.

Based on scientific evidence of the dispersion of CNS-based SRM in response to various stunning and pithing techniques, the EU has ranked the stunning methods used in ruminants, from the most hazardous to the least dangerous, as: (1) penetrative stunning with air injection; (2) penetrative stunning without air injection; (3) captive bolt stunner with pithing; (4) captive bolt stunner without pithing and free bullet; (5) non-penetrative stunner and electro-narcosis.[66] Two techniques, air-injection stunning and pithing are now prohibited worldwide. However, there is still a 50% probability that emboli will be deposited in the blood of an animal stunned with a non-air-injection stunner.[44]

3.3. Carcass dressing

During animal slaughtering, the general sequence followed during the dressing procedure is stunning followed by exsanguination, dehiding, evisceration, splitting, washing, and chilling.[42] Of these steps, stunning is not the only point at which the carcass can potentially be contaminated with CNS tissue; other steps in the slaughtering procedure also have a high probability of dispersing CNS-based SRM across the carcass (Table 1). CNS tissue has been found in samples collected at a number of points along the slaughter line, including on the captive bolt gun, on the aprons worn by the operators, on their hands and the saws used, and in the longissimus muscle.[67]

In a study tracking two marker microorganisms, *E. coli* and *P. fluorescens*, both bacteria were transferred from the first stunned sheep to the stun wounds of the next 10 sheep stunned using the same contaminated penetrative cartridge-operated pistol.[58] These bacteria were also detected in the sheep's blood and on their carcass surfaces.[58] Perhaps most worryingly, the marker organisms were found in the air, and on the hands and aprons of the operator.[58] In another similar study,[57] the marker organism *P. fluorescens* was detected in the slaughter environment immediately after stunning and at each subsequent stage of the slaughter-dressing process, including on the hands of the operators, the slaughter equipment (captive bolt gun and knife), the cattle hide, and the carcass splitting saw.

Spinal cord material is easily spread to bovine carcasses during carcass splitting. The majority of the CNS contamination was found on the internal surfaces of the carcass and along the cut vertebral surface, with lesser levels of contamination inside the body cavity.[68] CNS tissue was also detected in the surroundings during splitting, including on the hand held screen, the tray and apron, the captive bolt, the captive bolt aperture, and the floor of the receiving platform.[42, 69] Spinal cord tissue can also be transferred to subsequent bovine carcasses during carcass splitting. The main risk of subsequent carcass contamination comes from the splitting saw.[68] Overall, these studies demonstrate that the practice of carcass dressing, especially splitting, can lead to the extensive spread of SRM within the abattoir environment, contaminating equipment, surfaces, operators, and edible carcasses. There are also grounds for concern regarding operator safety when dealing with CNS tissue, particularly due to the aerosol of cerebrospinal fluid and spinal cord produced during the splitting process.

3.4. Advanced meat recovery

Mechanical systems have been developed to separate meat from bone by scraping, shaving, or pressing the meat from the bone, for example, via advanced meat recovery systems (AMRS). AMRS enables processors to remove any remaining muscle tissue from beef carcasses without breaking the bones.[70] About 70% of fed cattle and hogs, and 60% of cows slaughtered in the USA were processed using AMRS in 1998[71] and around 5,000 metric tons of mechanically recovered meat (MRM) were produced yearly during the period 1980–95 in the UK.[72] Although AMRS makes deboning more efficient and considerably faster than using a knife, if spinal cord material is attached to the spinal column as it enters these machines, it can be incorporated into the meat product produced.

A number of studies have detected CNS tissue in MRM products (Table 1).[73-76] In 2002, about 35% of US bovine MRM contained CNS tissue[77] and other studies detected spinal cord tissue in 29%[77] and 33.3%[75] of bovine MRM samples. Dorsal root ganglia tissue was found in 10% of the MRM samples.[77] Schmidt *et al.*[74] reported that 50% of the bovine MRM samples from 14 slaughter plants that they tested were positive for CNS protein. In a later study by the same group,[76] 17.2% of the MRM samples tested were contaminated by CNS tissue.

By the 1990s, about 90% of the beef MRM produced was being used in burgers in the UK.[78] Around 75–80% of individuals surveyed in France consumed MRM products from burgers, and the consumption of burgers increased by 40% over the period 1980–1995.[79] Frequent consumption of beef and beef products, including burgers and meat pies, containing the MRM or head meat has been linked to an increased risk for vCJD;[80] infection with vCJD is thought to be predominantly due to exposure to BSE in beef MRM and head meat products.[81]

Looking at the results of the present review (Table 1), every stunning method except for electrical stunning could potentially contaminate meat products by dispersing CNS-based SRM, including any PrP^{Sc} present in sub-clinically infected cattle. Although the use of air-injection stunning, pithing, and AMRS has now been prohibited during animal slaughtering and meat processing (Table 2), penetrative stunning without air-injection and non-penetrative stunning are still in widespread legal use. Slaughter conditions and procedures such as carcass dressing, especially the splitting process, can also result in widespread contamination within abattoirs, contaminating equipment, surfaces, operators, and carcasses destined for human consumption. CNS contamination of bovine carcasses simply cannot be eliminated by the slaughtering technology currently used.

4. The regulation of SRM

The contamination of animal carcass by SRM is not only a food and feed safety issue, but also of global economic importance, leading countries or areas such as the UK, the EU, Japan, Canada, and the USA to institute strict regulations to prevent SRM entering the human food/ animal feed chain (Table 2). Although SRM has different definitions in different countries and areas, bovine CNS tissue (brain and spinal cord) tops every list of prohibitive materials due to its high prion infectivity.

Prohibited activity	UK	EU	Japan	USA	Canada
Air-injection stunning	April 1, 2001	April 1, 2001	N/A*	January 12, 2004	July 24, 2003
Pithing	April 1, 2001	January 1, 2001	N/A	Never be used	July 24, 2003
AMRS	December 15, 1995	October 1, 2000	N/A	January 12, 2004	July 24, 2003
SRM in food	December 17, 1997	January 1, 1998	July 4, 2002	January 12, 2004	August 23, 2003
SRM in feed	January 1, 1998	January 1, 2001	October 2001	April 27, 2009	July 12, 2007

* N/A, not available

Table 2. Legislation related to prion disease control

Given that it was the first and most seriously area affected, it is not surprising that the early legislation to control the spread of prion diseases came from the UK. In 1995, the UK banned the use of MRM from bovine vertebral column for human consumption[82] and in 1996, the "over thirty months" rule came into force, which automatically banned older cattle from entering the human food chain.[83] In 1997, the UK enacted a comprehensive set of SRM regulations,[84] which classified SRM into specified sheep or goat material and specified bovine material. It also emphasized that no person should use or sell any SRM, or any food containing SRM, for human consumption. To further prevent SRM from entering the human food chain, the UK prohibited the practice of pithing in 2001.[85] The EU was also quick to take action; since the consumption of BSE-infected feedstuffs by ruminants was rapidly identified to be the main BSE transmission channel, in 1994 the EU banned the use of proteins originating from mammalian tissues for feeding ruminants.[86] Largely as a result of this feed ban, the UK's BSE risk status dropped from high to low between 1996 and 2012. Since 2013, the regulations have been relaxed somewhat and cattle aged over 72 months (O72M) are now permitted to enter the food chain if they have tested negative for BSE.[87]

In 2001, the World Health Organization recommended that all countries should introduce risk management procedures such as the identification and removal of entire bovine heads and/or prohibiting the harvesting of all MRM. All tissues that have been shown to be capable of carrying BSE infectivity should be removed and destroyed. If the risk is high, additional precautions should be taken such as prohibiting cattle over a certain age from entering the food chain.[88] The World Organization for Animal Health suggests that when countries import beef products from a BSE risk country or area, meat processing should not use air-injection stunning or pithing.[89]

In the EU, the use of air-injection stunning has been banned for slaughtering cattle since 2001;[90] it has also prohibited pithing since 2001.[65] In 1997, the EU enacted strict regulations prohibiting the use of SRM,[91] including the skull, brain, eyes, tonsils and spinal cord of cattle, sheep, and goats aged over 12 months and the spleens of sheep and goats. This was later extended to include bovine intestines in 2000[92] and vertebral columns in 2001.[93] Since 2000, all the member states of the EU, including the UK, are prohibited from producing MRM from bones of the head and vertebral columns of bovine, ovine, and caprine animals,[65] and this was extended to include all types of bones from these three species in 2001.[93] From 2001 onwards, SRM was excluded from the feed chain as a result of an EU-wide ban on the feeding of processed animal protein to all farmed animals.[94] In 2013, BSE testing in the EU has been changed to the O72M rule.[87]

Among the Asian countries, Japan has been the most severely affected by TSE, with about 36 cases of BSE and one case of vCJD being confirmed since 2001.[21, 95] In 2002, Japan announced that bovine MBM could not be used as an ingredient in animal feedstuffs and prohibited the use of specified materials from cattle.[96] However, the bovine vertebral column can still be consumed as food if it is derived from cattle originating from a BSE-free country or zone.[97] Because the BSE risk status has moderated somewhat, the age of cattle subject to inspection for BSE in Japan was revised upward from zero months to 21 months in 2005.[98] In 2013, it was further revised and is now only required for cattle over 48 months of age.[99]

In Canada, the first cases of BSE and vCJD were reported in 2003 and 2002, respectively, triggering a food directorate policy on SRM in the food supply that was implemented in 2003.[100] This required the removal of SRM from all cattle during the slaughtering process and prohibited its sale or import for human consumption.[101] The vertebral column from all cattle aged 30 months or older must now be removed as an inedible product and cannot be used in the preparation of MRM.[102] Canada has also prohibited the use of air-injection stunning and the pithing technique for cattle.[102]

The USA banned the use of proteins originating from ruminant tissues for feeding ruminants in 1997.[103] Later, the USA issued a prohibition of SRM consumption as food in January 2004, adopting a definition for SRM that is similar to that used in Canada.[104, 105] To ensure that AMRS do not become a means of spreading CNS tissue into meat products, the USA has also prohibited the use of brain, spinal cord, dorsal root ganglia, trigeminal ganglia, and significant amounts of bone solids or marrow of all cattle, as well as the skull and vertebral column of cattle 30 months of age and older, in AMRS.[106] The National Animal Identification System, an efficient and effective animal identification program, has been implemented in the USA since 2004 to support animal disease monitoring, surveillance, and eradication programs.[107] The USA has also prohibited the use of a number of cattle materials, including CNS-based SRM, in animal feedstuffs since 2009.[108] The USA currently prohibits the slaughter of cattle that are unable to stand or walk ("downer" cattle) when presented for pre-slaughter inspection[109] and in January 2004 also banned air-injection stunning of cattle; pithing has never been used.[52] From January 2004 to May 2005, the USDA performed inspection and verification procedures in about 6,000 meat and poultry establishments and found 1,036 procedures (< 1%) that were not in compliance with the regulations related to SRM.[110] In 2008, about 1.5% of the US companies handling materials prohibited from use in ruminant feed (n–7,876) committed technical violations, mostly consisting of minor recordkeeping lapses or conditions involving non-ruminant feeds.[111] However, in January 2006, Japan suspended all US beef imports after discovering SRM in beef products exported from the USA.[112] In South Korea, US beef imports did not resume until June 26, 2008, after having been suspended in 2003 after the first BSE case was reported in the USA.[113]

Based on statistical information related to MBM,[114] it is clear that the problems with SRM-contaminated feed should not be forgotten. In 2000, the total amount of mammalian MBM produced in the USA was about three million metric tons.[115, 116] Ruminant tissues have been prohibited in ruminant feed in the USA to prevent the spread of BSE since 1997[103] and about 41,520 metric tons of animal feed in the USA were recalled between 2006 and 2007 due to the omission of the cautionary BSE statement on the label or as a result of ruminant MBM contamination during the feed processing procedure.[117] As the data shown in Table 1, it is very easy for meat products to become contaminated by bovine CNS-based SRM when the animal is slaughtered and during meat processing. In 2012, about 53% of the red meat produced in the USA. was beef,[118] but fewer than 1.2% of the cattle slaughtered are tested for BSE each year.[119, 120] As part of the regulation enforcement, about 65,693 metric tons of ruminant SRM-contaminated cattle products in the USA have been recalled since 2003.[121]

5. Detection methods for surveillance of prion diseases

As international trade continues to increase, in order to successfully enforce food and feed safety regulations and avoid economic loss, many nonclinical methods for the detection of the contaminated products have been developed. Overall, these detection methods can be classified into two types: non-immunochemical detection (i.e., chromatography, spectroscopy, and polymerase chain reaction (PCR)) and immunochemical detection methods.

Among the non-immunochemical detection methods, chromatography is typically used to separate molecules based on differential absorption and elution, which involves the flow of a fluid carrier over an immobile absorbing phase. Using chromatography, brain-specific fatty acids such as lignoceric acid and the *cis/trans* isomers of nervonic acid[122, 123] had been used as the analyte for the detection of the presence of CNS tissue in meat products through their characteristic component patterns. Spectroscopic techniques are based on the unique absorbance profiles of the sample components at specific wavelengths of the electromagnetic spectrum.

Near-infrared spectroscopy[124] and attenuated total reflectance Fourier transform infrared spectroscopy[125] have both been used to analyze bovine spinal cord in ground beef. Some CNS fatty acids have also been used as markers in gas chromatography–mass spectrometry (GC–MS).[126, 127] However, these methods require expensive instruments and reagents and highly trained staff; their sophisticated nature and the laborious sample preparation involved have limited the utility of these instrumental methods for routine analysis. Alternatively, its ability to detect GFAP mRNA makes PCR a sensitive technique for the detection of bovine CNS tissue in meat products.[128-130] However, although PCR can achieve a very low detection limit, it suffers from serious drawbacks, once again requiring expensive instrumentation and reagents and expert technicians. PCR methods are also prone to contamination.[130-132]

Immunochemical detection methods based on the specific immunoreactions between an antibody and its target antigen have been extensively employed for the detection of CNS contamination in meat (Table 1), generally in the form of an enzyme-linked immunosorbent assay (ELISA),[76, 128, 133-143] immunoblotting[137-140, 144-149] or immunohistochemistry.[145, 146, 150-152] There are a number of advantages associated with using an immunoassay. For example, no serious instrumentation is required; it is easy to operate and it employs minimal reagents. In the case of ELISA, it not only has a large-scale screening and field test capability, but is also a rapid, specific, and sensitive technique. Several markers have been explored to detect bovine CNS tissue in meat products with immunoassays, either using monoclonal (mAb) or polyclonal (pAb) antibodies.

5.1. CNS markers for non-immunochemical detection methods

5.1.1. Cholesterol

Cholesterol (MW: 386.7 g/mol) is a sterol component of cell membranes, hormones, and bile acids. It has been reported that 85 g of brain tissue may contain as much as 2,640 mg of cholesterol, while the same amount of a meat sample will only contain up to 85 mg cholesterol.[153] Although cholesterol is not specific to CNS, it can serve as a useful marker for a screening test due to the low-cost and rapid procedures available for measuring it. Lucker *et al.*[146] pioneered the use of cholesterol to analyze CNS tissue in meat products, using it to detect the cholesterol content of 402 heat-treated meat products from different food markets in Germany. Sixteen field samples (4%) were identified as being possibly contaminated with CNS tissue using this procedure, but the presence of CNS tissue was confirmed by immunoblotting for both neuron-specific enolase (NSE) and GFAP in just 7 of these 16 meat products. This suggests that cholesterol is not a reliable marker for CNS tissue due to its non-tissue specificity. It can, however, be used to roughly screen for CNS residue in meat products even though the accuracy is much lower than that of markers such as GFAP and NSE.

5.1.2. Nervonic acid

Nervonic acid (15-tetracosenoic acid, MW: 366.6 g/mol), a monounsaturated long chain fatty acid (C 24:1), is enriched in nervous tissue and is mainly present in the sphingolipids of the brain. As it is seldom found in non-CNS tissue, nervonic acid has been used as a marker indicating the presence of CNS for the detection of SRM in meat products by chromatographic techniques. Based on the different ratios of *cis/trans* isomers of nervonic acid in different animal species, nervonic acid has also been used to differentiate CNS tissue from various animal species in meat products using an on-line liquid chromatography–gas chromatography (LC–GC) method developed by Barcarolo *et al.*[123] The authors reported that the ratio of the *cis/trans* isomers can be used to provide a rough estimate of the age of the animal but may not be an accurate method for the detection of CNS adulterant. Biedermann *et al.*[154] analyzed the concentrations of fatty acids typical in bovine CNS, including docosahexaenoic acid (C22:6), lignoceric acid (C24), nervonic acid, and cerebronic acid ($C_{24}OH$), using GC–MS to determine the CNS content The detection limit for CNS using this GC–MS method was 0.01%. In 2003, Agazzi *et al.*[122] reported an enhanced GC method for the detection of isomers of nervonic acid in meat products, reporting that the ratio of the *cis/trans* isomers of nervonic acid not only exhibit species specificity but also show a significant difference between cow and calf, and between pig and piglet.

Biedermann *et al.*[127] went on to study the structural characterization of nervonic acid as a marker for SRM using GC–MS and found that the ratio of *cis/trans* isomers used previously should be replaced by the ratio of positional isomers, i.e., the ratio of ω9/ω7-nervonic acid (15c-C24:1/17c-C24:1). Overall, although nervonic acid may be a specific and stable marker, the GC–MS-based CNS method is costly, requires a lengthy analysis time, and the data interpretation is highly complex. In addition, the lack of species and age specificity of nervonic acid for the detection of CNS disqualifies the GC–MS-based method as an effective method.

5.1.3. *Glial Fibrillary Acidic Protein (GFAP) mRNA*

In the CNS, GFAP (MW: about 50 kDa) is the major component of the filaments found in the astrocytes that support the functions of nerve cells.[155] In the peripheral nervous system (PNS), GFAP is mainly expressed by Schwann cells. The PCR method has been used in several studies to detect GFAP mRNA as a potential marker of CNS tissue contamination in meat products.

Seyboldt *et al.*[130] developed a reverse transcription-PCR (RT-PCR) assay coupled with restriction fragment length polymorphism for the detection of GFAP mRNA from bovine CNS tissue. Although this approach suffered from cross-reactions with unheated heart and skeletal muscle tissues, heat treatment (70 °C for 20 min) prior to RNA extraction was found to reduce the cross-reaction with a detection limit of 0.5% (g/g) heated bovine brain in bovine minced meat. In another study[143], a similar method was able to detect 0.25% (g/g) bovine CNS tissue in pork liver sausage (after a heat treatment of 80 °C for 80 min) that had been stored up to 28 days at 4 °C, although native peripheral nerves showed positive results.

Abdulmawjood and co-workers[129, 131, 156] developed a quantitative real-time RT-PCR to detect GFAP mRNA. Their system was capable of detecting down to 0.1% of CNS tissue in strongly heated (120 °C for 15 min) or medium-heated (80 °C for 90 min) meat samples (50% pork and 50% beef), and 0.2% of CNS tissue in pork liver sausages (80 °C for 1 h). However, low levels of GFAP could still be detected in peripheral nerves (sciatic and axillary nerves) and non-neuronal tissues (parotid gland, pancreas and adrenal gland).

5.2. CNS Markers for immunochemical detection

5.2.1. GFAP

The immunochemical differences between CNS-GFAP and PNS-GFAP are well known [157], and GFAP was first exploited to indicate the presence of brain or spinal cord tissues in meat as early as 1999 [134]. It has since been utilized as a marker for the detection of CNS tissue in meat products in a number of immunoassay systems [74, 76, 133, 141, 158]. Schmidt *et al.* were the first to develop a colorimetric sandwich ELISA in 1999 [134] and went on to develop an improved fluorescent sandwich ELISA in 2001 [135]. Although this fluorescent method was capable of detecting the presence of as little as 0.05% (g/g) bovine brain and spinal cord in beef, once again this assay suffers from a cross-reaction with PNS-GFAP, which is present in sciatic nerve and cervical ganglion tissue and is not considered an SRM [135]. These results suggest that the anti-GFAP pAb used in their assay was not actually specific to the CNS tissue.

GFAP has also been used as a CNS marker in a number of immunoblotting techniques [137, 145-147]. In a study by Lucker *et al.* [146], the immunoresponse of heat-treated samples was reduced significantly, however, and the sensitivity of the assay was inadequate when intensively heat-treated meat products were analyzed. Several other studies have also shown that the ability to detect GFAP is diminished when meat samples are subjected to severe heat treatment [133, 135].

5.2.2. Neuron-Specific Enolase (NSE)

NSE (MW: 48 kDa) is an enzyme that is found in both neurons and neuroendocrine cells. The immunological CNS specificity of $\gamma\gamma$-enolase has been used in clinical research for a long time [159], and in 1999 Lucker *et al.* were the first to adapt this clinical method to permit the use of NSE as a marker for the detection of CNS tissue in meat products using immunoblotting [148]. The detection limit of this assay was 1% (g/g) spiked brain in cooked (100 °C for 1 h) sausages of the emulsion type. They suggested that the sensitivity of the assay could be dramatically improved by removing the sample's fat content (30 to 40%) using Soxhlet extraction, although this is exceptionally time-consuming, requiring 8 h to extract the fat.

Several commercial immunochemical test kits have been developed [160]. Hughson *et al.* [140] compared the performance of two commercial kits, the RIDASCREEN GFAP kit and the Brainostic NSE kit (Table 3), for the detection of CNS tissue in meat products; these kits were based on the methods proposed by Schmidt *et al.* [134] and Lucker *et al.* [148], respectively. Hughson *et al.* [140] reported that levels down to 0.1% (g/g) of ovine spinal cord tissues could be detected in raw lamb meat using the GFAP kit, while 1.0% (g/g) was detected by the NSE kit. The detection limits of the GFAP and NSE methods for cooked samples (80 °C for 40 min) were 0.2% (g/g) bovine brain in pork and 0.1% (g/g) bovine spinal tissue in pork, respectively. They noted that the GFAP assay was easier to use and to interpret the data than the NSE assay. In a similar study by Agazzi *et al.* [138], the sensitivity of the NSE test kit was reported as being 0.5% (g/g) CNS tissue in raw and medium-heated (80 °C for 20 min) meat samples, rising to 2.0% (g/g) in strongly heated (120 °C for 20 min) meat materials. Another study [146] also found that the NSE immunoreactivity in samples of pure brain and brain muscle dropped significantly with time at 100 °C, almost disappearing after a 2 h heat treatment. The properties of these commercial kits are summarized in Table 3.

Company	ScheBo Biotech AG	R-Biopharm		ELISA Technologies	Neogen	
Commercial kit name	Brainostic Test (GFAP-ELISA)	RIDASCREEN Risk Material	RIDASCREEN Risk Material 10/5	MELISA-TEK Ruminant Kit	Reveal for Ruminant in Feed	Reveal for Ruminant in MBM
Detection method	sELISA*	sELISA	sELISA	sELISA	LFA*	LFA
LOD (%)	0.1	0.2	0.1	0.5	1	2
Target products	Raw, processed (heated) meat and on contaminated surfaces	Processed (heated) meat and sausages	Raw meat, meat products and on contaminated surfaces	MBM	Ruminant	Ruminant
Sample preparation	Homogenization, extraction, swabs, dilution	Homogenization, extraction, centrifugation, dilution	Place the sampling swab into the sample dilution buffer	Extraction	Extraction	Extraction
Species selectivity	Unknown	Cattle, veal, sheep, goat, horse, poultry and pig	Cattle, veal, sheep, goat, horse, poultry and pig	Beef, sheep and water buffalo	N/A*	N/A
Tissue specificity	Unknown	No	No	Muscle	Muscle	Muscle
Minimum sample amount	50 mg	2 g	50 mg	5 g	10 g	10 g
Antigen#	GFAP*	GFAP	GFAP	Troponin-I	Troponin-I	Troponin-I
Test implementation	45 min	1 h	15 min	1 h	10 min	15 min
Tests possible per kit	42 (each in duplicate)	96 (including standards)	96 (including standards)	16 (each in triplicate)	25	25
Kit store temperature	4–8 °C	2–8 °C	2–8 °C	4–8°C	18–30 °C	18 – 30°C
Standards provided	0.1, 0.2, 0.4 and 1.0%	0, 0.2, 1.0 and 2.0%	0, 0.1, 0.2 and 0.4%	No	No	No

*sELISA: sandwich ELISA; GFAP: glial fibrillary acidic protein; LFA: immunochromatographic lateral flow assay; N/A: not available

Table 3. Commercial immunoassay kits for the detection of prohibited CNS and ruminant tissue contaminants in meat and feed products

5.2.3. *Myelin proteolipid protein*

In the CNS, myelin proteolipid protein (MPP, MW: 30 kDa) makes up approximately 50% of the myelin proteins [161]. This protein can be readily extracted from brain white matter using organic solvents [162]. Although MPP is the most abundant myelin protein in the CNS, mRNA for PLP is also expressed in the PNS and small amounts of MPP are synthesized [162]. This protein has also been used as a marker for the detection of CNS tissue in meat products [163]. The detection limit for CNS in raw meat can be as low as 0.025% using Western blot [164], and 0.01% bovine brain in raw minced beef using dot blot [165]. However, the antibody used in these studies was not species-specific and was unable to differentiate between mammalian species. Another problem is that MPP is not thermostable; the reaction signal decreased significantly when the heating time was increased to 95 °C. MPP was not detectable after 3 h of heat treatment at this temperature using dot blot [164].

5.2.4. *Myelin Basic Protein (MBP)*

MBP is located in CNS on the cytoplasmic face of myelin membranes in the white matter corresponding to the major dense line. MBP makes up about 30% of the total protein in the mammalian CNS myelin. It is the only structural protein found so far that is essential for the formation of CNS myelin [166]. As a water-soluble protein, MBP can be extracted with either acid or salt solutions [162]. There are four isoforms of MBP, each with a different molecular weight: 21 kDa, 18.5 kDa, 17 kDa and 14 kDa [162]. Different animal species have different isoforms. In particular, 18.5-kDa MBP, the major isoform protein in the CNS, is highly unfolded with essentially no tertiary structure in solution and the amino acid sequence of this major basic protein is similar in a number of animal species [162]. The sequence identity of the 18.5-kDa MBP in bovines and pigs is 93% homologous [167]. It has been reported that bovine 18.5-kDa MBP is not only very stable in solution at neutral pH (pH 7.2), but also highly thermostable [168]. Its immunoreactivity has been shown to remain clearly detectable even after severe heat treatment at 133 °C for 30 min [168].

MBP has been used as a marker for the detection of CNS contamination using immunoassays in several studies.[136, 143, 145, 147, 151] Levels as low as 1% (g/g) of bovine brain have been detected in a heated luncheon meat-type product (115 °C for 1 h) using a pAb in immunohistochemistry[151] and a mAb in immunoblotting.[147] In another study,[136] a detection limit of 10% was recorded for CNS tissue using a mAb in indirect competitive ELISA. Using an improved MBP extraction method,[143] compared with the detection limit reported using the same antibody and the same ELISA method,[136] the detection limit for brain tissue could be enhanced from 10 to 0.05% (~200 times). This improved extraction method will provide a useful analytical basis for the subsequent development of a specific immunoassay for the detection of bovine CNT in processed meat and feed products.

5.3. Detection methods for the control of prion diseases in feed products

To safeguard livestock, a number of non-immunochemical and immunochemical detection methods have been developed to detect the presence of banned animal materials in feeds. A high-performance liquid chromatography method has been developed that utilizes dipeptide carnosine and related dipeptides as the markers,[169] and several other studies have used near-infrared reflectance spectroscopy (NIRS) to predict the amino acid content in animal feeds.[170-172] PCR has also been used for the detection of animal tissue in feeds. For example, Gao *et al.*[173] amplified a highly conserved eukaryotic DNA region of the 18S ribosomal gene using multiplex PCR (MPCR) and were able to detect levels as low as 0.02% bovine MBM in feedstuffs. Bellagamba *et al.*[174] detected 0.25% ruminant or pig adulterants in fish meal by MPCR.

A number of immunoassays have been developed to detect prohibited ruminant animal materials in feed. In our laboratory, a panel of species-specific mAbs have been produced for the detection of animal tissues in both food and feed products. [175-177] A skeletal muscle protein, troponin I (sTnI, 24 kDa), was first identified as a thermal-stable species marker protein; sTnI maintains its solubility and antigenicity even after undergoing severe heat treatment (126 °C for 120 min).[177] Given its even distribution in skeletal muscles, sTnI appears to be an ideal marker, allowing antibodies to be developed for the detection of animal tissues in severely heat-treated samples such as MBM. sTnI-specific mAbs have demonstrated their usefulness in an indirect non-competitive ELISA for the detection of rendered muscle tissues in animal feedstuffs, with detection limits of the mammalian and ruminant assays being between 0.3 and 2%.[175] In 2004, a sandwich ELISA was reported for the detection of ruminant proteins in feedstuffs.[176] This assay used a capture mAb specific to bovine and ovine sTnI and a biotin-conjugated detection mAb that reacts to all animal sTnI. The optimized assay achieved detection limits for bovine and ovine sTnI as low as 5.0 and 4.0 ng/ml, respectively.[176] Based on the detection of sTnI, several commercial immunoassay kits are now available for the enforcement of ruminant feed ban regulations (Table 3).

6. Conclusions

In order to maximize the economic profit from animal food production, humans made livestock animal's cannibals in the 1970s. Only a few years later, this backfired spectacularly, resulting in the rapid spread of infectious PrP^{Sc} through both the feed and food chains. These practices created a huge man-made disaster, mad cow disease, which went on to enter the human food chain through poor animal slaughtering and meat processing practices. The devastating impact of BSE on cattle has been called "a punishment from God" from which the global beef industry is only now beginning to recover. Although the negative impacts of the human version of mad cow disease, vCJD, are less significant, they continue to receive a great deal of publicity and the risk of recurrence of the disaster cannot be ignored. Most countries and areas have now enacted legislation to prohibit the use of SRM in the food and feed chain, and the goal of everyone concerned is to eventually eliminate the threat of BSE completely. Currently, through the strict implementation of feed bans and the enforcement of the regulations, both types of foodborne prion diseases appear to be under control. However, the prevention of prion diseases is not simply a matter of food or feed safety, but is also a significant factor affecting a number of political and economic issues. Effective detection methods for ruminant SRM in rendered feedstuffs are still lacking, as currently none of the available assays can effectively differentiate ruminant CNS tissue from that of non-ruminant animal species, or detect bovine CNS tissue in excessively processed (e.g., 133 °C for 20 min) meat and feed products. It is therefore vital to continue to conduct scientific research in this area if we are to gain a better understanding of these destructive prion diseases and develop more effective surveillance techniques for this disease in both humans and animals. We must take to heart this solemn lesson if we are to avoid another such punishment from God.

Author details

Qinchun Rao and Yun-Hwa Peggy Hsieh*

*Address all correspondence to: yhsieh@fsu.edu

Department of Nutrition, Food and Exercise Sciences, Florida State University, Tallahassee, FL, USA

References

[1] Doherr, M. G., Brief review on the epidemiology of transmissible spongiform encephalopathies transmissible (TSE). *Vaccine* 2007, *25*, 5619-5624.

[2] Prusiner, S. B., Novel proteinaceous infectious particles cause scrapie. *Science* 1982, *216*, 136-144.

[3] Prusiner, S. B., Molecular biology of prion diseases. *Science* 1991, *252*, 1515-1522.

[4] Collinge, J., Prion diseases of humans and animals: Their causes and molecular basis. *Annual Review of Neuroscience* 2001, *24*, 519-550.

[5] Prusiner, S. B., Molecular biology and pathogenesis of prion diseases. *Trends in Biochemical Sciences* 1996, *21*, 482-487.

[6] Mead, S., Prion disease genetics. *European Journal of Human Genetics* 2006, *14*, 273-281.

[7] Prusiner, S. B., Prion diseases and the BSE crisis. *Science* 1997, *278*, 245-251.

[8] Cobb, N. J.; Surewicz, W. K., Prion diseases and their biochemical mechanisms. *Biochemistry* 2009, *48*, 2574-2585.

[9] Cashman, N. R.; Caughey, B., Prion diseases - Close to effective therapy? *Nature Reviews Drug Discovery* 2004, *3*, 874-884.

[10] Safar, J.; Roller, P. P.; Gajdusek, D. C.; Gibbs, C. J., Thermal-stability and conformational transitions of scrapie amyloid (prion) protein correlate with infectivity. *Protein Science* 1993, *2*, 2206-2216.

[11] Appel, T.; Wolff, M.; von Rheinbaben, F.; Heinzel, M.; Riesner, D., Heat stability of prion rods and recombinant prion protein in water, lipid and lipid-water mixtures. *Journal of General Virology* 2001, *82*, 465-473.

[12] Glatzel, M.; Stoeck, K.; Seeger, H.; Luhrs, T.; Aguzzi, A., Human prion diseases: molecular and clinical aspects. *Archives of Neurology* 2005, *62*, 545-52.

[13] McKenzie, D.; Aiken, J., Prion diseases. In *Guide to Foodborne Pathogens*, John Wiley & Sons: Chichester, West Sussex, UK, 2013; pp 399-413.

[14] Comer, P. J.; Huntly, P. J., Exposure of the human population to BSE infectivity over the course of the BSE epidemic in Great Britain and the impact of changes to the Over Thirty Month Rule. *Journal of Risk Research* 2004, *7*, 523-543.

[15] European Commission's Scientific Steering Committee. Opinion of the scientific steering committee on the human exposure risk (HER) via food with respect to BSE. http://ec.europa.eu/food/fs/sc/ssc/out67_en.pdf. (Access: October 19, 2015).

[16] Wilesmith, J. W.; Ryan, J. B. M.; Atkinson, M. J., Bovine spongiform encephalopathy - epidemiologic studies on the origin. *Veterinary Record* 1991, *128*, 199-203.

[17] Smith, P. G.; Bradley, R., Bovine spongiform encephalopathy (BSE) and its epidemiology. *British Medical Bulletin* 2003, *66*, 185-198.

[18] Taylor, D. M.; Woodgate, S. L.; Fleetwood, A. J.; Cawthorne, R. J. G., Effect of rendering procedures on the scrapie agent. *Veterinary Record* 1997, *141*, 643-649.

[19] Collee, J. G.; Bradley, R., BSE: a decade on - part I. *Lancet* 1997, *349*, 636-641.

[20] Almond, J.; Pattison, J., Human BSE. *Nature* 1997, *389*, 437-438.

[21] World Organisation for Animal Health. Number of reported cases of bovine spongiform encephalopathy (BSE) in farmed cattle worldwide. http://www.oie.int/?id=505. (Access: May 17, 2015).

[22] World Organisation for Animal Health. Number of cases of bovine spongiform encephalopathy (BSE) reported in the United Kingdom. http://www.oie.int/en/animal-health-in-the-world/bse-specific-data/number-of-cases-in-the-united-kingdom/. (Access: May 17, 2015).

[23] Ferguson, N. M.; Donnelly, C. A.; Woolhouse, M. E. J.; Anderson, R. M., The epidemiology of BSE in cattle herds in Great Britain.2. Model construction and analysis of transmission dynamics. *Philosophical Transactions of the Royal Society of London Series B-Biological Sciences* 1997, *352*, 803-838.

[24] Anderson, R. M.; Donnelly, C. A.; Ferguson, N. M.; Woolhouse, M. E. J.; Watt, C. J.; Udy, H. J.; MaWhinney, S.; Dunstan, S. P.; Southwood, T. R. E.; Wilesmith, J. W.;

Ryan, J. B. M.; Hoinville, L. J.; Hillerton, J. E.; Austin, A. R.; Wells, G. A. H., Transmission dynamics and epidemiology of BSE in British cattle. *Nature* 1996, *382*, 779-788.

[25] Wigle, R.; Weerahewa, J.; Bredahl, M.; Samarajeewa, S., Impacts of BSE on world trade in cattle and beef: Implications for the Canadian economy. *Canadian Journal of Agricultural Economics* 2007, *55*, 535-549.

[26] Miran, B.; Akgungor, S., The effect of mad cow (BSE) scare on beef demand and sales loss: The case of Izmir. *Turkish Journal of Veterinary & Animal Sciences* 2005, *29*, 225-231.

[27] Jin, H. J.; Koo, W. W., The effect of the BSE outbreak in Japan on consumers' preferences. *European Review of Agricultural Economics* 2003, *30*, 173-192.

[28] Iraizoz, B.; Bardaji, I.; Rapun, M., The Spanish beef sector in the 1990s: impact of the BSE crisis on efficiency and profitability. *Applied Economics* 2005, *37*, 473-484.

[29] Herrmann, R., Economic consequences of BSE in Germany. *Deutsche Tierarztliche Wochenschrift* 2002, *109*, 358-361.

[30] CNN. Clashes over U.S. beef in S. Korea. http://www.cnn.com/2008/WORLD/asiapcf/06/01/skorea.usbeef/index.html?iref=newssearch#cnnSTCText. (Access: June 6, 2015).

[31] Coffey, B.; Mintert, J.; Fox, S.; Schroeder, T.; Valentin, L. The economic impact of BSE on the U.S. beef industry: product value losses, regulatory costs, and consumer reactions. http://www.ksre.ksu.edu/bookstore/pubs/MF2678.pdf. (Access: June 6, 2015).

[32] Brown, D. A.; Bruce, M. E.; Fraser, J. R., Comparison of the neuropathological characteristics of bovine spongiform encephalopathy (BSE) and variant Creutzfeldt-Jakob disease (vCJD) in mice. *Neuropathology and Applied Neurobiology* 2003, *29*, 262-272.

[33] Hill, A. F.; Desbruslais, M.; Joiner, S.; Sidle, K. C. L.; Gowland, I.; Collinge, J.; Doey, L. J.; Lantos, P., The same prion strain causes vCJD and BSE. *Nature* 1997, *389*, 448-450.

[34] U.S. Centers for Disease Control and Prevention. Fact sheet: variant Creutzfeldt-Jakob disease. http://www.cdc.gov/ncidod/dvrd/vcjd/factsheet_nvcjd.htm. (Access: May 17, 2015).

[35] Will, R. G.; Ironside, J. W.; Zeidler, M.; Cousens, S. N.; Estibeiro, K.; Alperovitch, A.; Poser, S.; Pocchiari, M.; Hofman, A.; Smith, P. G., A new variant of Creutzfeldt-Jakob disease in the UK. *Lancet* 1996, *347*, 921-925.

[36] UK National Creutzfeldt-Jakob Disease Surveillance Unit. Creutzfeldt-Jakob disease in the UK. http://www.cjd.ed.ac.uk/documents/figs.pdf. (Access: June 7, 2015).

[37] Munro, R., Neural tissue embolism in cattle. *Veterinary Record* 1997, *140*, 536.

[38] Coore, R. R.; Barr, F. J.; McKinstry, J. L.; Anil, M. H., Neural embolism and cerebral venous drainage at stunning and slaughter. *Veterinary Record* 2004, *155*, 86-87.

[39] Anil, M. H.; Harbour, D. A., Current stunning and slaughter methods in cattle and sheep - Potential for carcass contamination with central nervous tissue and microorganisms. *Fleischwirtschaft* 2001, *81*, 123-124.

[40] Love, S.; Helps, C. R.; Williams, S.; Shand, A.; McKinstry, J. L.; Brown, S. N.; Harbour, D. A.; Anil, M. H., Methods for detection of haematogenous dissemination of brain tissue after stunning of cattle with captive bolt guns. *Journal of Neuroscience Methods* 2000, *99*, 53-58.

[41] Gregory, N. G., Recent concerns about stunning and slaughter. *Meat Science* 2005, *70*, 481-491.

[42] Prendergast, D. M.; Sheridan, J. J.; Daly, D. J.; McDowell, D. A.; Blair, I. S., Dissemination of central nervous system tissue from the brain and spinal cord of cattle after captive bolt stunning and carcass splitting. *Meat Science* 2003, *65*, 1201-1209.

[43] Lim, D. G.; Erwanto, Y.; Lee, M., Comparison of stunning methods in the dissemination of central nervous system tissue on the beef carcass surface. *Meat Science* 2007, *75*, 622-627.

[44] Cohen, J. T.; Duggar, K.; Gray, G. M.; Kreindel, S. Evaluation of the potential for bovine spongiform encephalopathy in the United States. http://www.aphis.usda.gov/newsroom/hot_issues/bse/background/documents/appendix1.pdf. (Access: June 7, 2015).

[45] Miyaishi, S.; Moriya, F.; Yamamoto, Y.; Ishizu, H., Massive pulmonary embolization with cerebral tissue due to gunshot wound to the head. *Brain Injury* 1994, *8*, 559-564.

[46] Garland, T.; Bauer, N.; Bailey, M., Brain emboli in the lungs of cattle after stunning. *Lancet* 1996, *348*, 610-610.

[47] Coore, R. R.; Love, S.; McKinstry, J. L.; Weaver, H. R.; Phillips, A.; Hillman, T.; Hiles, M. J.; Shand, A.; Helps, C. R.; Anil, M. H., Dissemination of brain emboli following captive bolt stunning of sheep: Capacity for entry into the systemic arterial circulation. *Journal of Food Protection* 2004, *67*, 1050-1052.

[48] Anil, M. H.; Love, S.; Williams, S.; Shand, A.; McKinstry, J. L.; Helps, C. R.; Waterman-Pearson, A.; Seghatchian, J.; Harbour, D. A., Potential contamination of beef carcases with brain tissue at slaughter. *Veterinary Record* 1999, *145*, 460-462.

[49] Anil, M. H.; Love, S.; Helps, C. R.; McKinstry, J. L.; Brown, S. N.; Philips, A.; Williams, S.; Shand, A.; Bakirel, T.; Harbour, D., Jugular venous emboli of brain tissue induced in sheep by the use of captive bolt guns. *Veterinary Record* 2001, *148*, 619-620.

[50] Anil, M. H.; Love, S.; Helps, C. R.; Harbour, D. A., Potential for carcass contamination with brain tissue following stunning and slaughter in cattle and sheep. *Food Control* 2002, *13*, 431-436.

[51] Schmidt, G. R.; Hossner, K. L.; Yemm, R. S.; Gould, D. H., Potential for disruption of central nervous system tissue in beef cattle by different types of captive bolt stunners. *Journal of Food Protection* 1999, *62*, 390-393.

[52] U.S. Department of Agriculture., 9 CFR Parts 310 and 313 Prohibition of the use of certain stunning devices used to immobilize cattle during slaughter. *Federal Register* 2004, *69*, 1885-1891.

[53] Taylor, K. C., Brain emboli in the lungs of cattle. *Lancet* 1996, *348*, 749-749.

[54] EU Civic Consulting. Study on the stunning/killing practices in slaughterhouses and their economic, social and environmental consequences-final report part I: red meat. http://ec.europa.eu/food/animal/welfare/slaughter/docs/report_parti_en.pdf. (Access: June 7, 2015).

[55] Coore, R. R.; Love, S.; McKinstry, J. L.; Weaver, H. R.; Philips, A.; Hillman, T.; Hiles, M.; Helps, C. R.; Anil, M. H., Brain tissue fragments in jugular vein blood of cattle stunned by use of penetrating or nonpenetrating captive bolt guns. *Journal of Food Protection* 2005, *68*, 882-884.

[56] Lucker, E.; Schlottermuller, B.; Martin, A., Studies on contamination of beef with tissues of the central nervous system (CNS) as pertaining to slaughtering technology and human BSE-exposure risk. *Berliner und Munchener Tierarztliche Wochenschrift* 2002, *115*, 118-121.

[57] Daly, D. J.; Prendergast, D. M.; Sheridan, J. J.; Blair, I. S.; McDowell, D. A., Use of a marker organism to model the spread of central nervous system tissue in cattle and the abattoir environment during commercial stunning and carcass dressing. *Applied and Environmental Microbiology* 2002, *68*, 791-798.

[58] Buncic, S.; McKinstry, J.; Reid, C. A.; Anil, M. H., Spread of microbial contamination associated with penetrative captive bolt stunning of food animals. *Food Control* 2002, *13*, 425-430.

[59] Grandin, T. 2004 Restaurant animal welfare audits of stunning and handling in federally inspected U.S. and Canadian beef, veal, pork, lamb, and poultry slaughter plants. http://www.grandin.com/survey/2004.restaurant.audits.html. (Access: June 7, 2015).

[60] Grandin, T. 2005 Restaurant animal welfare audits of federally inspected beef and pork slaughter plants. http://www.grandin.com/survey/2005.restaurant.audits.html. (Access: June 7, 2015).

[61] Grandin, T. 2011 Restaurant animal welfare and humane slaughter audits in u.S. Federally inspected beef and pork slaughter plants. http://www.grandin.com/survey/2011.restaurant.audits.html. (Access: June 7, 2015).

[62] Gilbert, K. V.; Devine, C. E., Effect of electrical stunning methods on petechial haemorrhages and on the blood pressure of lambs. *Meat Science* 1982, *7*, 197-207.

[63] Gracey, J., *Meat plant operations*. Chalcombe Publications: Southampton, UK, 1998.

[64] Geering, W. A.; Penrith, M.-L.; Nyakahuma, D. Methods of destruction (Part 1: Destruction of animals). http://www.fao.org/docrep/004/Y0660E/Y0660E01.htm#ch1.3. (Access: 16 April 2009).

[65] European Commission, Commission Decision of 29 June 2000 regulating the use of material presenting risks as regards transmissible spongiform encephalopathies and amending Decision 94/474/EC. *Official Journal of the European Communities* 2000, *L158*, 76-82.

[66] European Commission. Scientific report on stunning methods and BSE risks. http://ec.europa.eu/food/fs/sc/ssc/out247_en.pdf. (Access: June 7, 2015).

[67] Prendergast, D. M.; Sheridan, J. J.; Daly, D. J.; McDowell, D. A.; Blair, I. S., Dissemination of central nervous system tissue during the slaughter of cattle in three Irish abattoirs. *Veterinary Record* 2004, *154*, 21-24.

[68] Helps, C. R.; Fisher, A. V.; Harbour, D. A.; O'Neill, D. H.; Knight, A. C., Transfer of spinal cord material to subsequent bovine carcasses at splitting. *Journal of Food Protection* 2004, *67*, 1921-1926.

[69] Helps, C. R.; Hindell, P.; Hillman, T. J.; Fisher, A. V.; Anil, H.; Knight, A. C.; Whyte, R. T.; O'Niell, D. H.; Knowles, T. G.; Harbour, D. A., Contamination of beef carcasses by spinal cord tissue during splitting. *Food Control* 2002, *13*, 417-423.

[70] U.S. Department of Agriculture. Revised directive for advanced meat recovery systems. http://www.fsis.usda.gov/Frame/FrameRedirect.asp?main=http://www.fsis.usda.gov/Oa/background/amrdirec.htm. (Access: June 7, 2015).

[71] Sparks Companies Inc. Advanced meat recovery systems - an economic analysis of the proposed USDA regulation. http://www.fsis.usda.gov/OPPDE/rdad/FRPubs/03-038IF/AMRSparksAnalysis.pdf. (Access: June 7, 2015).

[72] UK Spongiform Encephalopathy Advisory Committee. Sources of BSE infection: historical uses of mechanically recovered meat. http://www.seac.gov.uk/papers/seac76_4.pdf. (Access: 30 March 2009).

[73] Hajmeer, M. N.; Cliver, D. O.; Marsden, J. L., Central nervous system tissue detection in meat from advanced meat recovery systems. *Meat Science* 2006, *72*, 656-659.

[74] Schmidt, G. R.; Yemm, R. S.; Childs, K. D.; O'Callaghan, J. P.; Hossner, K. L., Verification of different glial fibrillary acidic protein (GFAP) analyses as accurate detectors of

central nervous system tissue in advanced meat recovery (AMR) products. *Meat Science* 2002, *62*, 79-84.

[75] Kelley, L. C.; Hafner, S.; McCaskey, P. C.; Sutton, M. T.; Langheinrich, K. A., An evaluation of methods for the detection of spinal cord in product derived from advanced meat recovery systems. *Journal of Food Protection* 2000, *63*, 1107-1112.

[76] Hossner, K. L.; Yemm, R. S.; Sonnenshein, S. E.; Mason, G. L.; Cummings, B. A.; Reddy, M. C. S.; Sofos, J. N.; Scanga, J. A.; Tatum, J. D.; Smith, G. C.; Belk, K. E., Comparison of immunochemical (enzyme-linked immunosorbent assay) and immunohistochemical methods for the detection of central nervous system tissue in meat products. *Journal of Food Protection* 2006, *69*, 644-650.

[77] U.S. Department of Agriculture. Analysis of 2002 FSIS bovine AMR products survey results. http://www.fsis.usda.gov/oa/topics/AMRAnalysis.pdf. (Access: June 7, 2015).

[78] Cooper, J. D.; Bird, S. M., UK dietary exposure to BSE in beef mechanically recovered meat: by birth cohort and gender. *Journal of Cancer Epidemiology and Prevention* 2002, *7*, 59-70.

[79] Chadeau-Hyam, M.; Tard, A.; Bird, S.; Le Guennec, S.; Bemrah, N.; Volatier, J. L.; Alperovitch, A., Estimation of the exposure of the French population to the BSE agent: comparison of the 1980-95 consumption of beef products containing mechanically recovered meat in France and the UK, by birth cohort and gender. *Statistical Methods in Medical Research* 2003, *12*, 247-260.

[80] Ward, H. J. T.; Everington, D.; Cousens, S. N.; Smith-Bathgate, B.; Leitch, M.; Cooper, S.; Heath, C.; Knight, R. S.; Smith, P. G.; Will, R. G., Risk factors for variant Creutzfeldt-Jakob disease: A case-control study. *Annals of Neurology* 2006, *59*, 111-120.

[81] Cooper, J. D.; Bird, S. M., Predicting incidence of variant Creutzfeldt-Jakob disease from UK dietary exposure to bovine spongiform encephalopathy for the 1940 to 1969 and post-1969 birth cohorts. *International Journal of Epidemiology* 2003, *32*, 784-791.

[82] UK Minister of Agriculture Fisheries and Food. Statutory Instrument 1995 No. 3246: The specified bovine offal (amendment) order 1995. http://www.opsi.gov.uk/si/si1995/Uksi_19953246_en_1.htm. (Access: June 7, 2015).

[83] UK Minister of Agriculture Fisheries and Food. Statutory Instrument 1996 No. 2097: The fresh meat (beef controls) (No. 2) regulations 1996. http://www.opsi.gov.uk/si/si1996/Uksi_19962097_en_1.htm. (Access: June 7, 2015).

[84] UK Minister of Agriculture Fisheries and Food. Statutory Instrument 1997 No. 2965: The specified risk material regulations 1997. http://www.opsi.gov.uk/si/si1997/19972965.htm. (Access: June 7, 2015).

[85] UK Minister of Agriculture Fisheries and Food. Statutory Instrument 2001 No. 447: The restriction on pithing (England) regulations 2001. http://www.opsi.gov.uk/si/si2001/20010447.htm. (Access: June 7, 2015).

[86] European Commission, 94/381/EC: Commission Decision of 27 June 1994 concerning certain protection measures with regard to bovine spongiform encephalopathy and the feeding of mammalian derived protein. *Official Journal of the European Communities* 1994, *L172*, 23-24.

[87] European Commission, Commission implementing decision of 4 February 2013 amending Decision 2009/719/EC authorising certain Member States to revise their annual BSE monitoring programmes. *Official Journal of the European Communities* 2013, *L35*, 6-7.

[88] World Organisation for Animal Health. Joint WHO/FAO/OIE technical consultation on BSE: public health, animal health and trade. http://whqlibdoc.who.int/publications/2001/9290445556.pdf. (Access: June 7, 2015).

[89] World Organisation for Animal Health. Terrestrial animal health code - 2014 Chapter 11.4. Bovine spongiform encephalopathy. http://www.oie.int/index.php?id=169&L=0&htmfile=chapitre_bse.htm. (Access: June 7, 2015).

[90] European Commission, Regulation (EC) No 999/2001 of the European Parliament and of the Council of 22 May 2001 laying down rules for the prevention, control and eradication of certain transmissible spongiform encephalopathies. *Official Journal of the European Communities* 2001, *L147*, 1-40.

[91] European Commission, Commission Decision of 30 July 1997 on the prohibition of the use of material presenting risks as regards transmissible spongiform encephalopathies. *Official Journal of the European Communities* 1997, *L216*, 95-98.

[92] European Commission, Commission Decision of 27 December 2000 amending Decision 2000/418/EC regulating the use of material presenting risks as regards transmissible spongiform encephalopathies. *Official Journal of the European Communities* 2001, *L1*, 21-22.

[93] European Commission, Commission Decision of 14 March 2001 amending Decision 2000/418/EC as regards mechanically recovered meat and bovine vertebral column. *Official Journal of the European Communities* 2001, *L84*, 59-61.

[94] European Commission, Council Decision of 4 December 2000 concerning certain protection measures with regard to transmissible spongiform encephalopathies and the feeding of animal protein. *Official Journal of the European Communities* 2000, *L306*, 32-33.

[95] World Health Organization. Variant Creutzfeldt-Jakob disease. http://www.who.int/mediacentre/factsheets/fs180/en/. (Access: June 6, 2015).

[96] Japan Ministry of Health Labour and Welfare. Law on special measures against bovine spongiform encephalopathy. http://www.mhlw.go.jp/english/topics/foodsafety/bse/dl/2-1.pdf. (Access: June 7, 2015).

[97] Japan Ministry of Health Labour and Welfare. Standards for foods, food additives, etc. under the Japanese Food Sanitation Law total ban on the food use of bovine ver-

tebral column from BSE countries or zones. http://www.mhlw.go.jp/english/topics/foodsafety/vertebral.html. (Access: June 7, 2015).

[98] Japan Ministry of Health Labour and Welfare. About the partial revision of the MHLW-related implementation rules for the law on special measures against bovine spongiform encephalopathy. http://www.mhlw.go.jp/english/topics/foodsafety/bse/dl/1-4.pdf. (Access: June 7, 2015).

[99] Ministry of Health Labour and Welfare. Measures to ensure food safety. http://www.mhlw.go.jp/english/policy/health-medical/food/dl/pamphlet.pdf. (Access: June 6, 2015).

[100] Health Canada. Food directorate policy on specified risk material (SRM) in the food supply. http://www.hc-sc.gc.ca/fn-an/securit/animal/bse-esb/policy_srm-politique_mrs-eng.php. (Access: June 7, 2015).

[101] Canada Gazette. Statutory Instruments 2003 SOR/2003-265 Regulations amending the food and drug regulations (1389 — specified risk material). http://publications.gc.ca/gazette/archives/p2/2003/2003-08-13/pdf/g2-13717.pdf. (Access: June 6, 2015).

[102] Canadian Food Inspection Agency. Meat hygiene manual of procedures, chapter 17 - ante and post-mortem procedures, dispositions, monitoring and controls - meat species, ostriches, rheas and emus, annex D - specified risk materials. http://www.inspection.gc.ca/food/meat-and-poultry-products/manual-of-procedures/chapter-17/annex-d/eng/1369768468665/1369768518427. (Access: June 6, 2015).

[103] U.S. Food and Drug Administration, 21 CFR Part 589 Substances prohibited from use in animal food or feed; animal proteins prohibited in ruminant feed; final rule. *Federal Register* 1997, *62*, 30936-30978.

[104] U.S. Department of Agriculture, 9 CFR Parts 309, 310, 311, 318, and 319 Prohibition of the use of specified risk materials for human food and requirements for the disposition of non-ambulatory disabled cattle. *Federal Register* 2004, *67*, 1862-1874.

[105] U.S. Department of Agriculture, 9 CFR Parts 310 and 318 Prohibition of the use of specified risk materials for human food and requirements for the disposition of non-ambulatory disabled cattle. *Federal Register* 2005, *70*, 53043-53050.

[106] U.S. Department of Agriculture, 9 CFR Parts 301, 318, and 320 Meat produced by advanced meat/bone separation machinery and meat recovery (AMR) systems. *Federal Register* 2004, *69*, 1874-1885.

[107] U.S. Department of Agriculture, National animal identification system; user guide and technical documents. *Federal Register* 2007, *72*, 4680-4681.

[108] U.S. Food and Drug Administration, 21 CFR Part 589 Substances prohibited from use in animal food or feed; final rule. *Federal Register* 2008, *73*, 22720-22758.

[109] USDA, 9 CFR Parts 309, 310 and 318 Prohibition of the use of specified risk materials for human food and requirements for the disposition of non-ambulatory disabled cattle; prohibition of the use of certain stunning devices used to immobilize cattle during slaughter; rule. *Federal Register* 2007, *72*, 38699-38730.

[110] U.S. Department of Agriculture. BSE rules being strictly enforced. http://www.fsis.usda.gov/wps/portal/fsis/topics/food-safety-education/get-answers/food-safety-fact-sheets/production-and-inspection/bse-rules-being-strictly-enforced/bse-rules-being-strictly-enforced. (Access: June 7, 2015).

[111] U.S. Food and Drug Administration. CVM reports BSE inspection figures as of November 15, 2008. http://madcowfeed.blogspot.com/2008_11_01_archive.html. (Access: June 8, 2015).

[112] Hanrahan, C. E.; Becker, G. S. Mad cow disease and U.S. beef trade. http://nationalaglawcenter.org/wp-content/uploads/assets/crs/RS21709.pdf. (Access: June 7, 2015).

[113] Kim, J. South Korea to allow return of US beef imports. http://www.reuters.com/article/latestCrisis/idUSSP212539. (Access: June 7, 2015).

[114] World Organisation for Animal Health. Terrestrial animal health code: glossary. http://www.oie.int/index.php?id=169&L=0&htmfile=glossaire.htm. (Access: June 7, 2015).

[115] Sparks Companies Inc. *The rendering industry: economic impact of future feeding regulations*; National Renderers Association: Alexandria, VA, 2001.

[116] Becker, G. S. Animal rendering: economics and policy. http://nationalaglawcenter.org/wp-content/uploads/assets/crs/RS21771.pdf. (Access: June 6, 2015).

[117] U.S. FDA Animal and veterinary recalls archive. http://www.fda.gov/AnimalVeterinary/SafetyHealth/RecallsWithdrawals/ucm393160.htm. (Access: October 19, 2015).

[118] U.S. Department of Agriculture. Agricultural statistics 2013. http://www.nass.usda.gov/Publications/Ag_Statistics/2013/Agricultural_Statistics_2013.pdf. (Access: June 7, 2015).

[119] U.S. Department of Agriculture. BSE surveillance in the United States. http://www.aphis.usda.gov/wps/portal/aphis/ourfocus/animalhealth/sa_animal_disease_information/sa_cattle_health/sa_bse/ct_surv_in_usa/!ut/p/a0/04_Sj9CPykssy0xPLMnMz0vMAfGj-zOK9_D2MDJ0MjDzdgy1dDTz9wtx8LXzMjf09TPQLsh0VAZdihIg!/. (Access: June 7, 2015).

[120] U.S. Department of Agriculture. Agricultural statistics annual. http://www.nass.usda.gov/Publications/Ag_Statistics/. (Access: June 7, 2015).

[121] U.S. Department of Agriculture. Recall case archive. http://www.fsis.usda.gov/wps/portal/fsis/topics/recalls-and-public-health-alerts/recall-case-archive. (Access: June 7, 2015).

[122] Agazzi, M. E.; Bau, A.; Barcarolo, R.; Lucker, E.; Barrero-Moreno, J.; Anklam, E., In-house validation of an improved sample extraction and clean-up method for GC determination of isomers of nervonic acid in meat products. *Analytical and Bioanalytical Chemistry* 2003, *376*, 360-365.

[123] Barcarolo, R.; Bau, A.; Moreno, J. B.; Dimitrova, B.; Anklam, E., On-line LC-GC method for determination of isomers of nervonic acid in meat-derived food. *Journal of Separation Science* 2003, *26*, 1347-1352.

[124] Gangidi, R. R.; Proctor, A.; Pohlman, F. W.; Meullenet, J. F., Rapid determination of spinal cord content in ground beef by near-infrared spectroscopy. *Journal of Food Science* 2005, *70*, C397-C400.

[125] Gangidi, R. R.; Proctor, A.; Pohlman, F. W., Rapid determination of spinal cord content in ground beef by attenuated total reflectance Fourier transform infrared spectroscopy. *Journal of Food Science* 2003, *68*, 124-127.

[126] Lucker, E.; Biedermann, W.; Lachhab, S.; Truyen, U.; Hensel, A., GC-MS detection of central nervous tissues as TSE risk material in meat products: analytical quality and strategy. *Analytical and Bioanalytical Chemistry* 2004, *380*, 866-870.

[127] Biedermann, W.; Lucker, E.; Porschmann, J.; Lachhab, S.; Truyen, U.; Hensel, A., Structural characterisation of some fatty acids from the brain as biomarkers of BSE risk material. *Analytical and Bioanalytical Chemistry* 2004, *379*, 1031-1038.

[128] Nowak, B.; von Mueffling, T.; Kuefen, A.; Ganseforth, K.; Seyboldt, C., Detection of bovine central nervous system tissue in liver sausages using a reverse transcriptase PCR technique and a commercial enzyme-linked immunosorbent assay. *Journal of Food Protection* 2005, *68*, 2178-2183.

[129] Abdulmawjood, A.; Schonenbrocher, H.; Bulte, M., Novel molecular method for detection of bovine-specific central nervous system tissues as bovine spongiform encephalopathy risk material in meat and meat products. *Journal of Molecular Diagnostics* 2005, *7*, 368-374.

[130] Seyboldt, C.; John, A.; von Mueffling, T.; Nowak, B.; Wenzel, S., Reverse transcription-polymerase chain reaction assay for species-specific detection of bovine central nervous system tissue in meat and meat products. *Journal of Food Protection* 2003, *66*, 644-651.

[131] Schoenbrucher, H.; Abdulmawjood, A.; Gobel, K. A.; Bulte, M., Detection of central nervous system. tissues in meat products: Validation and standardization of a real-time PCR-based detection system. *Veterinary Microbiology* 2007, *123*, 336-345.

[132] Shi, X. J.; Ma, G. P.; Li, B. L.; Yang, J. L.; Wang, Y.; Li, Y. X.; Liu, X. H.; Liu, Q. G., Detection of bovine central nervous system tissue as bovine spongiform encephalopathy risk material by real-time reverse transcriptase-PCR in raw and cooked beef products. *Animal Biotechnology* 2008, *19*, 225-230.

[133] Agazzi, M. E.; Moreno, J. B.; von Holst, C.; Lucker, E.; Anklam, E., Quantitative analysis of tissues of the central nervous system in food products by GFAP-ELISA test kit. Results of an interlaboratory study. *Food Control* 2004, *15*, 297-301.

[134] Schmidt, G. R.; Hossner, K. L.; Yemm, R. S.; Gould, D. H.; O'Callaghan, J. P., An enzyme-linked immunosorbent assay for glial fibrillary acidic protein as an indicator of the presence of brain or spinal cord in meat. *Journal of Food Protection* 1999, *62*, 394-397.

[135] Schmidt, G. R.; Yemm, R. S.; Childs, K. D.; O'Callaghan, J. P.; Hossner, K. L., The detection of central nervous system tissue on beef carcasses and in comminuted beef. *Journal of Food Protection* 2001, *64*, 2047-2052.

[136] Rencova, E., Comparison of commercially available antibodies for the detection of central nervous system tissue in meat products by enzyme-linked immunosorbent assay. *Journal of Food Protection* 2005, *68*, 630-632.

[137] Lucker, E.; Hardt, M.; Groschup, M. H., Detection of CNS and PrP^{Sc} in meat products. *Berliner und Munchener Tierarztliche Wochenschrift* 2002, *115*, 111-117.

[138] Agazzi, M. E.; Barrero-Moreno, J.; Lucker, E.; von Holst, C.; Anklam, E., Performance comparison of two analytical methods for the detection of tissues of the central nervous system in sausages: results of an interlaboratory study. *European Food Research and Technology* 2002, *215*, 334-339.

[139] Hajmeer, M.; Cliver, D. O.; Provost, R., Spinal cord tissue detection in comminuted beef: comparison of two immunological methods. *Meat Science* 2003, *65*, 757-763.

[140] Hughson, E.; Reece, P.; Dennis, M. J.; Oehlschlager, S., Comparative evaluation of the performance of two commercial kits for the detection of central nervous system tissue in meat. *Food Additives and Contaminants* 2003, *20*, 1034-1043.

[141] Yesilbag, K.; Kalkan, A., Detection of central nervous system tissues as BSE specified risk material in meat products in Turkey. *Food Control* 2005, *16*, 11-13.

[142] Reddy, M. C. S.; Hossner, K. L.; Belk, K. E.; Scanga, J. A.; Yemm, R. S.; Sofos, J. N.; Smith, G. C., Detection of central nervous system tissue on meat and carcass-splitting band saw blade surfaces using modified fluorescent glial fibrillary acidic protein enzyme-linked Immunosorbent assay sampling and extraction procedures. *Journal of Food Protection* 2006, *69*, 1966-1970.

[143] Rao, Q. C.; Hsieh, Y.-H. P., Enhanced immunodetection of bovine central nervous tissue using an improved extraction method. *Food Control* 2014, *46*, 282-290.

[144] Lucker, E.; Horlacher, S.; Eigenbrodt, E., Brain in human nutrition and variant Creutzfeldt-Jakob disease risk (vCJD): detection of brain in retail liver sausages using cholesterol and neuron specific enolase (NSE) as markers. *British Journal of Nutrition* 2001, *86*, S115-S119.

[145] Herde, K.; Bergmann, M.; Lang, C.; Leiser, R.; Wenisch, S., Glial fibrillary acidic protein and myelin basic protein as markers for the immunochemical detection of bovine central nervous tissue in heat-treated meat products. *Journal of Food Protection* 2005, *68*, 823-827.

[146] Lucker, E.; Eigenbrodt, E.; Wenisch, S.; Leiser, R.; Bulte, M., Identification of central nervous system tissue in retail meat products. *Journal of Food Protection* 2000, *63*, 258-263.

[147] Sultan, K. R.; Tersteeg, M. H. G.; Koolmees, P. A.; de Baaij, J. A.; Bergwerff, A. A.; Haagsman, H. P., Western blot detection of brain material in heated meat products using myelin basic protein and neuron-specific enolase as biomarkers. *Analytica Chimica Acta* 2004, *520*, 183-192.

[148] Lucker, E.; Eigenbrodt, E.; Wenisch, S.; Failing, K.; Leiser, R.; Bulte, M., Development of an integrated procedure for the detection of central nervous tissue in meat products using cholesterol and neuron-specific enolase as markers. *Journal of Food Protection* 1999, *62*, 268-276.

[149] Salman, M. D.; Jemmi, T.; Triantis, J.; Dewell, R. D., Assessment and modification of a Western blot assay for detection of central nervous system tissue in meat products in the United States. *Journal of Food Protection* 2005, *68*, 1706-1711.

[150] Kummer, V.; Maskova, J.; Rencova, E., Immunohistochemical detection of central nervous tissue in meat products. *Acta Veterinaria Brno* 2005, *74*, 449-454.

[151] Tersteeg, M. H. G.; Koolmees, P. A.; van Knapen, F., Immunohistochemical detection of brain tissue in heated meat products. *Meat Science* 2002, *61*, 67-72.

[152] Wenisch, S.; Lucker, E.; Eigenbrodt, E.; Leiser, R.; Bulte, M., Detection of central nervous tissue in meat products - An immunohistochemical approach. *Nutrition Research* 1999, *19*, 1165-1172.

[153] Wardlaw, G. M.; Kessel, M. W., Lipids. In *Perspectives in Nutrition*, 5 ed.; Meyers, L. M., Ed. McGraw-Hill: New York, 2002; pp 203-254.

[154] Biedermann, W.; Lucker, E.; Hensel, A., Detection of tissues of the central nervous system (CNS) as specified risk material (SRM) in meat products by means of gas chromatography mass spectrometry (GC-MS). *Berliner und Munchener Tierarztliche Wochenschrift* 2002, *115*, 131-133.

[155] Eng, L. F.; Ghirnikar, R. S.; Lee, Y. L., Glial fibrillary acidic protein: GFAP-thirty-one years (1969-2000). *Neurochemical Research* 2000, *25*, 1439-1451.

[156] Abdulmawjood, A.; Schonenbrucher, H.; Bulte, M., Collaborative trial for validation of a real-time reverse transcriptase-polymerase chain reaction assay for detection of

central nervous system tissues as bovine spongiform encephalopathy risk material: Part 1. *Journal of AOAC International* 2006, *89*, 1335-1340.

[157] Feinstein, D. L.; Weinmaster, G. A.; Milner, R. J., Isolation of cDNA clones encoding rat glial fibrillary acidic protein: expression in astrocytes and in Schwann cells. *Journal of Neuroscience Research* 1992, *32*, 1-14.

[158] Kale, M.; Kursun, O.; Pehlivanoglu, F., Detection of central nervous system tissues as bovine spongiform encephalopathy specified risk material in processed and raw meat products in Turkey. *Journal of Food Safety* 2007, *27*, 56-65.

[159] Sheppard, M. N.; Marangos, P. J.; Bloom, S. R.; Polak, J. M., Neuron specific enolase: a marker for the early development of nerves and endocrine cells in the human lung. *Life Sciences* 1984, *34*, 265-271.

[160] Weyandt, R. G., Detection of BSE-risk materials. *Fresenius' Journal of Analytical Chemistry* 2001, *371*, 574-575.

[161] Baumann, N.; Pham-Dinh, D., Biology of oligodendrocyte and myelin in the mammalian central nervous system. *Physiological Reviews* 2001, *81*, 871-927.

[162] Siegel, G. J.; Agranoff, B. W.; Fisher, S. K.; Albers, R. W.; Uhler, M. D., *Basic neurochemistry: molecular, cellular and medical aspects*. 6th ed.; Lippincott Williams & Wilkins: Philadelphia, 1999.

[163] Villmann, C.; Sandmeier, B.; Seeber, S.; Hannappel, E.; Pischetsrieder, M.; Becker, C. M., Myelin proteolipid protein (PLP) as a marker antigen of central nervous system contaminations for routine food control. *Journal of Agricultural and Food Chemistry* 2007, *55*, 7114-7123.

[164] Sandmeier, B.; Baeuerlein, R.; Villmann, C.; Duethorn, T.; Gareis, M.; Becker, C. M.; Pischetsrieder, M., Detection of central nervous system tissue in meat and meat products with a newly developed immunoassay selective for Myelin proteolipid protein. *Food Chemistry* 2007, *105*, 871-878.

[165] Bauerlein, R.; Sandmeier, B.; Villmann, C.; Hammon, A.; Gareis, M.; Becker, C. M.; Pischetsrieder, M., Development of a dot blot assay for the rapid detection of central nervous system tissue on meat and contact surfaces. *Journal of Agricultural and Food Chemistry* 2008, *56*, 44-49.

[166] Boggs, J. M., Myelin basic protein: a multifunctional protein. *Cellular and Molecular Life Sciences* 2006, *63*, 1945-1961.

[167] Gasteiger, E.; Gattiker, A.; Hoogland, C.; Ivanyi, I.; Appel, R. D.; Bairoch, A., ExPASy: the proteomics server for in-depth protein knowledge and analysis. *Nucleic Acids Research* 2003, *31*, 3784-3788.

[168] Rao, Q. C.; Hsieh, Y.-H. P., Effect of pH, temperature and storage time on the stability of bovine myelin basic protein. *Food Control* 2015, *50*, 166-172.

[169] Schonherr, J., Analysis of products of animal origin in feeds by determination of carnosine and related dipeptides by high-performance liquid chromatography. *Journal of Agricultural and Food Chemistry* 2002, *50*, 1945-1950.

[170] Fontaine, J.; Horr, J.; Schirmer, B., Near-infrared reflectance spectroscopy enables the fast and accurate prediction of the essential amino acid contents in soy, rapeseed meal, sunflower meal, peas, fishmeal, meat meal products, and poultry meal. *Journal of Agricultural and Food Chemistry* 2001, *49*, 57-66.

[171] Fontaine, J.; Horr, J.; Schirmer, B., Amino acid contents in raw materials can be precisely analyzed in a global network of near-infrared spectrometers: Collaborative trials prove the positive effects of instrument standardization and repeatability files. *Journal of Agricultural and Food Chemistry* 2004, *52*, 701-708.

[172] Fontaine, J.; Schirmer, B.; Horr, J., Near-infrared reflectance spectroscopy (NIRS) enables the fast and accurate prediction of essential amino acid contents. 2. Results for wheat, barley, corn, triticale, wheat bran/middlings, rice bran, and sorghum. *Journal of Agricultural and Food Chemistry* 2002, *50*, 3902-3911.

[173] Gao, H. W.; Zhang, D. B.; Pan, A. H.; Liang, W. Q.; Liang, C. Z., Multiplex polymerase chain reaction method for detection of bovine materials in foodstuffs. *Journal of AOAC International* 2003, *86*, 764-767.

[174] Bellagamba, F.; Valfre, F.; Panseri, S.; Moretti, V. M., Polymerase chain reaction-based analysis to detect terrestrial animal protein in fish meal. *Journal of Food Protection* 2003, *66*, 682-685.

[175] Chen, F. C.; Hsieh, Y.-H. P.; Bridgman, R. C., Monoclonal antibodies against troponin I for the detection of rendered muscle tissues in animal feedstuffs. *Meat Science* 2002, *62*, 405-412.

[176] Chen, F. C.; Hsieh, Y.-H. P.; Bridgman, R. C., Monoclonal antibody-based sandwich enzyme-linked immunosorbent assay for sensitive detection of prohibited ruminant proteins in feedstuffs. *Journal of Food Protection* 2004, *67*, 544-549.

[177] Chen, F. C.; Hsieh, Y.-H. P., Porcine troponin I: a thermostable species marker protein. *Meat Science* 2002, *61*, 55-60.

[178] Rovira, P. J.; Scanga, J. A.; Grandin, T.; Hossner, K. L.; Yemm, R. S.; Belk, K. E.; Tatum, J. D.; Sofos, J. N.; Smith, G. C., Central nervous system tissue contamination of the circulatory system following humane cattle stunning procedures. *Food Protection Trends* 2007, *27*, 524–529.

[179] Rotterud, O. J.; Helps, C. R.; Hillman, T. J.; Fisher, A. V.; Harbour, D.; Anil, H.; Nesbakken, T., Hot boning of intact carcasses: a procedure to avoid central nervous system self-contamination in beef and beef products. *Journal of Food Protection* 2006, *69*, 405-411.

[180] [Anon] Glossary. http://www.nature.com/nrm/journal/v5/n8/glossary/nrm1438_glossary.html. (Access: June 6, 2015).

[181] Allore, R.; Ohanlon, D.; Price, R.; Neilson, K.; Willard, H. F.; Cox, D. R.; Marks, A.; Dunn, R. J., Gene encoding the beta subunit of S100 protein is on chromosome 21: implications for Down syndrome. Science 1988, 239, 1311-1313.

[182] Bennett, M. K.; Calakos, N.; Scheller, R. H., Syntaxin: a synaptic protein implicated in docking of synaptic vesicles at presynaptic active zones. Science 1992, 257, 255-259.

3

Risk Factors Influencing Microbial Contamination in Food Service Centers

Antonio Valero, Magdevis-Yanet Rodríguez,
Guiomar Denisse Posada-Izquierdo, Fernando Pérez-Rodríguez,
Elena Carrasco and Rosa Maria García-Gimeno

Abstract

An improvement of food service centers in recent years has been made based on the implementation of the principles of the Hazard Analysis and Critical Control Points (HACCP) system. Food safety preventive measures have been focused on training of handlers in hygiene practices and on improving the sanitary quality of meals. In Europe, an increasing trend in foodborne outbreaks has been attributed to catering businesses. This fact highlights that the impact of preventive measures in the past few years has not been sufficiently effective as expected. Special attention should be paid to food services destined to susceptible population, such as hospitals, long-term care facilities, or school canteens, because people could be more susceptible to become ill when exposed to foodborne agents. There are numerous relevant factors influencing microbial contamination of foods, according to the preparation method, hygienic sanitary conditions of catering facilities, or food handling, storage, and distribution. In the present chapter, a review of the most significant risk factors influencing microbial contamination of foods in food service centers are described with special focus on those establishments where susceptible population (i.e., children, elderly, immunocompromised people) is present. Besides, potential preventive measures to be considered in that establishments and correct implementation of food safety actions are given to provide useful recommendations to food handlers, food operators, and risk managers.

Keywords: Food safety, catering establishments, risk factors, handlers, microbial indicators

1. Introduction

Some individuals tend to eat out of the home, often at food service operations, such as cafeterias, canteens, fast food outlets, bars, and restaurants [1]. In recent years, the catering sector has been experiencing an increase in technological innovation in correspondence with changes in consumer habits of the population, transformed by numerous factors and changing lifestyles, demographic trends, and so on, which have increased consumer preferences for healthy, safe, and convenient foods.

Food service or catering industry defines those businesses, institutions, and companies responsible for any meal prepared outside home. These industries include restaurants, school and hospital cafeterias, catering operations, and some other small- and big-scale establishments. A catering food establishment also means an approved food establishment that is serving or preparing food at a location other than its permitted location for a contracted food service event. During these operations, foods are often transported, distributed, handled, and consumed in a short-time framework (often less than 1 week).

The catering sector can be divided into three groups based on the population they serve, the way of working, the technology used, or the food types served (Table 1): commercial (residential and non-residential) catering sector, and non-commercial residential establishments.

Residential	Hotels, guest houses, holiday parks, farmhouses, public houses, bed, and breakfasts
Non-residential	Restaurants, cafes, fast-food outlets, wine bars, delicatessen and salads, bars, take-away outlets, schools catering, and burger vans
Non-commercial residential establishments	Hospitals, residential homes, prisons, and armed services

Table 1. Types of catering establishments

The Hazard Analysis and Critical Control Points (HACCPs) system has been recognized as a useful tool to prevent food contamination in food service establishments. Control measures are mainly focused on training of food handlers in hygiene practices to improve safety of served meals.

Unquestionably, consumption outside home is linked to a fact of modern life that causes some concern among health professionals and food authorities because of the potential adverse effects that may entail for human health. Food poisoning poses a serious problem for public health worldwide, and the most vulnerable population (children, elderly, pregnant, and sick) counts among the most affected.

Data from the World Health Organization assert that about 2.2 million people become daily ill in the world for more than 200 foodborne diseases and about two-thirds of the outbreaks that occur originate in their homes and in catering establishments. Given the higher volume of meals prepared in the later ones together with meals exposure to handling and environmental factor, the risk of becoming ill is expected to be considerably higher [2]. In Europe, an increasing trend in foodborne diseases outbreaks has been attributed to catering businesses [3]. This fact highlights that the impact of prevention measures in the past few years has not been sufficiently effective as expected. Important aspects such as the size of establishments and heterogeneity of foods served justify the creation of specific regulations on food safety management, to reduce the risks of foodborne illness.

Moreover, it seems to be necessary to strengthen food hygiene and compliance with HACCP system to prevent food outbreaks. These outbreaks are mainly caused by foodborne pathogens and norovirus. Special attention should be paid to food services destined to susceptible population, such as hospitals, long-term care facilities, or school canteens, because people could be more susceptible to become ill when exposed to foodborne agents. Besides this, the large number of meals served per day could have an impact on public health if prepared foods are contaminated.

In food service environments, various factors may be related to foodborne diseases. According to the Food and Drug Administration (FDA), these factors are the food served coming from unsafe sources, poor personal hygiene, inadequate cooking, improper holding temperatures, or utilization of contaminated equipment. On the other hand, the World Health Organization (WHO) [4] suggested that drinking water could be an important factor related to foodborne diseases. It was already demonstrated that training of food handlers and knowledge acquisition in hygienic food preparation, processing, and distribution of meals is crucial in the prevention of most types of foodborne diseases [5]. The use of normalized questionnaires on good manufacturing practices (GMPs) was achieved to evaluate training of food handlers. It was found that, after implemented, there is a substantial improvement in GMP [6–8]. However, it is recognized that these training activities should be repeated over time to minimize the reluctance of food handlers to apply the acquired knowledge [9].

Legislation in food hygiene at EU level prioritizes control measures to protect public health, making food operators responsible to assure product safety [10]. Implementation of HACCP system in food service operations may increase food safety management. However, given the complexity of the food chain and variety of menus and meals prepared, simplified and flexible self-control measures must be required in most cases to increase efficiency and homogeneity of implemented food safety management systems [7]. Evaluation of microbial risks is crucial to determine food safety of prepared meals [11].

In the present chapter, a review of the most significant risk factors influencing microbial contamination of foods in food service centers will be described with special focus on those establishments where susceptible population (i.e., children, elderly, immunocompromised people) is present. Besides, potential preventive measures to be considered in that establishments and correct implementation of food safety actions will be given to provide useful recommendations to food handlers, food operators, and risk managers.

2. Factors enhancing microbial foodborne outbreaks worldwide

The increase in the global burden of foodborne diseases constitutes a concern to governments and food operators today than a few years ago. Identification of emerging pathogens (or environmentally adapted) causing life-threatening conditions, introduction of ethnic foods, environmental changes, food security aspects, migrant populations, the ease of worldwide shipment of fresh and frozen food, and the development of new food industries including aquaculture are some of the reasons for this concern. Over the past years, foodborne diseases caused by bacteria, viruses, parasites, and prions have been prioritized by governments of industrialized countries generating substantial media attention.

However, to monitor foodborne diseases, an effective surveillance system at the local, national, and international levels should be implemented. The appearance of multistate outbreaks (i.e., contamination in a commercial product occurring in one country and affecting persons in several other countries, or tourists being infected abroad and possibly transmitting the pathogen to others at home) contributes to the increase in the number of illnesses at international level. Currently, funding sources are not easily available for non-developed countries and regions where most of the information comes from passive reporting mechanisms [12]. In the industrialized countries, a need for funding sources allocated to the improvement of foodborne disease surveillance, and control is required.

Several factors (such as environmental, socioeconomic, chemical, physical, and biological) are influencing on such public concern.

Environmental factors such as weather influence the transport and dissemination of microbial hazards via rainfall and runoff and their survival and/or growth through according to temperature or humidity conditions [13]. It is shown that the increase in climate variability influences on current and future deficiencies in areas, such as watershed protection, infrastructure, and storm drainage systems, thus enhancing the risk of food contamination events. More knowledge is needed about transport processes and fate of microbial hazards to predict risks associated to weather variability. In this sense, application of existing technologies such as molecular fingerprinting to track contaminant sources could be expanded.

On the other hand, despite of the development of novel traceability systems and inspection controls, infectious diseases still remain a leading cause of global disease burden with high morbidity and mortality in non-developed countries. The emerging and re-emerging diseases have been a big impact at socioeconomic and public health levels. Their control requires continuing surveillance, research and training, better diagnostic facilities, and improved public health system. Food safety is of particular concern in a developing country context given the higher incidence of foodborne illnesses and their associated economic and social cost [14]. Furthermore, the economic and social changes associated with development (i.e., urbanization, changes in food production systems, and consumption patterns) could increase the risk of emerging foodborne illnesses. This recognizes the connections between disease and socioeconomic factors such as poverty and malnutrition and the wider economic, social, physical, and cultural environment in which people live [15]. Promotion of trade together with improvement of agricultural and animal practices at primary production will definitely help developing countries to better manage food safety. Improving food safety in international trade would also require numerous policy and technical interventions that include an effective market access; a better analysis of the costs and benefits of global trade rules for developing countries or integration of developing countries into the global economy [16].

Besides environmental and socioeconomic factors, according to the Center for Disease Control and Prevention (CDC), the absence of control, a key elementary factor along the production chain, contributes to the extension of foodborne outbreaks, some of them affecting to food service centers:

- Producing and harvesting food: lack of quality assurance programs and unsafe agricultural practices.
- Processing foods: absence or undefined inspection systems at industrial facilities; poor processing preservation technologies (i.e., thermal treatments, addition of preservatives, and unsafe food formulations).
- Distributing and preparing foods: absence of food purchasing specifications, untrained food handlers, improper hand-washing procedures and facilities, and lack of food safety education programs for consumers.

3. Residential catering

Institutional food establishments have a key role to public health. Consumers that are generally more likely to suffer from foodborne diseases occupy them, such as children, elderly, sick, or immunocompromised individuals. Their physiological characteristics often require high degree of food safety in the meals served.

According to the European Federation of Contract Catering Organizations (FERCO), approximately 33% of firms or collective organizations currently have a contract with a Contract Catering company. Indeed, this is a sector that represents an alternative to meet the basic food needs of a group of very important people. Among institutional catering companies stand Eurest SA authorities; Serunión S.A; Sodexho S.A. Spain; and Aramark catering services, SL, among others, that allocate their production mainly to hospitals, nursing geriatric, dining, and study centers.

The fact of preparing and serving large volumes of food in a relatively short-time framework involves the use of new technologies for conservation and/or optimization of existing technologies, to ensure the hygienic and sanitary quality and shelf life of food.

Studies related to systems development, maintenance, and transportation of prepared meals cite the refrigerated cold chain, frozen cold chain, hot chain, or vacuum cooking [17]:

- In the refrigerated cold chain, food is abated once drawn up, from 65°C to 10°C in a time not exceeding 90 min, before being stored at 0–3°C. Afterward, food is regenerated until reaching the appropriate temperature before serving.
- In the frozen cold chain, food is placed in a blast chiller given the same procedures above and subsequent cold storage at −18° C until consumption, at which time must be regenerated to its proper serving temperature (65°C).

- In the hot chain, foods are maintained at temperatures above 65°C and should be consumed within a maximum of 12 h.
- In the vacuum cooking, the food is prepared at low temperatures and for a long time. The food is then packaged in a waterproof material, which is not affected by high temperatures. Before cooking, air is extracted and sealed. After cooking, the product is cooled quickly and is cooled, and then regenerated (65°C) before consumption.

Utilization of more traditional food preservation techniques is also common [17] such as industrial pasteurization or chemical preservation (salting, smoking, marinating, and pickling).

Lastly, other preservation technologies are used to reduce the amount of food handling in the kitchen, such as high hydrostatic pressure, the light pulses, dehydration, irradiation, and modified atmosphere preservation, among others [17–19].

4. Health and socioeconomic implications of microbial contamination and its effects on international trade

Provision of safe foods supports national economies, international trade, and consumer confidence, thus underpinning sustainable development. However, globalization and changes in consumers' habits to a more convenient and healthy foods led to increase the awareness of potential and/or emerging hazards for public health. This also triggered a growing consumer demand for a wider variety of foods, thus leading to a more complex food chain.

The consequence of the population growth is an intensification of agriculture and animal production to meet consumer demands. Food safety challenges should consider the potential effect of climate change because temperature changes can modify the risk profile of a given food commodity during the whole production chain [20]. These effects produce a greater degree of responsibility to food producers and handlers to ensure food safety. It should be highlighted that the spread of a localized outbreak can increase largely due to the globalized food chain and international trade. Examples include the contamination of infant formula with melamine in 2008, and the 2011 Enterohemorrhagic *Escherichia coli* outbreak in Germany linked to contaminated sprouts. This multistate outbreak was reported in eight countries in Europe and North America, leading to 53 deaths. Losses caused to farmers and food industries were quantified in US$ 1.3 billion, while emergency aid payments to 22 EU member states were around US$ 236 million [21].

According to data reported by USDA, foodborne illnesses are annually costing the economy more than $15.6 billion. Each year, more than 8.9 million Americans will be sickened by one of the 15 pathogens, with more than 5.4 million of those illnesses due to the stomach churning, but usually short-lived, Norovirus. In EU countries, 5196 foodborne outbreaks, such as waterborne outbreaks, were reported [22]. Governmental authorities should make food safety a public health priority, as they play key role in developing policies and regulatory frameworks. They also are in charge of establishing and implementing effective food safety systems that ensure that food producers and suppliers along the whole food chain operate responsibly and supply safe food to consumers.

Among the food commodities causing outbreaks, meat products are the most frequently reported, given the high consumption associated to these products. Vegetable salads are recognized as potential vehicles for enteric pathogens, as they are not subjected to any heat treatment before consumption. In Table 2, notified outbreaks in EU and the United States for produce commodities are reported. Scientific studies highlight the importance of an adequate training of food handlers and implementation of GMPs when elaborating vegetable salads in catering establishments.

Regarding cooked meat products, they are susceptible to be contaminated after heat treatment, during storage, and distribution. This is because they can be subjected to poor handling practices (i.e., slicing, packaging) during preparation in catering establishments. As an example, in Table 3, notified outbreaks in EU countries and the United States associated to consumption of meat products are reported. It is concluded that problems in kitchen design, inadequate handling and disinfection practices, and lack of knowledge on food safety by handlers are the main risk factors influencing microbial contamination.

Year	Vehicle	Microbial hazard involved	Place	Country	Number of cases	Reference
2012	Romaine lettuce	*E. coli* O157:H7	Retail outlets	The United States	58	[23]
2011	Bean sprouts	*E. coli* O104:H4	Multiple places	Multistate EU outbreak	3910	[24]
2011	Romaine lettuce	*E. coli* O157:H7	Retail outlets	The United States	60	[25]
2011	Fresh basil	*Shigella sonnei*	Not available	Norway	46	[26]
2010	Lettuce	*E. coli* enterotoxigénica and norovirus	Catering establishments	Denmark	260	[27]
2007	Shredded lettuce	*E. coli* O157	Processing industries	The Netherlands, Iceland	50	[28]
2007	Alfalfa sprouts	*Salmonella stanley*	Domestic homes	Sweden	51	[29]
2006	Shredded romaine lettuce	*E. coli* O145	Processing industries	The United States	26	[30]
2004	Salad lettuce	*Salmonella newport*	–	The United Kingdom	375	[31]
1994	Iceberg lettuce	*S. sonnei*	Domestic homes	The United Kingdom, Norway	–	[32]

Table 2. Notified foodborne outbreaks in Europe and the United States by the consumption of produce commodities

Year	Vehicle	Microbial hazard involved	Place	Country	Number of cases	Reference
2011	Turkey meat	*Salmonella* Heidelberg	–	The United States	77	[33]
2009	Cured RTE meat	*Salmonella* montevideo	–	The United States	272	[34]
2009	Cooked meat	*E. coli* O157	School canteens	Wales	150	[35]
2008	Not available	*Listeria monocytogenes*	Processing meat industry	Canada	22	[36]
2006	RTE pork meat	*Yersinia enterocolítica* O:9	–	Norway	11	[37]
2006	Fermented sausages	*E. coli* O103:H25	–	Norway	17	[38]
2005	Minced beef	*E. coli* O157:H7	–	France	69	[39]
2002	Turkey meat	*L. monocytogenes*	Processing industry	The United States	54	[40]
2001	Cooked meat	*E. coli*	Butchery	The United Kingdom	30	[41]
–	Minced beef	*E. coli* O157	–	–	732	[42]
2000	Ham	*L. monocytogenes*	–	New Zealand	28	[43]
2000	Turkey meat slices	*L. monocytogenes*	Processing industry	The United States	11	[44]

Table 3. Notified foodborne outbreaks in Europe and the United States by consumption of meat commodities

5. Microbial contaminants of prepared meals in catering establishments

During the whole production chain, there is constant exposure of food to microbial contamination. Therefore, a strict quality and safety food control should be promoted with a view to minimize the incidence of food poisoning.

Undoubtedly, for catering establishments, the HACCP system assesses the condition under which the product was elaborated, determines the main risk factors of food contamination, and manages effective measures to reduce contamination by pathogenic and spoilage microorganisms.

Microbial indicators are able to highlight deficiencies in the hygienic and sanitary food quality. Indeed, their presence at high levels leads to a reduction of shelf life and is probably related to the presence of pathogenic microorganisms.

According to the International Commission on Microbiological Specifications for Foods (ICMSF), microbial indicators do not offer a direct risk to human health. These groups are mainly aerobic mesophilic bacteria, lactic acid bacteria, total coliforms, fecal coliforms, enterococci, enterobacteriaceae, *Staphylococcus aureus* and *E. coli* [45]. In the next subsections, the most representative microbial indicators will be described. Besides, *Listeria monocytogenes* is included due to its relevance and presence in a wide range of food commodities as well as for the current EU regulation (No. 1441/2007) [46] where it is included as safety criteria for ready to eat foods.

5.1. Aerobic mesophilic microorganisms

Microbial species belonging to this group are quite heterogeneous and include all bacteria, fungi, and yeasts growing at aerobic conditions. The presence of aerobic mesophilic microorganisms in fresh foods demonstrates the effectiveness of sanitary procedures during processing, handling, and storage before [47].

Ready to eat foods (apart from fermented foods, cheeses, and dairy products) with significant concentration levels of aerobic mesophilic microorganisms should not be considered suitable for human consumption, even if microbial species are not pathogenic.

Generally, contamination occurs because of the use of contaminated raw materials or inefficient health treatments as well as inadequate conditions of storage time and temperature [48].

In general, high levels pose a greater risk of pathogen contamination. Several authors agree that the recommended concentrations for ready to eat foods should be less than 5.0 log cfu/g [49]. However, other guidelines for ready to eat foods such as those proposed by the Health Protection Agency (UK) [50] establish acceptable limits between 6 and 8 log cfu/g, depending on the food type.

5.2. Lactic acid bacteria

Lactic acid bacteria comprise a wide range of microorganisms with common morphological, metabolic, and physiological characteristics. Some of the most representative species are *Streptococcus* spp., *Pediococcus* spp., *Leuconostoc* spp., *Lactobacillus* spp., and *Lactococcus* spp. [51].

In the food industry, they have multiple uses as starter cultures in the manufacture of cheese, yogurt, and fermented meats. They are also recognized as natural antimicrobial agents against foodborne pathogens in biopreservation processes [52]. They represent the predominant group in fermented meat products reaching levels between 8 and 9 log cfu/g during the maturation processes. The most common species are *Lactobacillus sakei, Lactobacillus curvatus, Lactobacillus plantarum, Lactobacillus pentosus, Pediococcus acidilactici,* and *Pediococcus pentasaceus* [53].

Despite its protective function, they are able to produce end metabolites that lead to food spoilage and thus shortening its shelf life. Their final levels depend largely on the storage temperature and packaging methods [54]. Deterioration caused by the growth of lactic acid bacteria is shown by undesirable changes in smell, taste, color, and gas production. Some studies have found these changes in vacuum-packed meat products or modified atmosphere products [55].

5.3. Enterobacteriaceae

Enterobacteriaceae are considered as food quality indicators including *E. coli* being mainly related to fecal contamination. Generally, the presence of these microorganisms in foods is closely linked with the implementation of inadequate handling practices, inefficient cooking processes, cross-contamination, inadequate personal hygiene of food handlers, equipment and food-contact surfaces as well as inadequate holding time and temperature conditions [56].

Enterobacteriaceae species are Gram-negative bacteria, aerobic or facultative anaerobic, non-sporulated, mobile or immobile, and being able to ferment glucose and to reduce nitrate to nitrite. Some of the most representative species include *Salmonella enterica*, *Shigella* spp., *Yersinia* spp. (intestinal pathogens in humans), *Edwarsiella* spp., *Hafnia* spp., *Proteus* spp., *Morganella* spp., *Erwinia* spp., *E. coli*, *Enterobacter* spp., *Citrobacter* spp., *Serratia* spp., or *Klebsiella* spp. Most of them produce endotoxins and thermolabile and/or thermostable exotoxins. Some *E. coli* serotypes are producing verotoxins and shigatoxins, which are linked to a high rate of morbidity and mortality in humans [57].

Food commodities where Enterobacteriaceae can be found are processed meat products [58], nutritional formulas for infants [59], mixed salads, raw vegetables, and milk/dairy products, among others [60].

5.4. Total and fecal coliforms

Total and fecal coliforms are specific groups within the Enterobacteriaceae family, including species, such as *E. coli*, *Klebsiella* spp., *Enterobacter* spp., or *Citrobacter* spp. These are Gram-negative bacteria, aerobic or facultative anaerobes, non-sporulated, whose optimal growth temperature is around 35–40°C. These food quality indicators are taking part of the intestines of humans and warm-blooded animals and other organisms often located on the ground or plant.

The main difference between total and fecal coliforms is that the latter group ferments lactose at temperatures between 44 and 45°C. The group includes primarily *E. coli* (~90%) with certain *Klebsiella* and *Citrobacter* species. Coliforms are considered a reliable indicator of fecal contamination and are sometimes found in contaminated equipment and utensils, as well as in a wide variety of foods.

Contamination of ready to eat foods by coliforms is commonly attributed to environmental contamination, the use of inadequate hygiene practices, and/or insufficient control of the storage temperature. In the case of thermally treated food, the presence of coliforms is indicative of inadequate treatment or post-processing contamination as they are thermolabile microorganisms [47].

5.5. *Escherichia coli*

Enteropathogenic *E. coli* comprise different serotypes that can be present in contaminated foods. Most of them are able to produce Shiga-like toxins and/or other heat-labile or heat-stable toxins that can potentially cause diarrheagenic diseases in humans [61]. Besides, some serotypes of enterotoxigenic *E. coli* can also produce a cytotoxin to Vero cells (VTEC *E. coli*).

Normally, outbreaks caused by VTEC serotypes are of low prevalence (1.2 cases per 100,000 people in the EU) [39]; however, the high infectivity and severity of the disease increase the importance of performing novel research on this pathogen. It is reported that human outbreaks attributed to *E. coli* serotypes were mostly originated from catering services or restaurants [62].

Generally, *E. coli* can be present in animal origin foods (pork, beef, and poultry), water sources, or produce such as cabbage, lettuce, or spinach. They can enter the food chain through cross-contamination or recontamination phenomena [63] or through the irrigation with contaminated water, which may result in the internalization of certain *E. coli* serotypes in vegetables [64].

E. coli O157:H7 was the most studied serotype due to the severity of the illnesses caused and its low infective dose, around 100 cells [65]. However, other non-O157 serotypes have been associated to human infections through the ingestion of risk food products, such as fermented and minced meats or raw milk [66].

5.6. *Staphylococcus aureus*

S. aureus has been reported as a microbial indicator most likely associated to reduced water activity (a_w) foods, such as ready to eat cooked or cured meats [67, 68]. The presence of *S. aureus* is often associated to contamination of raw material, such as poultry carcasses or raw chicken samples [69] or cross-contamination events occurring because of mishandling during processing and storage [63].

Food poisoning is attributed to the ingestion of foods that contain thermotolerant Staphylococcal Enterotoxins (SEs) in doses around 20–100 ng [70]. The staphylococcal enterotoxin A (SEA) is the one most frequently reported. A wide range of environmental factors, such as pH, a_w, temperature, food type, and processing conditions, have been suggested to play an important role on SEs production. Generally, growth of *S. aureus* is necessary for SE production, although this phenomenon does not always accompany growth [67]. Indeed, some published studies consider hazardous *S. aureus* levels from 6 log cfu/g in contaminated foods for SE production [71].

5.7. *Listeria monocytogenes*

L. monocytogenes is a foodborne pathogen causing listeriosis, with high mortality rates between 20% and 30% [72]. It is mainly distributed in the field, soil, contaminated water sources, and decaying vegetation. It is also categorized as a psychrotrophic microorganism, being ubiquitous in food-processing environments. Consequently, *L. monocytogenes* is often found as a post-contamination pathogen in food products like sliced cooked meat products, smoked fish, cut vegetables, or ready-to-eat (RTE) products. Raw chicken, milk, and raw meat are frequently implicated in foodborne outbreaks [73]. The associated high mortality rates to pregnant women and their unborn child, neonates, elderly people, and immunocompromised people makes that its level in food products should remain low. The Commission Regulation No. 1441/2007 *on microbiological criteria for foodstuffs* states that, for *L. monocytogenes*, in the food category *RTE foods able to support the growth of L. monocytogenes other than those intended for infants and for special medical purposes, two different microbiological criteria are proposed: (i) L. monocytogenes levels should not exceed 100 cfu/g throughout the shelf life of the product and (ii) absence in 25 g of the product before the food has left the immediate control of the food business operator*, who has produced it. Their application depends on the ability of the food operator to demonstrate that the targeted food is able or not to support the growth of *L. monocytogenes* up to the end of the shelf life. Also in the United States, the limit of 100 cfu/g for *L. monocytogenes* that does not support growth of the microorganism in foods is being considered [74].

6. Risk factors associated to microbiological contamination and foodborne outbreaks

Foods can become contaminated during growth and harvesting of raw materials, storage and transport to the industry, and processing into finished products. Recontamination can also occur during transport to retail outlets and before consumption at domestic homes and/or in catering establishments. Contamination vectors are mainly animals, surfaces, environment (air, water), and people in contact with foods (food handlers). Processing conditions, packaging materials, and equipment used can also be contamination sources. Survival of microorganisms on contaminated surfaces could lead to their multiplication at high levels, thus compromising food quality and safety [75].

Animals are important reservoirs of microorganisms, and slaughter of animals could introduce high concentration of microorganisms in food industries. Zoonotic pathogens are normally present on the skin and in the gastrointestinal and respiratory tracts. Pathogens carried on the hands are also a major source of contamination [76].

Airborne contamination represents a significant medium for the microbial transfer to food products. [77] Installation of proper air filters is recommended; otherwise, microorganisms can be present together with dust, debris, or insects.

Water sources are used in the food industry as an ingredient, a processing aid, and for cleaning. Therefore, not to increase both microbial and chemical contaminations, it is important to use decontaminated water (i.e., chlorinated and electrolyzed). Water used in hand-washing facilities can pose a potential risk because of the presence of condensations, leaking pipes, or aerosols. Microorganisms colonizing these surfaces can multiply rapidly if conditions are favorable. Thus, checking the microbiological quality of water is essential to guarantee food safety.

Food handlers can act as vectors for food contamination leading to the transmission of enteric and respiratory pathogens to food, e.g., through aerosol droplets from coughing near the processing line [78]. They can also favor cross-contamination through the skin if hand-washing is not properly done.

Finally, pests, such as birds, insects, and rodents, are potentially a major contamination problem. Therefore, care must be taken to avoid their entrance into food factories. Facilities should be designed in such a way they cannot live and breed. To do so, appropriate disinsectization fumigation methods must be achieved.

The above factors when combined together may increase the risk of food contamination. According to CDC data, 1527 foodborne disease outbreaks, resulting in 29,444 cases of illness, 1184 hospitalizations, and 23 deaths were reported within 2009–2010. [79] Among the 790 outbreaks with a laboratory-confirmed illness, norovirus was the most commonly reported infection, accounting for 42% of outbreaks, followed by *Salmonella,* with 30% of outbreaks. Outbreaks caused by some pathogens were particularly severe. For example, *Listeria* outbreaks resulted in the highest proportion of persons hospitalized (82%), followed by *Clostridium botulinum* (67%). Among the 23 deaths, 22 were linked to bacteria (9 *Listeria,* 5 *Salmonella, E. coli* O157, 3 *Clostridium perfringens,* and 1 *Shigella*), and 1 was linked to norovirus. Regarding European data [22], in 2013, *Campylobacter* continued to be the most commonly reported gastrointestinal bacterial pathogen in humans in the European Union (EU) and has been so since 2005 (214,779 cases) with an EU notification rate of 64.8 per 100,000 population. However, the high mortality rate associated to *L. monocytogenes* was confirmed since 191 deaths were reported in 2013, much higher than deaths associated to *Campylobacter* or *Salmonella* (59 or 56, respectively).

6.1. Risk factors affecting microbial safety of foods in catering establishments

6.1.1. Hygienic food handlers` practices

In production processes, storage, and distribution of prepared foods, the role of food handlers seems essential to ensure food safety, supported mainly on good hygienic practices and implementation of improved self-control measures. The food handler is defined as "anyone who by their work have direct contact with food during preparation, manufacture, processing, manufacturing, packaging, storage, transport, distribution, sale, supply and service."

Then existing laws applied to food handlers are cited as follows:

- Regulation (EC) 852/2004 of the European Parliament and of the Council of 29 April 2004 on the hygiene of foodstuffs [10].
- Regulation (EC) 882/2004 of the European Parliament and of the Council of 29 April 2004 on official controls performed to ensure the verification of compliance with the legislation on feed and food law, animal health, and animal welfare [80].

In a food catering environment, hygiene procedures may be improved as part of food poisoning occurs as a result of the risk factors associated with food handling, related to poor hygiene, improper cooking procedures, cross-contamination, or improper storage of food [81].

Handlers sometimes act as vehicles for the spread of indicators or pathogens directly and indirectly through the hands to other food-contact surfaces and handlers. In the United States, the hygienic practices of food handlers are one of the five most important risk factors of food poisoning and about 89% of the outbreaks occur mainly by inadequate hygiene [78].

In certain circumstances, the hands may represent the most important vehicle of fecal and respiratory microorganisms [82]. It has been shown that microorganisms, such as *S. aureus, E. coli,* and *S. enterica.,* can survive on the hands if hygiene measures are not sufficiently appropriate. Several studies describe outbreaks of food poisoning associated with catering establishments [83]; these studies indicate that sanitary measures may be insufficient when dealing with consumer food safety.

Ayçiçek et al. [84] evaluated the sanitary measures of food handlers in a hospital central kitchen. They concluded that these measures were insufficient as significant counts of *S. aureus* and *E. coli* in both gloved hands and bare hands were presented when handling food. Specifically, *S. aureus* positive samples were obtained in 70% of the isolates. Other microorganisms, such as *Bacillus* spp. or *Staphylococcus* coagulase negative, were also isolated. Lues and Van Torden [85] attempted to relate the microbiological contamination found in the hands of food handlers and that presented in cloths and aprons. To do this, they visited several retail points in South Africa dedicated to the selling of RTE meat. In the study, it was found aerobic mesophilic counts (hands and aprons, respectively) in 98% and 8% of analyzed samples, total coliforms (40% and 26%), Enterobacteriaceae (44% and 16%), and *S. aureus* (88% and 40%). However, they did not find any significant correlation between the microbial counts, so potential cross-contamination could not be concluded. Besides, they considered that inadequate hygiene could be a potential risk factor in the microbiological contamination of food, as 32% of the hands analyzed presented high counts for total coliforms. Fecal coliforms also prevailed in more than half of the samples from the hands (55.6%) of food handlers of several school canteens in Brazil [8]. The lack of the annual medical examination (51.9%), lack of regular training for handlers (74.1%), and poor hygiene practices (100%) could have an influence of these results, as stated by other authors [86]. Other microbial agents, such as norovirus and hepatitis A virus in humans, can survive in the hands of food handlers when they do not follow good hygienic practices [2]. In this case, contamination is often associated with asymptomatic carriers and direct hand contact with contaminated food. At the same time, contamination of food-contact surfaces is also promoted.

While hand-washing is a quick and simple method, it is also considered by many authors as the most convenient and effective way to reduce foodborne pathogens [82]. If done correctly, it prevents the risk of cross-contamination and the presence of high microbial loads in foods not submitted to intense inactivation treatments.

The principles for hand-washing are universal, though effective reduction of microorganisms depends on the following considerations [82]:

- Origin and level of organic and microbial contamination.
- The use of water power.
- Washing time (15–30 s).
- Type of soap and amount used.
- Degree of exposure to the washing process of the fingers, palms, back, wrists, nails, and subungual region.

- Frequency and intensity of rubbing fingers and palms during rinsing.

In the light of the importance of maintaining adequate hygiene habits and attitudes of food handlers to ensure food safety, various methods have been used according to legal regulations for guidance handlers' hygiene. The Theory of Planned Behavior (TPB) has been advocated by many researchers to predict determinants of food handler's behavior [87].

However, some studies warn of possible deficiencies in hygiene by food handlers. For example, Green et al. [88] conducted a study to identify factors related to the hygienic practices of food handlers. To do this, they performed an observational study that evaluated the hand-washing and use of gloves of 321 food handlers. The results showed that washing hands and wearing gloves were most frequent in food preparation activities than when handling soiled equipment and direct contact with body parts. They also found that the use of gloves resulted in a decrease in the frequency of washes hands. In Clayton and Griffith [89], the habits of 115 handlers in 29 establishments during food preparation together with their corresponding hygiene measures were analyzed. The results indicated that only 9% of handlers washed their hands after touching their hair or face and 25% washed their hands after handling of contaminated equipment. Lubran et al. [90] examined the frequency of contact of the hands of food handlers with objects and food ready for consumption before the sale, washing hands and changing gloves during food preparation, as well as hygienic measures used for the equipment, utensils, and food-contact surfaces. The study was conducted in nine retail stores of RTE food and found a high percentage (60–80%) of handlers that did not wash their hands when handling food while maintaining contact with other surfaces. Likewise, the hygiene of the hands of food handlers in 15 retail shops was evaluated in another study [91]. Out of the 29 food handlers examined, only 48.3% used soap to wash their hands properly and rubbed hands including washing between fingers, fingertips, and wrists. They also found that most handlers who washed their hands in less than 10 s (41.4%) had slightly higher levels of contamination of aerobic mesophilic bacteria and *S. aureus* than those food handlers who took a longer time for hand-washing. To this end, some studies [82] highlight the importance of rubbing hands and increase in the frequency of hand-washing to have higher efficiency to remove microbial load.

On the other hand, hand-drying is also a very important in the hand-washing stage. Some authors consider most critical is the last washing stage and needs to be implemented correctly to ensure proper hygiene, thus reducing the risk of cross-contamination [92]. According to the study conducted by Michaels et al. [93], effective hand-drying may reduce microbial population up to 90%. As result of inappropriate drying practices, residual moisture of hands, drying hands with sheets of cloth, or inefficient air dryers that lengthen the drying time are encountered [94]. It should be pointed out that rubbing hands during drying could promote skin contamination after washing. In an observational study by Clayton and Griffith [89], it was found that the effectiveness of hand-washing was affected by a high percentage of inappropriate drying practices (61%).

6.1.2. Availability of health resources

Obviously, the availability of resources and the functioning of health facilities (sink, hot water, soap, etc.) in the catering establishments constitute also another factor that may adversely affect the personal hygiene practices of food handlers [94]. Another study aimed at evaluating the implementation of the HACCP system and knowledge of hygiene and food safety as well as food handlers' attitudes and practices in 20 establishments in Spain. [95] revealed that the difficulties encountered were improper maintenance of sinks and showers in 95% of the establishments visited, as well as the availability of hot water in bathrooms and changing rooms (50%). Likewise, another study evaluated 123 food retail outlets in the United States to investigate the knowledge and hygienic practices of food handlers, as well as the availability of resources for hand-washing [96]. The results indicated that the main constraints were linked to the absence of brushes to wash the nails (38%) and insufficient sinks (24%) concluding that only the fully equipped establishments (55%) had a properly trained supervisory staff.

6.1.3. Gloves

It is already demonstrated that hand-washing does not always guarantee complete removal of microorganisms. Thus, the use of gloves is necessary to prevent direct hand contact with food and food-contact surfaces as a measure of increasing food safety and minimizing risks of cross-contamination in the food industry. The most important issue is that sometimes the use of gloves can create a false sense of hygiene among handlers [97] and jeopardize the food safety.

It should be noted that the irregular change of gloves as well as their improper use could enhance cross-contamination. In the United States, this is attributed as one of the main reasons that favor the occurrence of food poisoning [98]. In another study [99], it was found that the use of the same pair of gloves for handling different foods increases the risk of transfer of pathogens. In this case, they found an increased risk of transfer of *L. monocytogenes* from contaminated raw chicken to cooked meat slices. In turn, if gloves are changed, this risk was minimized. Besides, not only the frequency of gloves changing but also their integrity is highly important to avoid cross-contamination. Some authors state that washing hands before and after the change of gloves is highly desirable to reduce this microbial contamination [81]. Even the process of changing gloves is fraught with hazards, because many glove materials cause excess moisture build-up, causing difficulty in disinfecting contamination from the nail region to spread all over the hand. However, gloves should be changed regularly because the risk of transmission of foodborne pathogens could be higher from dirty, unchanged gloves than from bare hands.

The pros and cons of using gloves are extensively explained [97]. However, it should be highlighted that a proper glove change must be done because their efficiency as physical barrier for microorganisms is limited over working time.

Besides this, physical properties of gloves (tensile strength, flexibility, resistance to puncture, and tears), material used (natural rubber gloves, vinyl, nitrile, polyurethane), and glove features (single use vs. multiuse, powder-free, allergenic reactions, etc.) should be primarily considered in the quality control system as they have a great impact on pathogens transmission from food handlers to prepared meals.

Gloves can be made of different types of material, according to their use and physical properties. Polyethylene copolymer gloves could be the least expensive of all glove types. Density materials are quite variable, and they usually have a loose fit. They are built for using in short-time periods, and some glove types contain antibacterial compounds. However, their use is quite limited in comparison to other gloves.

Vinyl gloves (PVC) can be used as an alternative to latex gloves being more resistant to heat damage. However, they are susceptible to being torn with snag on nail edges. They also have a limited use in food industries because of their short shelf life.

Nitrile gloves were also developed to replace latex gloves. They are less elastic but have a longer shelf life. They are more resistant to chemicals but sensitive to alcohol degradation. Although they are sensible to be torn, their garish color helps them to be distinguished within a food lot.

Natural rubber latex gloves are most commonly used because they are most cost-efficient and comfortable. They provide good tactile sensitivity and good dexterity. However, latex and chemical compounds added to the gloves can produce allergies and migration of particles to food, especially in the presence of bleach.

In summary, according to the intended use, convenience, and cost-effectiveness, glove material should be carefully chosen. However, it has been demonstrated that regardless of the gloves used, handling practices and gloves changing are critical steps that influence microbial transfer to foods.

6.1.4. Cross-contamination

As mentioned above, the cross-contamination is a major cause of food poisoning worldwide. Cross-contamination phenomena arise as a consequence of the application of inadequate hygiene practices, contact with contaminated equipment and utensils, by direct hand contact with foods prepared by the improper storage of food, bad processing food, by direct food contact with air or contaminated environments, and so on [94]. Recontamination routes and sources (e.g., raw materials, food contact surfaces, food handlers) were revised [100] demonstrating their relevance to foodborne disease outbreaks. These information sources should also be incorporated in Quantitative Microbiological Risk Assessments (QMRAs) to perform mitigation strategies and reduce foodborne disease [101].

Particularly, RTE foods are highly susceptible to be contaminated during handling. Some studies refer to the direct contact of the RTE food (e.g., meat) with food-contact surfaces and contaminated cutting utensils [58]. It is also emphasized its importance as a source of transmission of enteric pathogens, *L. monocytogenes*, or *S. aureus* [102].

6.1.5. Food storage

Temperature is the most important factor that governs microbial growth in food. Most microorganisms grow at temperatures between 5°C and 60°C (called danger zone), being the optimum growth temperature at 37°C. Thus, maintaining the cold chain and a correct heat treatment for hot foods are essential measures to maintain food safety, where food handlers play a key role throughout the production chain.

Table 4 presents the recommended temperature storage/transport and freezing of raw materials and finished products in catering establishments.

Maintaining food to inadequate time intervals/storage temperature often constitutes one of the most common risk factors for food poisoning. Previous studies indicate that in school canteens and other related catering establishments exposure to abuse refrigerated temperature for extended periods of time could lead to an increase of pathogens at hazardous levels for human health from preparation to the distribution thereof [103]. Other possible deficiencies are related to temperature control storage of raw and processed ready for consumption, lack of knowledge of food handlers about cooking and refrigeration temperatures suitable to prevent the growth and survival of microorganisms, inadequate cooling and warming food and preparing several hours before consumption [104], or joint cooling of raw and cooked foods [105].

Food product	**Temperatures**	
	Storage/ Transport	**Freezing**
Meat	≤7°C	≤–18°C
Meat wastes	≤3°C	≤–18°C
Broiler meat	≤4°C	≤–18°C
Minced and mechanically separated meat	≤2°C	≤–18°C
Meat preparations	≤4°C	≤–18°C
Ham, cooked meat, deli meats	0–5°C	≤–18°C
Prepared meals to be consumed within 24 h from preparation	≤8°C	
Prepared meals to be consumed after 24 h from preparation	≤4°C	
Prepared meals (frozen)	≤–18°C	
Hot meals	≥65°C	
Frozen fruits and vegetables	≤–18°C	

Table 4. Recommended storage temperatures of raw ingredients and prepared meals

6.1.6. Training of food handlers

Training of food handlers has been considered an important measure as a part of the HACCP systems, given that it helps to prevent most foodborne diseases. Although knowledge alone is not enough to change practices, food handlers with adequate knowledge can change their practices easier if they are closely supervised and supported by their onsite managers. In addition, guidance and supervision by their managers during work improve attitudes and practices [106].

For some time ago, questionnaires or "checklists" have represented an effective tool to evaluate the level of knowledge and skills on food hygiene and safety of food handlers [98]. Nevertheless, further studies detailing more sophisticated methods to obtain a greater quantity and reliability of information to improve the training of food handlers are needed.

Additionally, novel strategies leading to more effective training methods have been performed. For example, some studies proposed to strengthen the training of less experienced food handlers and validate the knowledge of those more experienced in a period not exceeding 10 years [107]. In this sense, a better knowledge on food safety by food handlers ensures better performance and motivation [108]. In this sense, it is highly important to food handlers in the HACCP systems companies to correct their attitudes and behavior at work. Also, a periodic training is found as an effective way to raise awareness of food handlers [9].

Training of food handlers in food hygiene is a mandatory requirement for the food industry. At EU level, requirements on food safety and hygiene procedures are stated in Regulation (EC) 852/2004 [10].

Specifically, in its Annex II (Chapter XII), the food business operators must ensure the following:

- "Supervision and instruction or training of food handlers in food hygiene matters, according to their work."
- "That those who are in charge of the development and maintenance of the procedure referred to in paragraph 1 of Article 5 of this Regulation or the application of relevant guides have received adequate training in regard to the application of the HACCP principles."
- "Compliance with all requirements of national legislation concerning training programs for persons working in certain food sectors."

Recently, the European Federation of Food, Agriculture and Tourism (EFFAT) and the FERCO launched the project "Food hygiene training for all" [109]. The project is based on the development of a software tool for basic training of food handlers of the contract-catering sector across Europe. This tool is available online free of charge (www.contract-catering-guide.org/food-hygiene-training-for-all) and aims at obtaining a better qualification of workers in catering establishments and also offers the opportunity for training staff in those companies that do not have sufficient resources to invest in training.

6.1.7. Intervention strategies against microbial foodborne outbreaks

The burden reduction of foodborne diseases is a major goal of societies. The strategies developed by countries to achieve this goal are numerous and very different depending on issues, such as political and socioeconomic status, actual or emerging pathogens, resources, trade (import/exports), temporal limitations, and inter-regional cooperation.

Woteki and Kineman [110] described different approaches to reducing foodborne illness and grouped them into four categories: (i) population surveillance and better outbreak detection, (ii) prevention-based regulatory approaches, (iii) information and education, and (iv) risk-based system.

Early identification of foodborne outbreaks and the implicated organism should be directed to controlling the outbreak, stopping exposure, and perhaps more importantly, preventing future outbreaks [111]. Also, a rapid and coordinated response is needed among state officials and federal agencies. Some authors [111, 112] pointed out that surveillance based on molecular analysis of foodborne pathogens involved in outbreaks and sporadic cases together with the creation of a platform to share this information would allow for anticipation of potential future episodes. In this sense, Fisher et al. [113] reported the creation in the EU of platforms where data and information on potential outbreaks of foodborne pathogens are available and can be disseminated rapidly to those who need to know; the Enter-net is a surveillance network database of bacterial enteric pathogens, while Salm-gene is a molecular typing network. In the United States, similar platforms are available [110], i.e., FoodNet, a system of disease surveillance that provides information on the incidence of foodborne illness, and Pulse-net, a common name for National Molecular Subtyping Network for Foodborne Disease Surveillance.

Regulatory agencies allow food business operators to set up performance standards in the industry through the well-known HAZARD plan. The HAZARD plan should be designed upon the analysis of the likely hazards in the food and the strategies put in place to eliminate them or to reduce them to acceptable levels [114]. This system has shown to have a very positive and crucial impact on food safety, thus on public health, as has been recognized by key organizations like the World Health Organization [115]. Special attention deserves establishments, which deliver meals to a large number of people, and even more, in those centers where there is an important proportion of consumers with a weak or impaired immune system like hospitals or nurseries. Unfortunately, in the past years it has been reported some cases where the HACCP plan was not fully implemented. This is the case of the study by Kokkinakis et al. [116], who reported that only two out of the seven major hospitals interviewed in Crete (Greece) had implemented the HACCP plan during the period of 2004–2009. These authors identified 14 crucial elements for HACCP implementation in hospitals. Shih and Wang [117], in their study on factors influencing HACCP implementation in 23 public hospitals in Taiwan, revealed that the most important concern perceived by managers was related to economic issues, i.e., "getting funds from the hospital" and "difficulty of allocation funds for facility improvement." In addition, it was shown that more support, HACCP training, and coordination with other hospitals were necessary to avoid staff reluctance to implement the HACCP plan. Shih and Wang [117] also pointed out that kitchen design and flow charts of food production are the first two issues to consider before the HACCP implementation. The lack of financial support and poor HACCP training were also reported by Garayoa et al. [95] in their survey directed to staff from 20 contract catering companies throughout Spain.

In the food industry, emerging and existing technologies should be assessed in terms of food safety [118]. However, a new concept of food safety arisen in the early 2000s, with Regulation 178/2002, *laying down the general principles and requirements of food law, establishing the European Food Safety Authority and laying down procedures in matters of food safety*. In this document, risk analysis was introduced as the pillar on which food law should be based on the aim to achieve the general objective of a high level of protection of human health and life. This risk-based approach would enable the setting of national and international targets for disease reduction as well as provide the basis for such reduction efforts [110, 119].

Education and information of consumers are highly relevant to prevent outbreaks, and above all, sporadic cases. The World Health Organization launched in 2001 an educational campaign called "Five keys to Safer Foods," where five brief and clear messages were given to food handlers: (i) keep clean, (2) separate raw and cooked, (3) cook thoroughly, (4) keep food at safe temperatures, and (5) use safe water and raw materials (WHO, 2006). Other campaigns like Fight bac™ [120] or Thermy™ [121] aimed at getting consumers informed about hygienic food handling practices the former, and the use of thermometers in the cooking of food products the latter. Other most recent campaigns like "The chicken challenge" clearly show short messages with the objective of cutting *Campylobacter* food poisoning in half by the end of 2015 [122].

The reduction of foodborne illness incidence is a challenge for governments, which should manage the different strategies to lower the risk posed by food hazards up to acceptable levels. Current knowledge and tools on risk assessment allow for science-based decision-making.

7. Conclusions and recommendations

By following a systematic approach in assessing risks from production to serving food safety managers will better define the control measures to be adopted in catering settings to prevent foodborne infections. GMPs and HACCP principles should be followed together with special training of food handlers. Although microbiological quality of prepared meals is often satisfactory, special care should be taken regarding indicator microorganisms or prevalence of pathogens such as *L. monocytogenes*. Control of time and temperature along the food chain might prevent microbial growth until risk levels. Other measures such us excluding key high-risk foods to the most susceptible population (i.e., children, elderly, immunocompromised people) would also be advisable.

Acknowledgements

AGR-1906 Project of Excellence from the Andalusia Government, RTA-2013-00055-C02-02 project from the Spanish government, and European ERDF funding are greatly acknowledged for providing material and especially human resources, making possible the continuation of risk assessment and management activities at national and European levels by our research group AGR170 "HIBRO."

Author details

Antonio Valero*, Magdevis-Yanet Rodríguez, Guiomar Denisse Posada-Izquierdo, Fernando Pérez-Rodríguez, Elena Carrasco and Rosa Maria García-Gimeno

*Address all correspondence to: avalero@uco.es

Department of Food Science and Technology, University of Cordoba, Cordoba, Spain

References

[1] Nyachuba DG. Foodborne illness: is it on the rise? *Nutrition Reviews*. 2002; 68: 257–269. ISSN 1753-4887. DOI: 10.1111/j.1753-4887.2010.00286.x.

[2] Hedberg WC, Smith SJ, Kirkland E, Radke V, Jones TF, Selman CA. Systematic environmental evaluations to identify food safety differences between outbreak and non-outbreak restaurants. The Ehs-Net Working Group. *Journal of Food Protection*. 2006; 69(11): 2697–2702.

[3] Chapman B, Eversley T, Fillion K, MaClaurin T, Powell D. Assessment of food safety practices of food service food handlers (risk assessment data): testing a communication intervention (evaluation of tools). *Journal of Food Protection*. 2010; 73(6): 1101–1107.

[4] World Health Organization. Food Safety and Foodborne Illness. Fact Sheet 237; 2007. Available at: http://www.who.int/mediacentre/factsheets/fs237/en/index.html [Accessed August 12, 2015].

[5] Dharod JM, Paciello S, Bermúdez-Millán A, Venkitanarayanan K, Damio G, Pérez-Escamilla R. Bacterial contamination of hands increases risk of cross-contamination among low-income Puerto Rican meal preparers. *Journal of Nutrition Education and Behavior*. 2009; 41(6): 389–397. DOI: 10.1016/j.jneb.2008.11.001 ISSN 1499-4046.

[6] Acikel CH, Ogur R, Yaren H, Gocgeldi E, Ucar M Kir T. The hygiene training of food handlers at a teaching hospital. *Food Control*. 2008; 19(2): 186–190. DOI: 10.1016/j.foodcont.2007.03.008.

[7] Veiros MB, Proença RPC, Santos MCT, Kent-Smith L, Rocha A. Food safety practices in a Portuguese canteen. *Food Control*. 2009; 20(10): 936–941. DOI: 10.1016/j.foodcont.2009.02.002.

[8] Santana NG, Almeida RCC, Ferreira JS, Almeida PF. Microbiological quality and safety of meals served to children and adoption of good manufacturing practices in public school catering in Brazil. *Food Control*. 2009; 20(3): 255–261. DOI: 10.1016/j.foodcont.2008.05.004.

[9] Furnari G, Molino N, Bruno S, Quaranta G, Laurenti P, Ricciardi G. Efficacy and critical implications of food handlers' professional training: analysis of an experience. *Annals of Hygiene*. 2002; 14(5): 419–426. PMID: 12508450.

[10] Regulation (EC) No. 852/2004, of 29 April 2004, on the hygiene of foodstuffs. Official Journal of the European Union 139: 1–54.

[11] Lahou E, Jacxsens L, Daelman J, Van Landeghem F, Uyttendaele M. Microbiological performance of a food safety management system in a food service operation. *Journal of Food Protection*. 2012; 75(4): 706–716. DOI: 10.4315/0362-028X.JFP-11-260.

[12] Rose J, Epstein PR, Lipp EK, Sherman BH, Bernard SM, Patz A. Climate variability and change in the United States: potential impacts on water- and foodborne diseases caused by microbiologic agents. *Environmental Health Perspective*. 2001; 109: 211–221.

[13] Newell DG, Koopmans M, Verhoef L, Duizer E, Aidara-Kane A, Sprong H, Opsteegh M, Langelaar M, Threfall J, Scheutz F, van der Giessen J, Kruse H. Food-borne diseases—the challenges of 20 years ago still persist while new ones continue to emerge. *International Journal of Food Microbiology*. 2010; 139: S3–S15. DOI: 10.1016/j.ijfoodmicro.2010.01.021.

[14] Unnevehr L. Food safety in developing countries: moving beyond exports. *Global Food Security*. 2015; 4: 24–29. DOI: 10.1016/j.gfs.2014.12.001.

[15] Forget G, Lebel J. An ecosystem approach to human health. *International Journal of Occupational and Environmental Health*. 2001; 7: S3–38.

[16] Schillhorn van Veen T. International trade and food safety in developing countries. *Food Control*. 2005; 16: 491–496. DOI: 10.1016/j.foodcont.2003.10.014.

[17] Arévalo CA. Update in hospital food science. In: Ed. Glosa SL (Ed.), Chapter 6: Novel technologies associated to hospital foodservice Spain, 2009, 95–117. ISBN: 978-84-7429-429-3.

[18] Nissen H, Rosnes JT, Brendehaug J, Kleiberg GH. Safety evaluation of sous vide-processed ready meals. *Letters in Applied Microbiology*. 2002; 35(5): 433–438. DOI: 10.1046/j.1472-765X.2002.01218.x.

[19] Devlieghere F, Vermeiren L, Debevere J. New preservation technologies: possibilities and limitations. *International Dairy Journal*. 2004; 14(4): 273–285. DOI: 10.1016/j.idairyj.2003.07.002.

[20] Uyttendaele M, Liu C, Hofstra N. Special issue on the impacts of climate change on food safety. *Food Research International*. 2015; 68: 1–6. DOI: 10.1016/j.foodres.2014.09.001.

[21] World Health Organization. Global Health Observatory Data; 2014. Available at: http://www.who.int/gho/publications/world_health_statistics/2014/en/ [Accessed October 31, 2015].

[22] European Food Safety Authority (EFSA) and European Centre for Disease Prevention and Control (ECDC). The European Union Summary Report on Trends and Sources of Zoonoses, Zoonotic Agents and Food-borne Outbreaks in 2013. *EFSA Journal*. 2015; 13(1): 3991, 162 pp. DOI: 10.2903/j.efsa.2015.3991.

[23] Centers for Disease Control and Prevention (CDC). Investigation update: multistate outbreak of *E. coli* O157:H7 infections linked to romaine lettuce; 2012. Available at: http://www.cdc.gov/ecoli/2011/ecoliO157/romainelettuce/032312/index.html [Accessed October 31, 2015].

[24] European Food Safety Authority (EFSA) and European Centre for Disease Prevention and Control (ECDC). The European Union Summary Report on Trends and Sources of Zoonoses, Zoonotic Agents and Food-borne Outbreaks in 2011. *EFSA Journal*. 2013; 11(4): 3129, 250 pp. DOI: 10.2903/j.efsa.2013.3129.

[25] Centers for Disease Control and Prevention (CDC). Investigation announcement: multistate outbreak of *E. coli* O157:H7 infections linked to romaine lettuce; 2011. Available at: http://www.cdc.gov/ecoli/2011/ecoliO157/romainelettuce/120711/index.html [Accessed October 31, 2015].

[26] Guzman-Herrador B, Vold L, Comelli H, MacDonald E, Heier BT, Wester AL, Stavnes TL, Jensvoll L, Lindegård AA, Severinsen G, Aasgaard GJ, Werner JO, Cudjoe K, Nygard K. Rapid communications. Outbreak of *Shigella sonnei* infection in Norway linked to consumption of fresh basil, October, 2011. *Eurosurveillance*. 2011; 16(44): 20007.

[27] Ethelberg S, Lisby M, Bottiger B, Schultz AC, Villif A, Jensen T, Olsen KE, Scheutz F, Kjelso C, Muller L. Outbreaks of gastroenteritis linked to lettuce, Denmark, January 2010. *Eurosurveillance*. 2010; 15(6): 19484.

[28] Friesema I, Sigmundsdottir G, van der Zwaluw K, Heuvelink A, Schimmer B, de Jager C, Rump B, Briem H, Hardardottir H, Atladottir A, Gudmundsdottir E, van Pelt W. An international outbreak of shiga toxin-producing *Escherichia coli* O157 infection due to lettuce, September–October 2007. *Eurosurveillance*. 2008; 13(50): 19065.

[29] Werner S, Boman K, Einemo I, Erntell M, Helisola R, de Jong B, Lindqvist A, Löfdahl M, Löfdahl S, Meeuwisse A, Ohlen G, Olsson G, Persson I, Runehagen A, Rydevik G, Stamer U, Sellström E, Andersson Y. Outbreak of *Salmonella stanley* in Sweden associated with alfalfa sprouts, July–August 2007. *Eurosurveillance*. 2007; 12(42): E071018.2.

[30] Centers for Disease Control and Prevention (CDC). Investigation update: multistate outbreak of human *E. coli* O145 infections linked to shredded romaine lettuce from a single processing facility; 2010. Available at: http://www.cdc.gov/ecoli/2010/ecoli_O145/index.html [Accessed October 31, 2015].

[31] Health Protection Agency (HPA). Update – outbreak of *Salmonella newport* infection in England, Scotland, and Northern Ireland: association with the consumption of lettuce. Communicable Disease Report (CDR) Weekly [serial online] 14, *News*; 2004. Available at: http://www.hpa.org.uk/cdr/archives/2004/cdr4104.pdf [Accessed October 31, 2015].

[32] Kapperud G, Rørvik LM, Hasseltvedt V, Høiby EA, Iversen BG, Staveland K, Johnsen G, Leitao J, Herikstad H, Anderson Y. Outbreak of *Shigella sonnei* infection traced to imported iceberg lettuce. *Journal of Clinical Microbiology*. 1995; 33(3): 609–614. PMCID: PMC227998.

[33] Centers for Disease Control and Prevention (CDC). Multistate outbreak of human *Salmonella* Heidelberg infections linked to ground Turkey (final update); 2011. Avail-

able at: http://www.cdc.gov/salmonella/2011/ground-turkey-11-10-2011.html [Accessed November 5, 2015].

[34] Centers for Disease Control and Prevention (CDC). Multistate outbreak of human *Salmonella* montevideo infections (final update); 2010. Available at: http://www.cdc.gov/salmonella/2010/montevideo-5-4-2010.html [Accessed November 5, 2015].

[35] Meldrum RJ, Mannion PT, Garside J. Microbiological quality of ready-to-eat food served in schools in Wales, United Kingdom. *Journal of Food Protection*. 2009; 72(1): 197–201. PMID: 19205487.

[36] Public Health Agency of Canada. *Listeria monocytogenes* outbreak update; 2009. Available at: http://www.phac-aspc.gc.ca/alert-alerte/listeria/listeria_2008-eng.php [Accessed October 31, 2015].

[37] Grahek-Ogden D, Schimmer B, Cudjoe KS, Nygård K, Kapperud G. Outbreak of *Yersinia enterocolitica* serogroup O:9 infection and processed pork, Norway. *Emerging Infectious Diseases*. 2007; 13(5): 754–756. DOI: 10.3201/eid1305.061062.

[38] Sekse C, O'Sullivan K, Granum PE, Rørvik LM, Wasteson Y, Jørgensen HJ. An outbreak of *Escherichia coli* O103:H25-Bacteriological investigations and genotyping of isolates from food. *International Journal of Food Microbiology*. 2009; 133(3): 259–264. DOI: 10.1016/j.ijfoodmicro.2009.05.026.

[39] European Food Safety Authority (EFSA). The community summary report on trends and sources of zoonoses, zoonotic agents and antimicrobial resistance and foodborne outbreaks in the European Union in 2005. *The EFSA Journal*. 2006; 94: 2–288. ISBN: 978-92-9199-046-7.

[40] Gottlieb SL, Newbern EC, Griffin PM, Graves LM, Hoekstra RM, Baker NL, Hunter SB, Holt KG, Ramsey F, Head M, Levine P, Johnson G, Schoonmaker-Bopp D, Reddy V, Kornstein L, Gerwel M, Nsubuga J, Edwards L, Stonecipher S, Hurd S, Austin D, Jefferson MA, Young SD, Hise K, Chernak ED, Sobel J, Listeriosis Outbreak Working Group. Multistate outbreak of listeriosis linked to turkey deli meat and subsequent changes in US regulatory policy. *Clinical Infection Diseases*. 2006; 42(1): 29–36.

[41] Rajpura A, Lamden K, Forster S, Clarke S, Cheesbrough J, Gornall S, Waterworth S. Large outbreak of infection with *Escherichia coli* O157 PT21/28 in Eccleston, Lancashire, due to cross contamination at a butcher's counter. *Communicable Disease and Public Health*. 2003; 6(4): 279–284. PMID: 15067851.

[42] Meng J, Doyle MP, Zhao T, Zhao S. Enterohemorrhagic *Escherichia coli*. In: Doyle MP, Beuchat LR, Montville TJ (Eds.), Food Microbiology: Fundamentals and Frontiers. Washington, DC: ASM Press, 2001, 193–213.

[43] Gilbert S, Lake R, Hudson A, Cressey P. Risk profile: listeria monocytogenes in processed ready-to-eat meats; 2009. Available at: http://foodsafety.govt.nz/elibrary/indus-

try/Risk_Profile_Listeria_Monocytogenes_Processed-Science_Research.pdf [Accessed October 31, 2015].

[44] Olsen SJ, Patrick M, Hunter SB, Reddy V, Kornstein L, MacKenzie WR, Lane K, Bidol S, Stoltman GA, Frye DM, Lee I, Hurd S, Jones TF, LaPorte TN, Dewitt W, Graves L, Wiedmann M, Schoonmaker-Bopp DJ, Huang AJ, Vincent C, Bugenhagen A, Corby J, Carloni ER, Holcomb ME, Woron RF, Zansky SM, Dowdle G, Smith F, Ahrabi-Fard S, Ong AR, Tucker N, Hynes NA, Mead P. Multi-state outbreak of *Listeria monocytogenes* infection linked to delicatessen turkey meat. *Clinical Infectious Diseases*. 2005; 40(7): 962–967. PMID: 15824987.

[45] Silva Jr, EA. Manual of hygienic control in food service systems (Sixth edition). São Paulo: Varela, 2007, Vol. 623. ISBN: 8585519533.

[46] Commission Regulation (EC) No 1441/2007 of 5 December 2007 amending Regulation (EC) No 2073/2005 on microbiological criteria for foodstuffs. *Official Journal of the European Union*. 322: 12–29.

[47] Ray B. Fundamental Food Microbiology (3rd edition). CRC Press, 2004, 608. LLC. N.W. Boca Raton, FL: Corporate Blvd., 2000, 33431. ISBN: 0-8493-1610-3.

[48] Caballero-Torres AE. Lessons in food hygiene. 2008. Medical Sciences (Ed.), página 379. National Center of information on medical sciences La Habana, Cuba. ISBN: 978-959-212-363-2.

[49] Jay JM. Modern Food Microbiology (5th edition). New York: Chapman and Hall, 1996. ISBN: 0412076918.

[50] Health Protection Agency. Guidelines for Assessing the Microbiological Safety of Ready-to-Eat Foods. London: Health Protection Agency, 2009.

[51] Parra-Huertas RA. Review: lactic acid bacteria: functional role in foods. *Agrofisheries Faculty*. 2010; 8(1). ISSN: 1692-3561.

[52] Vermeiren V, Devlieghere F, Debevere J. Co-culture experiments demonstrate the usefulness of *Lactobacillus sakei* 10A to prolong the shelf-life of a model cooked ham. *International Journal of Food Microbiology*. 2006; 108(1): 68–77. DOI: 10.1016/j.ijfoodmicro.2005.11.001.

[53] Hammes WP, Bantleón A, Min S. Lactic acid bacteria in meat fermentation. *FEMS Microbiology Letters*. 1990; 87(1–2): 165–174. DOI: 10.1111/j.1574-6968.1990.tb04886.x.

[54] Borch E, Kant-Muermans ML, Blixt Y. Bacterial spoilage of meat and cured meat products. *International Journal of Food Microbiology*. 1996; 33(1): 103–120.

[55] Nychas GJE, Skandamis PN, Tassou CC, Koutsoumanis KP. Meat spoilage during distribution. *Meat Science*. 2008; 78(1–2): 77–89. DOI: 10.1016/j.meatsci.2007.06.020.

[56] Rodríguez-Caturla M, Valero A, Carrasco E, Pérez-Rodríguez F, Posada-Izquierdo GD, Zurera G. Hygienic conditions and microbiological status of chilled ready-to-eat

products served in Southern Spanish Hospitals. *Food Control*. 2011; 22(6): 874–882. DOI: 10.1016/j.foodcont.2010.11.015.

[57] Valero A, Rodríguez-Caturla M, Carrasco E, Pérez-Rodríguez F, García-Gimeno RM, Zurera G. Studying the growth boundary and subsequent time to growth of pathogenic *Escherichia coli* serotypes by turbidity measurements. *Food Microbiology*. 2010; 27(6): 819–828. DOI: 10.1016/j.fm.2010.04.016.

[58] Al-Mutairi MF. The incidence of Enterobacteriaceae causing food poisoning in some meat products. *Advanced Journal of Food Science and Technology*. 2011; 3(2): 116–121. ISSN: 2042-4876.

[59] Shaker R, Osailia T, Al-Omary W, Jaradata Z, Al-Zuby M. Isolation of *Enterobacter sakazakii* and other *Enterobacter* sp. from food and food production environments. *Food Control*. 2007; 18(10): 1241–1245. DOI: 10.1016/j.foodcont.2006.07.020.

[60] Warren BR, Parish ME, Schneider KR. *Shigella* as a foodborne pathogen and current methods for detection in food. *Critical Reviews in Food Science and Nutrition*. 2006; 46(7): 551–567. DOI: 10.1080/10408390500295458.

[61] Chu P, Hemphill RR. Acquired hemolytic anemia. In: Tintinalli JE, Kelen GD, Stapczynski JS (Eds.), Emergency Medicine: A Comprehensive Study Guide (6th edition). New York, NY: McGraw-Hill, 2004. ISBN: 0-07-138875-3.

[62] European Food Safety Authority (EFSA). The European Union summary report on trends and sources of zoonoses, zoonotic agents and foodborne outbreaks in 2009. *The EFSA Journal*. 2010; 8(1): 1496. DOI: 10.2903/j.efsa.2010.1496.

[63] Pérez-Rodríguez F, Valero A, Todd ECD, Carrasco E, García-Gimeno RM, Zurera G. Modeling transfer of *Escherichia coli* O157:H7 and *Staphylococcus aureus* during slicing of a cooked meat product. *Meat Science*. 2007; 76(4): 692–699. DOI: 10.1016/j.meatsci.2007.02.011.

[64] Sivapalasingam S, Friedman CR, Cohen L, Tauxe RV. Fresh produce: a growing cause of outbreaks of foodborne illness in the United States, 1973 through 1997. *Journal of Food Protection*. 2004; 67: 2342–2353.

[65] Strachan NJC, MacRae M, Ogden ID. Quantifying the *Escherichia coli* O157 reservoir in Grampian, Scotland. *Veterinary Record*. 2005; 156: 282–283. DOI: 10.1136/vr.156.9.282.

[66] Scavia G, Botta A, Ciofi degli Atti ML, Di Fluri G, Ferretti A, Galero G, Marziano ML, Merla R, Minelli F, Montini G, Pecoraio C, Pizzuti, R, Tozzi AE, Trani AM, Caprioli A. Reported outbreak of Hemolytic Uremic Syndrome (HUS) associated to *Escherichia coli* O26 infection, in Salerno province (Italy); 2005. Available at: http://www.epicentro.iss.it/problemi/seu/seu.asp [Accessed August 20, 2015].

[67] Wallin-Carlquist N, Márta D, Borch E, Radström P. Prolonged expression and production of *Staphylococcus aureus* enterotoxin A in processed pork meat. *International*

Journal of Food Microbiology. 2010; 141: S69–S74. DOI: 10.1016/j.ijfoodmicro.2010.03.028.

[68] Khakhria R, Woodward D, Johnson WM, Poppe C. *Salmonella* isolated from humans, animals, and other sources in Canada. *Epidemiology and Infection*. 1997; 119: 15–23. PMCID: PMC2808817.

[69] Pesavento G, Ducci B, Comodo N, Lo Nostro A. Antimicrobial resistance profile of *Staphylococcus aureus* isolated from raw meat: a research for methicillin resistant *Staphylococcus aureus* (MRSA). *Food Control*. 2007; 18: 196–200. DOI: 10.1016/j.foodcont.2005.09.013.

[70] Asao T, Kumeda Y, Kawai T, Shibata T, Oda H, Haruki K, Nakazawa H, Kozaki S. An extensive outbreak of staphylococcal food poisoning due to low-fat milk in Japan: estimation of enterotoxin A in the incriminated milk and powdered skim milk. *Epidemiology and Infection*. 2003; 130: 33–40. DOI: 10.1017/S0950268802007951.

[71] Lindqvist R, Sylvén S, Vågsholm I. Quantitative microbial risk assessment exemplified by *Staphylococcus aureus* in unripened cheese made from raw milk. *International Journal of Food Microbiology*. 2002; 78: 155–170. DOI: 10.1016/S0168-1605(02)00237-4.

[72] Rocourt J. Risk factors for listeriosis. *Food Control*. 1996; 7(4/5): 195–202. DOI: 10.1016/S0956-7135(96)00035-7.

[73] Bell C, Kyriakides A. Listeria Monocytogenes. In: Blackburn CdW, McClure PJ (Eds.), Foodborne Pathogens—Hazards, Risk Analysis and Control. Cambridge, England: Woodhead Publishing Limited, 2002, 337–361. ISBN: 9788420009117.

[74] Federal Register. The Daily Journal of the United States Government. Best Practices Guidance for Controlling *Listeria monocytogenes* in Retail Delicatessens. Available at: https://www.federalregister.gov/articles/2015/06/11/2015-14330/best-practices-guidance-for-controlling-listeria-monocytogenes-in-retail-delicatessens [Accessed August 10, 2015].

[75] Bridier A, Sanchez-Vizuete P, Guilbaud M, Piard JC, Naïtali M, Briandet R. Biofilm-associated persistence of food-borne pathogens. *Food Microbiology*. 2015; 45: 167–178. DOI: 10.1016/j.fm.2014.04.015.

[76] Taylor JH, Holah JT. Hand Hygiene in the Food Industry: A Review. Chipping Campden: Review 18, Campden & Chorleywood Food Research Association; 2000.

[77] Brown KL, Wray S. Control of Airborne Contamination in Food Processing. In: Hygiene in Food Processing. Principles and Practice (2nd edition),. Woodhead Publishing Series in Food Science and Nutrition, 2014, 174–202.

[78] Guzewich J, Ross P. Evaluation of risks related to microbiological contamination of ready-to-eat food by food preparation workers and the effectiveness of interventions to minimise those risks. Food and Drug Administration White Paper, FDA, CFSAN, 1999. Available at: http://cfsan.fda.gov/~ear/. [Accessed January 31, 2015].

[79] Centers for Disease Control and Prevention (CDC). Tracking and Reporting Foodborne Disease Outbreaks; 2015. Available at: http://www.cdc.gov/features/dsfoodborneoutbreaks/ [Accessed October 31, 2015].

[80] Regulation (EC) No. 882/2004 of the European Parliament and of the Council of 29 April 2004 on official controls performed to ensure the verification of compliance with feed and food law, animal health and animal welfare rules. *Official Journal of the European Union* 165: 1–139.

[81] Michaels B, Keller C, Blevins M, Paoli G, Ruthman T, Todd E, Griffith CJ. Prevention of food worker transmission of foodborne pathogens: risk assessment and evaluation of effective hygiene intervention strategies. *Food Service Technology*. 2004; 4(1): 31–49. DOI: 10.1111/j.1471-5740.2004.00088.x.

[82] Todd E, Michaels B, Greig J, Smith D, Bartleson C. Outbreaks where food workers have been implicated in the spread of foodborne disease. Part 9. Washing and drying of hands to reduce microbial contamination. *Journal of Food Protection*. 2010; 73(10): 1937–1955.

[83] Lynch RA, Elledge BL, Charles PH, Griffith CC, Boatright DT. A comparison of food safety knowledge among restaurant managers, by source of training and experience, in Oklahoma County, Oklahoma. *Journal of Environmental Health*. 2003; 66(2): 9–14. PMID: 12971041.

[84] Ayçiçek H, Aydoğan H, Küçükkaraaslan A, Baysallar M, Başustaoğlu AC. Assessment of the bacterial contamination on hands of hospital food handlers. *Food Control*. 2004; 15(4): 253–259. DOI: 10.1016/S0956-7135(03)00064-1.

[85] Lues JFR, Van Tonder I. The occurrence of indicator bacteria on hands and aprons of food handlers in the delicatessen sections of a retail group. *Food Control*. 2007; 18(4): 326–332. DOI: 10.1016/j.foodcont.2005.10.010.

[86] Couto-Campos AK, Soares Cardonha AM, Galvão Pinheiro LB, Rocha Ferreira N, Medeiros de Azevedo PR, Montenegro Stamford TL. Assessment of personal hygiene and practices of food handlers in municipal public schools of Natal, Brazil. *Food Control*. 2009; 20(9): 807–810. DOI: 10.1016/j.foodcont.2008.10.010.

[87] Seaman P, Eves A. Efficacy of the theory of planned behaviour model in predicting safe food handling practices. *Food Control*. 2010; 21(7): 983–987. DOI: 10.1016/j.foodcont.2009.12.012.

[88] Green LR, Radke V, Mason R, Bushnell L, Reimann DW, Mack JC, Motsinger MD, Stigger T, Selman CA. Factors related to food worker hand hygiene practices. *Journal of Food Protection*. 2007; 70(3): 661–666.

[89] Clayton DA, Griffith CJ. Observations of food safety practices in catering using notational analysis. *British Food Journal*. 2004; 106(3): 211–227. DOI: 10.1108/00070700410528790.

[90] Lubran MB, Pouillot R, Bohm S, Calvey EM, Meng J, Dennis S. Observational study of food safety practices in retail deli departments. *Journal of Food Protection*. 2010; 73(10): 1849–1857. PMID: 21067673.

[91] Fawzi M, Gomaa NF, Bakr WM. Assessment of hand washing facilities, personal hygiene and the bacteriological quality of hand washes in some grocery and dairy shops in Alexandria, Egypt. *Egypt Public Health Association*. 2009; 84(1): 71–93.

[92] Michaels B. Are gloves the answer? *Dairy, Food and Environmental Sanitation*. 2001; 21: 489–492.

[93] Michaels B, Gangar V, Meyers E, Johnson H, Curiale M. The significance of hand drying after handwashing. *Journal of Food Protection*. 2000; 63(Suppl. A): 106.

[94] Todd E, Michaels B, Greig, J, Smith D, Holah J, Bartleson C. Outbreaks where food workers have been implicated in the spread of foodborne disease. Part 7. Barriers to reduce contamination of food by workers. *Journal of Food Protection*. 2010; 73(8): 1552–1565.

[95] Garayoa R, Vitas AI, Díez-Leturia M, García-Jalón I. Food safety and the contract catering companies: food handlers, facilities and HACCP evaluation. *Food Control*. 2011; 22(12): 2006–2012. DOI: 10.1016/j.foodcont.2011.05.021.

[96] Allwood PB, Jenkins T, Paulus C, Johnson L, Hedberg CW. Hand washing compliance among retail food establishment workers in Minnesota. *Journal of Food Protection*. 2004; 67(12): 2825–2828. PMID: 15633696.

[97] Todd E, Michaels B, Greig J, Smith D, Bartleson C. Outbreaks where food workers have been implicated in the spread of foodborne disease. Part 8. Gloves as barriers to prevent contamination of food by workers. *Journal of Food Protection*. 2010; 73(9): 1762–1773.

[98] McCarthy PV, Guzewich JJ, Braden CR, Klontz KC, Hedberg CW, Fullerton KE, Bogard A, Dreyfuss M, Larson K, Vugia D, Nichols DC, Radke VJ, Shakir FK, Jones TF. Contributing factors (CFs) identified in produce-associated outbreaks from CDC's National Electronic Foodborne Outbreak Reporting System (eFORS), FoodNet sites, 1999–2002. Presented at the International Conference on Emerging Infectious Diseases, Atlanta, March 19–22, 2006.

[99] Pérez-Rodríguez F, Todd ECD, Valero A, Carrasco E, García RM, Zurera G. Linking quantitative exposure assessment and risk management using the food safety objective concept: an example with *Listeria monocytogenes* in different cross-contamination scenarios. *Journal of Food Protection*. 2006; 69(10): 2384–2394.

[100] Reij MW, den Aantrekker WD, ILSI Europe Risk Analysis in Microbiology Task Force. Recontamination as a source of pathogens in processed foods. *International Journal of Food Microbiology*. 2004; 91: 1–11. DOI: 10.1016/S0168-1605(03)00295-2.

[101] Pérez-Rodríguez F, Valero A, Carrasco E, García R, Zurera G. Understanding and modelling bacterial transfer to foods: a review. *Trends in Food Science & Technology*. 2008; 19: 130–143. DOI: 10.1016/j.tifs.2007.08.003.

[102] Malheiros P, Passos C, Casarin L, Serraglio L, Tondo EC. Evaluation of growth and transfer of *Staphylococcus aureus* from poultry meat to surfaces of stainless steel and polyethylene and their disinfection. *Food Control*. 2010; 21(3): 298–301. DOI: 10.1016/j.foodcont.2009.06.008.

[103] Rodríguez-Caturla M, Valero A, Carrasco E, Posada-Izquierdo GD, García-Gimeno RM, Zurera G. Evaluation of hygienic practices and microbiological status of ready-to-eat vegetable salads in Spanish school canteens. *Journal of the Science Food of Agriculture*. 2012; 92(11): 2332–2340. DOI: 10.1002/jsfa.5634.

[104] Jianu C, Chis A. Study on the hygiene knowledge of food handlers working in small and medium-sized companies in western Romania. *Food Control*. 2012; 26: 151–156. DOI: 10.1016/j.foodcont.2012.01.023.

[105] Gomes-Neves E, Araujo AC, Ramos E, Cardoso CS. Food handling: comparative analysis of general knowledge and practice in three relevant groups in Portugal. *Food Control*. 2007; 18(6): 707–712. DOI: 10.1016/j.foodcont.2006.03.005.

[106] Vo TH, Lea NH, Lea ATN, Minh NNT, Nuorti JP. Knowledge, attitudes, practices and training needs of food-handlers in large canteens in Southern Vietnam. *Food Control*. 2015; 57: 190–194. DOI: 10.1016/j.foodcont.2015.03.042.

[107] Hislop N, Shaw K. Food safety knowledge retention study. *Journal of Food Protection*. 2009; 72(2): 431–435. PMID: 19350994.

[108] Gomes-Neves E, Cardoso CS, Araújo AC, Correia da Costa JM. Meat handlers training in Portugal: a survey on knowledge and practice. *Food Control*. 2011; 22(3–4): 501–507. DOI: 10.1016/j.foodcont.2010.09.036.

[109] European Federation of Trade Unions in Food, Agriculture and Tourism (EFFAT)-European Federation of Contract Catering Organisations (FERCO). Food Hygiene Training for All; 2011. Available at: www.contract-catering-guide.org/food-hygiene-training-for-all [Accessed August 15, 2015].

[110] Woteki CE, Kineman BD. Challenges and approaches to reducing foodborne illness. *Annual Review of Nutrition*. 2003; 23: 315–344. DOI: 10.1146/annurev.nutr.23.011702.073327.

[111] Majkowski J. Strategies for rapid response to emerging foodborne microbial hazards. *Emerging Infectious Diseases*. 1997; 3: 551–554. DOI: 10.3201/eid0304.970420.

[112] Tauxe R. Emerging foodborne diseases: an evolving public health challenge. *Emerging Infectious Diseases*. 1997; 3: 425–434. DOI: 10.3201/eid0304.970403.

[113] Fisher IST, Threlfall EJ, Enter-net Salm-gene P. The Enter-net and Salm-gene databases of foodborne bacterial pathogens that cause human infections in Europe and be-

yond: an international collaboration in surveillance and the development of intervention strategies. *Epidemiology and Infection.* 2005; 133: 1–7.

[114] Woteki CE, Glavin MO, Kineman BD. HACCP as a model for improving food safety. In: *Perspectives in World Food and Agriculture.* 2003 (eds C.G. Scanes and J.A. Miranowski), Iowa State Press (USA). DOI: 10.1002/9780470290187.ch7

[115] Motarjemi Y, Kaferstein F, Moy G, Miyagawa S, Miyagishima K. Importance of HACCP for public health and development – the role of the World Health Organization. *Food Control.* 1996; 7: 77–85. DOI: 10.1016/0956-7135(96)00003-5.

[116] Kokkinakis E, Kokkinaki A, Kyriakidis G, Markaki A, Fragkiadakis GA. HACCP implementation in public hospitals: a survey in Crete, Greece. *11th International Congress on Engineering and Food (Icef11).* 2011; 1: 1073–1078.

[117] Shih KM, Wang WK. Factors influencing HACCP implementation in Taiwanese public hospital kitchens. *Food Control.* 2011; 22: 496–500. DOI: 10.1016/j.foodcont.2010.09.034.

[118] Juneja VK. Intervention strategies for control of foodborne pathogens. Conference on Monitoring Food Safety, Agriculture and Plant Health. Spie-Int Soc Optical Engineering, Providence, RI, 2003, 147–160.

[119] Schlundt J. New directions in foodborne disease prevention. *International Journal of Food Microbiology.* 2002; 78: 3–17. DOI: 10.1016/S0168-1605(02)00234-9.

[120] U.S. Food & Drug Administration. Fight bac! Keep Food Safe from Bacteria; 2002. Available at: http://www.fightbac.org/ [Accessed January 13, 2016].

[121] U.S. Food & Drug Administration. Thermy; 2002. Available at: http://www.fsis.usda.gov/wps/portal/fsis/topics/food-safety-education/teach-others/fsis-educational-campaigns/thermy [Accessed January 13, 2016].

[122] FSA. 2015. The chicken challenge; 2015. Available at: https://www.food.gov.uk/news-updates/campaigns/chicken-challenge-2015 [Accessed January 13, 2016].

4

Herbal Extracts – Possibility of Preventing Food-Borne Infection

Biljana Damjanović-Vratnica

Abstract

Despite the high degree of awareness of food preservation methods, there is increasing occurrence of disease outbreaks caused by pathogenic and spoilage microorganisms in foods. Due to consumer awareness and negative perception of artificial preservatives in food, in recent years attention is shifting toward alternatives that the consumers recognize as natural. Thus, herbal extracts are now getting more space in food industry to prevent the propagation of bacteria that affect the spoilage of food or for the spread of so-called food-borne diseases. Herbal extracts, particularly essential oils (EOs), have complex composition that quality and composition depend on the method of extraction. There are now numerous reports of the in vitro antimicrobial activity of EOs in the scientific and medical literature: EOs are found to have broad-spectrum inhibitory activities against various food-borne Gram-positive and Gram-negative pathogenic bacteria. In this chapter, definition, history, and economic importance of aromatic herbs and herbal extracts, particularly EOs, are described. Also, attention has been paid to techniques for extraction, as well as chemical composition and antimicrobial activity of herbal extracts. This chapter demonstrates the possibility of usage of herbal extracts in preventing food-borne infection through literature survey and original results.

Keywords: Aromatic herbs, herbal extracts, food-borne disease, food preservation

1. Introduction

In recent years, there has been a dramatic increase of reported cases of food-borne illness because of consumption of foods contaminated with pathogenic bacteria. The presence of various microorganisms in food followed by inappropriate storage results not only in a reduction of food quality but also in spoilage of food.

Food contaminants, such as harmful parasites, bacteria, viruses, prions, and chemical or radioactive substances, cause more than 200 diseases – ranging from infectious diseases to cancers [1]. Consequently, there is considerable interest in ways to stop this upward trend and reduce the incidence of food poisoning. The development of new and improved methods of food infectious intestinal disease preservation is of utmost importance [2]. The microbiological safety in ready-to-eat products is a cause of big concern not only for the consumers and food industries but also for the regulatory agencies. It can be said that the food industry faces a constant problem in providing the food where there are no food-borne pathogens present [3, 4]. The number of documented outbreaks of food-borne diseases has increased in the last decade, with *Salmonella* spp., *Listeria monocytogenes*, and *Escherichia coli* being responsible for the largest number of outbreaks and deaths [5].

Due to consumer awareness and negative perception of artificial preservatives in food, in recent years attention is shifting toward alternatives that the consumers recognize as natural, for example, herbal extracts. It can be said that herbal extracts have been relatively neglected as soon as modern antibiotics were discovered; they became the primary means of treating bacterial infections. Antibiotics have selective toxicity and that many important bacteria were exquisitely susceptible to them. However, bacterial infections have gone up even after the discovery of many antibiotics, mainly due to the reduced susceptibility to conventional antimicrobial agents shown by many important pathogens. Because of that, studies of the antimicrobial activity of plant extracts are intensified [6] and became prominent in science, this area has suddenly become more important in the scientific literature, and this issue is currently most attractive. Therefore, at the beginning of the 21st century, herbal extracts are getting more space in food industry to prevent the propagation of bacteria that are responsible for the spoilage of food or the spread of so-called food-borne diseases [7].

Herbal extracts, particularly essential oils (EOs) have complex composition, containing from a few dozen to several hundred constituents, especially hydrocarbons (terpenes and sesquiterpenes) and oxygenated compounds (alcohols, aldehydes, ketones, acids, phenols, oxides, lactones, ethers, and esters). The greatest use of EOs in the European Union (EU) is in food (as flavorings), perfumes (as fragrances and aftershaves), and pharmaceuticals (for their functional properties). The aroma oil is the result of the combination of the main components in the oil, but trace components are also important. Thus, it is significant that the quality and composition of an EO are maintained during isolation from herb matrix, where the method of extraction plays a crucial role [8]. There are now numerous reports of the in vitro antimicrobial activity of EOs in the scientific and medical literature: EOs are found to have broad spectrum inhibitory activities against various food-borne Gram-positive and Gram-negative pathogenic bacteria.

Hydrodistillation has traditionally been applied for EO recovery from plant materials, but, during isolation, heat-sensitive compounds can easily be destroyed and quality is extremely impaired. Thus, the extraction of EO components using solvents at high pressure, or supercritical fluids, has received much attention in the recent years, especially in food, pharmaceutical, and cosmetic industries, because it presents an alternative to conventional processes such as organic solvent extraction and steam distillation [9].

In this study, the antimicrobial properties of herbal extracts obtained by hydrodistillation and supercritical carbon dioxide (CO_2) extraction from plants originated from Montenegro, sage (*Salvia officinalis*), rosemary (*Rosmarinus officinalis*), oregano (*Origanum vulgare*), and savory (*Satureja montana*), were investigated against five pathogenic bacteria important in food industry.

2. Aromatic plants and herbal extracts

Aromatic plants and herbal extracts have been used for thousands of years as incense, perfumes, and cosmetics. After the discovery of plants' medicinal properties, natural flora became a valuable source of health improvement in the ancient civilizations.

Literature dating from around 2000 BC lists over 700 substances, including ginger, myrrh, and coriander, used for therapeutic purposes in India. Chinese ancient herbal tradition was recorded in Yellow Emperor's *Book of Internal Medicine*, dating more than 2000 BC. However, the most famous and richest associations concerning the first aromatic materials are those surrounding the ancient Egyptian civilization. Through thousands of recipes, Egyptian papyruses showed that, for example, coriander and castor oils were used as cosmetics, for medicinal applications, and as preservatives [10].

Therapeutic uses of herbal plants were described by several scholars, during Greek and Roman period, namely Hippocrates, Galen, Dioscorides, and many others [11]. Romanians are known for their use of medicinal herbs since very long; thus, Herodotus was the first to mention the method of distillation of turpentine, in about 425 BC. These great Graeco-Roman works were translated into Persian, Arabic, and other languages, and at the end of Byzantine Empire, their knowledge was passed to the Arab world. However, in 1975, archeological expedition in Pakistan found perfectly preserved perfume containers and terracotta apparatus similar to distillation apparatus, from about 3000 BC. This discovery suggests that the Arabs simply revived or improved upon a process that had been known for over 4000 years. In 19th century, herbal products were introduced by Romanian pharmacopoeia, whereas in 1904 the first institute of medicinal herbs was established in Cluj city [10].

2.1. History and definition of bioactive compounds

The use of aromatic plants and herbal extracts in the past demonstrates the history of bioactive compounds usage. Typically, bioactive compounds of plants are produced as secondary metabolites [12]. Traditionally, secondary plant metabolites have been defined as all compounds synthesized by the plant that do not appear to be essential for plant growth and development and/or those compounds without an obvious function [13]. In different words, secondary metabolites are those metabolites that are often produced in a phase of subsequent to growth, have no function in growth (although they may have survival function), have unusual chemical structures, and are often formed as mixtures of closely related members of a chemical family [14]. They are not universally synthesized in all plants. In contrast, primary metabolites are produced by all plants, usually are part of essential metabolic processes of growth and development, such as proteins, carbohydrates, lipids, and amino acids.

In recent years, as interest as well as investigation of secondary metabolites grow, this artificial and naive definition is changing. Natural functions of many secondary metabolites are unknown because they have never been investigated; this lack of evidence or knowledge is then interpreted as lack of function [15]. Recently, it was discovered that they can help plant to increase their overall ability to survive and overcome local challenges by allowing them to interact with their surroundings [16, 17]. Thus, floral species synthesize aroma to attract insect for their pollination and fertilization while synthesis of toxic chemical has evolved to pathogens as well as to herbivores for suppressing the growth of neighboring plants [18].

Among secondary metabolites, some of these substances have effect on biological systems, which are considered as bioactive. Thus, bioactive compounds in plants can be defined as follows: secondary plant metabolites eliciting pharmacological or toxicological effects in humans and animals [19]. Bioactive compounds of plants could be divided into three main categories: (1) terpenes and terpenoids (approximately 25,000 types), (2) alkaloids (approximately 12,000 types), and (3) phenolic compounds (approximately 8000 types) [13].

2.1.1. EOs as bioactive compounds

EOs are natural, volatile, complex plant compounds, oily or lipid-like in nature, and frequently characterized by strong fragrance [20, 21]. They have been known as herbal extracts to mankind for hundreds of years, even millenniums, but there are different opinions about historical origin of EO production [22]. According to some authors, China has been the cradle of hydrodistillation [23], whereas others claim, as already mentioned, that Arabs invented this process about 3000 BC. Also, there are other opinions that distillation as a method of producing EOs was first used in the East (Egypt, India, and Persia) more than 2000 years ago and improved in the 9th century by the Arabs [23, 24]. However, Villanova (ca. 1235–1311), a Catalan physician, was the first scientist who performed distillation of EOs and left authentic written account [23].

By the 13th century, EOs were being made by pharmacies and their pharmacological effects were described in pharmacopoeias [24], but their use does not appear to have been widespread in Europe until the 16th century when they were traded in the City of London [20]. In "Grete Herbal," published in 1526, some of the illustrations of the retorts and stills used for the extraction of volatile oils are presented [25].

While knowledge of the science of EOs did not increase during 17th century when pharmacies generally used 15–20 different oils, 18th century brought about only small progress in the design of equipment and in refinements of the techniques used [23]. At the end of the 18th century, the use of tea tree oil for medical purposes has been documented in Australia [26]. In 19th century, the first movable apparatus appeared, cooling methods were improved, and double-wall distillation plants appeared. The medical properties and application of increasing number of new EOs were analyzed and recorded by the pharmacists, and the first experimental measurement of the bactericidal properties of EOs were performed [23]. With the scientific revolution in this century, chemists were able to identify for the first time the various constituents of EOs, and the most important investigation was performed by O. Wallach, an assistant of Kekule [27]. In 1876, Haarman and Reimer started the first production of synthetic aroma chemicals: vanillin, coumarin, anisaldehyde, heliotropin, and terpineol. These researches, partly, laid the ground for the development of the EOs' synthetic counterparts and the growth of the modern drug industry. By the middle of 20th century, the use of EOs in medicine gradually became secondary to their use for flavor and aroma in perfumes, cosmetics, and foodstuffs [28].

2.1.2. Taxonomy of EO-producing plants

EOs are complex mixture of volatile compounds produced by living organisms, isolated by steam distillation, hydrodistillation, or solvent extraction from a whole plant or plant part of known taxonomic origin [29]. Around 3000 EOs have been produced from around 2000 plant species that belong to various genera distributed to around 60 families. It is well known that all plants possess the ability to produce volatile compounds, however, quite often only in traces. Only 300 EOs are important from the commercial point of view, and "EO-bearing plants" in particular are those plant species delivering an EO of commercial interest [20]. Some of those plants have specialized anatomic structures (secretory cells, glandular trichomes, and cavities/ducts), which leads to accumulation of volatiles and higher concentration in the plant [30]. Others, such as rose (*Rosa* spp.), jasmine (*Jasminum sambac*), or tuberose (*Polyanthes tuberosa*), produce and emit the volatiles by the epidermal layers of their petals [31, 32]. Their EO yield is exceptionally small so special techniques have to be applied to recover volatile fragrance compound.

The term "EO" groups together a wide range of chemical compounds on the basis of their historic use and method of isolation and belies the variety and complexity of compounds found within them [33]. Some plant families are particularly well known for their oil-bearing species and ability to produce EOs of medicinal and industrial value. These include *Alliaceae, Apiaceae, Asteraceae, Cupressaceae, Lamiaceae, Lauraceae, Myrtaceae, Piperaceae, Poaceae, Rosaceae, Rutaceae, Santalaceae, Zingiberaceae*, and *Zygophyllaceae* [30, 34]. Those plant families are not restricted to one specialized taxonomic group but are distributed among all plant classes. Gymnosperms, for example, the families Cupressaceae (cedarwood, cedar leaf, juniper oil, etc.) and Pinaceae (pine and fir oils, etc.), produce EOs with significant biological activities [34] as well as angiosperms (*Magnoliopsida, Rosospida*, and *Liliospida* [35–37]).

2.1.3. Composition of EOs

EOs as herbal extracts are not simple compounds or even simple mixtures of several individual compounds. They may comprise up to approximately 100 components, but usually between 20 and 60 components [38–40]. Major components can constitute up to 85% of the EO, whereas other components are present only as a trace [24, 41]. Numerous publications have presented data on the composition of the various EOs [20, 21, 42–48].

The major volatile constituents are hydrocarbons (e.g. pinene, limonene, and bisabolene), alcohols (e.g. linalol and santalol), acids (e.g. benzoic acid and geranic acid), aldehydes (e.g. citral), cyclic aldehydes (e.g. cuminal), ketones (e.g. camphor), lactones (e.g. bergaptene), phenols (e.g. eugenol), phenolic ethers (e.g. anethole), oxides (e.g. 1,8 cineole), and esters (e.g. geranyl acetate) [45]. They are defined as substances composed of isoprene (2-methylbutadiene) hydrocarbons joined together in a repetitive head-to-tail manner (known as the isoprene rule) [46, 49]. Leopold Ruzicka, recipient of the 1939 Nobel Prize in Chemistry, proposed biogenetic isoprene rule [50], which emphasizes the single biochemical origin of terpenes.

The composition of EOs are influenced by many factors such as plant ecotype or variety, genetic variation, plant nutrition, application of fertilizers, geographic location of plants, surrounding climate, seasonal variation, stress during growth or maturity, and postharvest handling, and also by isolation technique applied [51–53].

The major components of EOs isolated from aromatic herbs examined in this chapter are presented in Table 1.

Common name of EO	Latin name of plant source	Major components	Approximate % composition	References
Sage	*Salvia officinalis* L.	Camphor	6-23%	[55-57]
		B-pinene	2-10%	
		1,8-cineole	6-14%	
		α-thujone	20-42%	
Rosemary	*Rosmarinus officinalis*	α-pinene	2-51%	[57-59]
		camphor	2-32%	
		Bornyl-acetate	0-17%	
		1,8 cineole	3-89%	
		Limonene	4-15%	
Oregano	*Origanum vulgare*	Carvacrol	2-80%	[55, 60-62]
		Thymol	1-64%	
		γ-terpinene	2-52%	
		p-cymene	Trace-52%	
		sabinene	1-48%	
Savory	*Satureja montana*	Thymol	4-38%	[54, 63,64]
		Carvacrol	5-96%	
		γ-terpinene	1-31%	
		p-cymene	3-27%	

Table 1. Major components of selected EOs

2.1.4. Economic importance of EOs

Use of herbal extracts, particularly EOs, for perfumery, additives in food/confectionary as well as for pharmaceuticals and cosmetics is a growing market trend. The huge production of EOs (>70,000 tons per annum) with estimated market value of more than 700 million $US indicates that production and consumption of EOs is increasing all over the world [45]. This production is achieved mainly by major cultivators and producers like the United States, Brazil, India, and China; although some other countries are important contributors of EOs; for example, vetiver/khus, clove, lemongrass, basil, and celery oils are mainly produced in India, whereas Spain and France are major producers of rosemary oil [52, 60].

Today, aromatherapy and use of herbal extracts as "natural" products are fast developing segment of the industry, and this is a return to what was common practice in ancient and medieval times [64].

The largest world consumer of EOs is the flavor industry, especially for soft drinks. However, this is limited to a few EOs, mainly citrus (orange, lemon, grapefruit, and lime), ginger, cinnamon, clove, and peppermint. Similar oils are used in confectionery, bakery, desserts, and dairy products, besides some fruity products and spices. Alcoholic beverage industry is the one of the largest consumers of various EOs as well as dairy, desserts, sweet bakery, confectionery, and cream manufacturers. The fast food and processed food industries use EOs with spicy and herbal flavors: oregano, basil, fennel, pepper, dill, etc [65].

3. Techniques for extraction of bioactive compounds from aromatic plants

When considering quality and composition of an EO, the method of extraction plays a crucial role. EO are extracted from different part of the plants – flowers, buds, seeds, leaves, bark, herbs, wood, fruits, and roots – and different extraction techniques should be used in different conditions for understanding the extraction selectivity.

Furthermore, separation, identification, and characterization of bioactive compounds are only possible by an appropriate extraction process of plant matrix.

Different techniques could be used to extract bioactive compounds; many of them remain almost same through hundreds of years. All these techniques have some common objectives: (a) to extract targeted bioactive compounds from complex plant sample, (b) to increase selectivity of analytical methods, (c) to increase sensitivity of bioassay by increasing the concentration of targeted compounds, (d) to convert the bioactive compounds into a more suitable form for detection and separation, and (e) to provide a strong and reproducible method that is independent of variations in the sample matrix [66].

3.1. Conventional extraction techniques

Bioactive compounds from plant materials can be extracted by various classical extraction techniques. Most of these techniques are based on the extracting power of different solvents in use and the application of heat and/or mixing. To obtain bioactive compounds from plants, the existing classical techniques are maceration [19, 67], expression [25], solvent extraction [19], and hydrodistillation.

Hydrodistillation is a traditional method for extraction of bioactive compounds and EOs from plants. Organic solvents are not involved, and it can be performed before dehydration of plant materials. There are three types of hydrodistillation: water distillation, water and steam distillation, and direct steam distillation [68]. In hydrodistillation, first, the plant materials are packed in a still compartment; second, water is added in sufficient amount and then brought to boil. Alternatively, direct steam is injected into the plant sample. Constituents that are insoluble in the water but volatile enough to be driven off by the steam come over and are cooled, condensed, and collected in the receiving vessel. The condensed mixture flows from the condenser to a separator, where oil and bioactive compounds separate automatically from the water [69]. At a high extraction temperature, some volatile components may be lost. This drawback limits its use for thermolabile compound extraction. EOs extracted by hydrodistillation need further purification, especially drying, to remove water.

3.2. Nonconventional extraction techniques

It was found that the major drawbacks of conventional extraction are necessity of costly and high purity solvent, solvent or water removal, low extraction selectivity, thermal decomposition of thermolabile compounds and long extraction time [70]. In recent years, to overcome these limitations, so-called nonconventional extraction techniques are presented: ultrasound-assisted extraction [71], pulsed electric field–assisted extraction [19, 72], enzyme-assisted extraction [19, 73], microwave-assisted extraction [19, 74], and supercritical fluid extraction (SFE).

SFE was systematically investigated over the past decades [48]. SFE is performed by using fluids in supercritical state at temperature higher than their critical temperature and under the pressure higher than their critical pressure. Supercritical fluid possesses gas-like properties of diffusion, viscosity, and surface tension and liquid-like density and solvation power [7]. These characteristics enable easy penetration of the fluid in supercritical state into herbal material and the extraction of secondary herbal metabolites.

For several practical reasons, more than 90% of all analytical SFE is performed with carbon dioxide (CO_2). Apart from having relatively low critical pressure (74 bar) and temperature (32°C), CO_2 is relatively nontoxic, nonflammable, available in high purity at relatively low cost, and easily removed from the extract. The main drawback of CO_2 is its lack of polarity for the extraction of polar compounds [75]. The limitation of low polarity of carbon dioxide has been successfully overcome by the use of chemical modifier [76].

A basic SFE system consists of the following parts: a tank of mobile phase, usually CO_2; a pump to pressurize the gas; co-solvent vessel and pump; an oven that contains the extraction vessel; a controller to maintain the high pressure inside the system; and a trapping vessel. Usually, different type of meters such as flow meter and dry/wet gas meters could be attached to the system. The major variables influencing the extraction efficiency are temperature, pressure, particle size, packing density and moisture content of feed material, extraction time, flow rate of CO_2, and solvent-to-feed-ratio [9, 77]. By varying the extracting conditions, it is possible to obtain extract with the maximum content of desired active substances. The only disadvantage of industrial application of SFE, as opposed to conventional methods, is the larger investment in equipment due to working in conditions of elevated pressure. The costs of production, however, are significantly reduced; the process is simpler and more cost-efficient; and quality of the final product good [7].

Supercritical carbon dioxide (SC_{CO2}) is known as a good solvent for a wide range of natural bioactive principles. In the last 10 years, studies on the extraction of classical compounds like essential and seed oils from various sources: seeds, fruits, leaves, flowers, rhizomes, etc., with or without the addition of a co-solvent, have been published [78]. Extraction of bioactive compounds from plant material using SC_{CO2} has been indicated as a favorable technique for producing solvent-free extracts suitable for wide use in pharmaceutical, biomedical, cosmetic, and food industries [79].

4. Herbal extracts as antimicrobial agent

Even before the role of microorganisms in disease pathogenesis was understood, plant-based medicines were used for treating such illnesses. It is recognized that plant molecules have antimicrobial properties; especially, EOs exhibit broad spectrum inhibitory activities against various Gram-positive and Gram-negative bacteria pathogens [20, 38, 57].

However, herbal extract medicine diminished as soon as modern antibiotics were discovered. Renewed recent interest in their use has been attributed to several factors, including the desire for antimicrobial compounds with even better safety and toxicity profiles [80]. Also, severity of bacterial infections has gone up even after the discovery of many antibiotics, mainly due to the reduced susceptibility to conventional antimicrobial agents shown by many important pathogens. Therefore, infectious diseases caused by bacteria are still one of the leading causes of deaths [57, 81]. In addition, toxicity due to side effects limits the prolonged use of high concentrations of available antibacterial drugs [57].

Most EOs possess at least some degree of antibacterial activity. Generally, EOs with phenolics and aldehydes exhibit better antibacterial efficacies [38]. In few cases, a main component of the oil has been observed to possess activity better than the EO. For example, carvacrol and eugenol from clove oil, terpinen-4-ol in *Melaleuca alternifolia* (tea tree oil), or thymol from oregano oil displays greater efficacy than specific oil. Many of the plant molecules are effective against drug-sensitive as well as drug-resistant strains [58, 82].

The methods used for establishing antibacterial activities are, usually, disc diffusion methods or agar or broth dilution methods. Although disc diffusion methods are popular, the data they offer are less useful than others. Agar and broth dilution methods, in which serial dilutions of the test oil in agar or broth media are inoculated with a known concentration of test organism, allow minimum inhibitory concentrations (MICs) to be determined [83]. The MIC is generally defined as the lowest concentration of EO that inhibits growth of the test organism. Although solubilization or dispersion in these systems may be problematic, MICs can help establish safe and effective final concentrations in formulated products.

Two principal reasons for performing the in vitro tests are as follows:

- Identification of antimicrobially active compounds

- Control of microbial susceptibilities toward approved antibiotics and antimicotics

The procedure from identification of antimicrobially active compounds for their use in humans to treat infectious disease is a multistep pathway, which includes pharmacological (concentration of the active compound at the site of action, half-life time, serum levels, dose–response relationship, etc.) and toxicological (e.g. toxicity, allergic responses, and interactions) aspects [84].

4.1. Mode of action of bioactive compounds

The health benefits of medicinal plants are ascribed to their bioactive compounds, known as phytochemicals. It has been estimated that 74% of pharmacologically active plant-derived compounds were discovered after following up on ethnomedicinal use of the plants [85]. Various phytochemicals are recognized to possess antimicrobial, anti-inflammatory, analgesic, anesthetic, antioxidant, neuroprotective, and antitumor activity, thus providing medicinal plants with great therapeutic as well disease-preventive potential [86].

However, detailed knowledge about the mode of action of EOs and other bioactive compounds is still lacking. In general, the mechanism of action of EOs is to alter the structure of the cytoplasmic membrane, disrupting the proton motive force, electron flow, active transport, and coagulation of cell contents [87, 88]. The mode of action of phytochemicals depends on the type of microorganism and is generally related to their outer membrane arrangement as well as cell wall structure. For example, antimicrobial action of EOs depends on their hydrophilic or lipophilic character. Certain components of EOs can act as uncouplers, which interfere with proton translocation over a membrane vesicle and subsequently interrupt ADP phosphorylation [89]. Phytochemicals may also modulate transcription factors, redox-sensitive transcription factors, redox signaling, and inflammation [90].

4.1.1. Terpenoids

Specific terpenoids with functional groups, such as phenolic alcohols or aldehydes, interfere with membrane-integrated or associated enzyme proteins stopping their production or activity. It was found that some phenolic alcohols (e.g. carvacrol and thymol) cause a disruption of the lipopolysaccharide outer layer followed by partial disintegration of the outer membrane [89]. The interaction of thymol with the membrane affects membrane permeability and results in the release of K+ ions and ATP [91]. In some cases, thymol can induce the release of lipopolysaccharides, but it does not affect chelating cations [92]. Thymol integrates within the polar head groups of the lipid bilayer, inducing alterations of the cell membrane. Zengin and Baysal [93] found that α-terpineol 1,8-cineole and linalool alter the function of cell membranes and the permeability of outer membranes of *Staphylococcus aureus* and *E. coli*. Bard *et al.* [94] reported that geraniol enhances the permeability of whole cells of *Candida albicans* and also increases the fluidity of both *C. albicans* membranes and dipalmitoyl phosphatidylcholine DPPC liposomal membranes. Mendanha *et al.* [95] revealed that all the tested terpenoids (nerolidol, menthol, pulegone, carvone, (+)-limonene, α-terpineol, and 1,8-cineole) increase the fluidity of cell membrane and exert cytotoxic effects on fibroblast cells. Yin *et al.*

[96] found that borneol increases the fluidity of DPPC bilayer membranes. Terpenoid, isolated from purple prairie clover, petalostemumol, has significant activity against *Bacillus subtilis* and *S. aureus* [97]. Carvone is capable to disrupt pH gradient and membrane potential of cells. With increasing amount of carvone, Oosterhaven *et al.* [98] reported a decrease in the growth rate of *E. coli, Streptococcus thermophilus,* and *Lactococcus lactis* was caused by disturbing the metabolic energy status of cells.

4.1.2. Phenolics

Phenolic toxicity to microorganism is often explained by enzyme inhibition with the oxidized compounds, probably through reaction with sulfhydryl groups or in more nonspecific interactions with the proteins. The induced defense response includes formation of a lesion that limits the growth of the pathogen, where polyphenols and other antibiotic compounds accumulate [85].

Hydroxyl groups number as well sites in phenolics are thought to be related to their relative toxicity to microorganisms (increased hydroxylation results in increased toxicity) [85]. Thus, eugenol is considered bacteriostatic against both fungi and bacteria [88]. It was found that eugenol alters the membrane, affects the transport of ions and ATP, and changes the fatty acid profile of different bacteria [99]. It also acts against different bacterial enzymes, including ATPase, histidine carboxylase, amylase, and protease [100, 101]. Cinnamaldehyde is usually less powerful than eugenol [102], but extremely effective against *E. coli* and Salmonella *typhimurium* [92]. Catechin acts on different bacterial strains belonging to different species (*E. coli, Salmonella choleraesis, Klebsiella pneumonie, Serratia marcescens, B. subtilis, Pseudomonas aeruginosa,* and *S. aureus*) by generating hydrogen peroxide and by altering the permeability of the microbial membrane [85, 103]. It was found that polyphenols obstruct bacterial quorum sensing, that is, the production of small signal molecules by *E. coli, Pseudomonas putida,* and *Burkholderia cepacia* cells that trigger the exponential growth of a bacterial population [104, 105].

4.1.3. Other compounds

Flavonoids have been shown in vitro to be effective antimicrobial substances against a wide array of microorganisms [106]. Their activity is probably due to their ability to disrupt microbial membranes or to complex with extracellular and soluble proteins in bacterial cell walls [107, 108]. Moreover, Arora *et al.* [109] demonstrated that some flavonoids (naringenin and rutin) and isoflavonoids (genistein) decrease the membrane fluidity. Selvaray *et al.* [110] correlated bioactivities of flavonoids to their membrane localization and their induced changes in membrane fluidity. Thus, catechins inhibited in vitro *Vibrio cholerae, Streptococcus mutans, Shigella,* and other bacteria and microorganisms [97]. Several studies have documented the effectiveness of flavonoids such as swertifrancheside, glycyrrhizin (from licorice), and chrysin against multiple viruses [97, 106, 111]. It was found that membrane-interacting properties of flavonoids to modify permeability of cellular membranes could inhibit *E. coli* growth [112].

The mode of antimicrobial action of tannins could be related to their capability to inactivate enzymes, cell envelope, transport proteins, microbial adhesins, etc. Previous studies have shown that tannins can be toxic to filamentous fungi, yeasts, and bacteria. It was found that condensed tannins prevent growth and protease activity by binding the cell walls of ruminal bacteria [97].

The potential range of quinones' antimicrobial effects is great, due to its ability to complex irreversibly with nucleophilic amino acids in proteins, often leading to inactivation of the protein and loss of function. Possible targets in the microbial cell are cell wall polypeptides, surface-exposed adhesins, and membrane-bound enzymes. Quinones may also render substrates unavailable to the microorganism [85]. Hypericin, an anthraquinone from *Hypericum perforatum*, has received much attention lately as an antimicrobial agent [97].

Several alkaloids have the properties to interact with artificial and biological membranes to change the fluidity in association with their pharmacological effects [113]. Berberine is one of isoquinoline alkaloids, which have been considered to possess anti-inflammatory and antimicrobial effects [114]. Also, it has been found that alkaloid sanguinarine possesses antibacterial activity against bacteria and pathogenic fungi [115].

The bacterial resistance is conferred by multidrug resistance pumps (MDRs), membrane translocases that extrude structurally unrelated toxins from the cell. These protect microbial cells from both synthetic and natural antimicrobials [116]. The mechanism of action of EOs depends on their chemical composition, amount of the single components, and their antimicrobial activity is not attributable to a unique mechanism but is instead a cascade of reactions involving the entire bacterial cell [20]. Also, the synergistic effects of antibiotics and herbal extracts can provide successful therapy against drug-resistant bacteria. The use of herbal extracts and phytochemicals can be of great significance in therapeutic applications and could help control the problem of multidrug-resistant organisms [85].

4.2. Effect of some herbal extracts on selected food-borne microorganisms

Here, some original results of antibacterial activity of EO and supercritical extracts from aromatic herbs (from Montenegro) – sage (*Salvia officinalis*), rosemary (*Rosmarinus officinalis*), oregano (*Origanum vulgare*), and savory (*Satureja montana*) – against some pathogenic bacteria important in food industry are presented.

Leaves from selected herbs were collected in the central southern part of Montenegro, air-dried and stored in double-layer paper bags at the room temperature, until further analysis.

The selected test organisms, used to evaluate the antimicrobial activity of the herbal extracts, were as follows: Gram-positive (*Bacillus cereus* ATCC 11778, *S. aureus* ATCC 25923, and *Listeria innocua* ATCC 33090) and Gram-negative (*Salmonella enteritidis* ATCC 13076 and *E. coli* ATCC 25922).

To obtain EOs, herb material was subjected to hydrodistillation in a Clevenger-type apparatus for 2 hours according to Yugoslav Pharmacopoeia IV. Supercritical CO_2 extraction (SCE) procedure is previously described in detail [117], while the extraction conditions were: temperature 40 °C, pressure 100 bar, extraction time 360 min; and CO_2 flow rate 0.3 kg CO_2/h.

MIC values were determined for extracts displaying antimicrobial properties in screening studies, using a modified microdilution broth method [118]. Briefly, the extracts were first dissolved in DMSO, then diluted in sterile water and tested over a range of concentrations from 0.09 to 25 mg/ml against overnight broth cultures of selected bacteria grown to a population of 10^6 CFU/ml in tryptic soy broth (TSB). Microplates were incubated at optimum growth temperature for each bacterial strain, and growth was monitored by measuring absorbance at 600 nm every 45 min over 18 hours, using a microplate reader.

The antibacterial activity, for investigated herbal extracts against some pathogenic bacteria important in food industry, are summarized in Table 2.

Herb material	*Salvia officinalis*	*Rosmarinus officinalis*	*Origanum vulgare*	*Satureja montana*
Method of extraction	HD/SCE	HD/SCE	HD/SCE	HD/SCE
Bacterial strain	**MIC (mg/ml)**			
Bacillus cereus ATCC 11778	0.09/0.09	0.36/0.72	0.09/0.18	0.18/0.36
Staphylococcus aureus ATCC 25923	0.09/0.09	0.36/0.72	0.36/0.72	0.18/0.36
Listeria innocua ATCC 33090	0.18/0.18	3.13/6.25	0.18/0.36	1.57/3.13
Salmonella enteridis ATCC 13076	3.13/3.13	3.13/6.25	3.13/3.13	3.13/6.25
Eschericia coli ATCC 25922	0.18/0.18	0.36/0.72	0.36/0.36	1.57/3.13

Table 2. Antibacterial activity of isolates obtained by hydrodistillation and SCE from selected herbs against some pathogenic bacteria important in food industry

According to the results presented in Table 2, sage extracts had the highest antibacterial efficiency against tested bacteria strains (MIC=0.09–3.13 mg/ml) followed by oregano (MIC=0.09–3.13 mg/ml) extracts. In this study, the carriers of antimicrobial activity of the sage oil were probably α-thujone and camphor. High efficiency of oregano extracts could be attributed to the high content of compounds with known antimicrobial activity in examinated samples, such as phenolic components, thymol and its isomer carvacrol, as well as its precursors, γ-terpinene and p-cymene. Rosemary extracts showed somewhat smaller activity than expected, probably because rosemary is a cultivated herb, whereas all other examined herbs were wild growing.

Among tested bacteria, *B. cereus* and *S. aureus* were the most sensitive to presence of all tested extracts, especially to presence of sage extracts. Thus, Gram-positive bacteria seemed to be more susceptible to all tested herb extracts.

The results of the bioassays show that tested extracts obtained by SC-CO_2 extraction from different pretreated herb matrices exhibited the same or weaker antimicrobial activity when compared to the EO obtained by hydrodistillation.

The presented results in this study confirm the facts that plant molecules have significant antibacterial activity and therefore can be used as a strong antimicrobial agent. The use of EOs in foods as preservatives is limited due to toxicological aspects, but also possible reasons for this limitation may be the strong smell and taste of these substances when used at effective doses. The SC-CO_2 extracts bear the closest natural smell and taste of original material, thus further investigation should point to various combinations of investigated extracts, which should improve the level of inhibition due to synergistic effects.

5. Potential application of herbal extracts in foods

The microbiological safety in ready to eat products is a cause of big concern not only for the consumers and food industries but also for the regulatory agencies. It can be said that the food industry faces a constant problem in providing the food where there are no food-borne pathogens present [3, 4]. The number of documented outbreaks of food-borne diseases has increased in the last decade with *Salmonella* spp., *L. monocytogenes*, and *E. coli* being responsible for the largest number of outbreaks and deaths [5]. The different diseases, caused by food-related pathogenic bacteria, such as listeriosis, hemorrhagic colitis, campylobacteriosis, and salmonellosis, are still reported. For example, in meat and meat products, spoilage bacteria could shorten the shelf life by causing off-odors, discoloration, gas production, etc. Recently, we have trend in reducing the level of synthetic antimicrobial agents, as well as salt levels, in ready meals because of the proven development of hypertension and increased risk of cardiovascular disease. In recent years, consumers prefer fewer chemicals and more natural foods. Therefore, there is growing interest in using natural antimicrobial compounds, including extracts of herbs and spices, as salt replacers or alternatives to synthetic compounds for food preservation [119].

As already emphasized, in addition to providing flavor and fragrance, spices and herbs also have antimicrobial potential and thus can be used for preventing food deterioration and shelf life extension. It was found that, however, well herbal extracts perform in antibacterial assays in vitro, generally, a higher concentration is required to obtain the same efficacy in foods. [120]. The basic properties of the food (fat/protein/water content, antioxidants, preservatives, pH, salt, and other additives) are the most relevant in this respect, although, the extrinsic elements (temperature, packaging in vacuum/gas/air, and characteristics of microorganisms) can also influence bacterial sensitivity [20, 121].

Utilization of packaging materials containing these herbal extracts as antimicrobial compounds is also becoming an attractive option in the food industry. If herbal extracts were to be more widely applied as antibacterials in foods, the organoleptic impact would be important.

5.1. Meat, meat products, and fish dishes

Activity of oregano, thyme, basil, marjoram, lemongrass, ginger, and clove EOs against bacterial strains inoculated experimentally in irradiated minced meat and against natural microbiota found in minced meat samples was tested [122]. $MIC_{90\%}$ values ranged from 0.05%v/v (lemongrass oil) to 0.46%v/v (marjoram oil) to Gram-positive bacteria and from 0.10%v/v (clove oil) to 0.56%v/v (ginger oil) to Gram-negative strains.

Eugenol and coriander, clove, oregano, and thyme oils were found to be in inhibiting *L. monocytogenes, Aeromonas hydrophila,* and autochthonous spoilage flora in meat products, sometimes causing a marked initial reduction in the number of recoverable cells [123] whereas mustard, cilantro, mint, and sage oils were less effective or ineffective [124].

Carvacrol vapor antimicrobial activity was established against *S. enteritidis* on pieces of raw chicken [125]. The effectiveness of oils and vapors of lemon, sweet orange, and bergamot against *L. monocytogenes, S. aureus, B. cereus, E. coli* O157, and *Campylobacter jejuni* was investigated on chicken [126]. The results indicate that bergamot was the most inhibitory EO due to high content of citral and linalool.

The antimicrobial effect of extracts from *S. officinalis* L. and berries of *Schinus molle* L. against *Salmonella anatum* or *S. enteritidis* inoculated on minced beef meat was studied [127]. It was found that use of 0.1% or 1.5% *S. officinalis* with 6% or 4% NaCl or 0.1% or 1.5% *S. molle* with 4% or 8% NaCl could effectively eliminate *S. anatum* from refrigerated raw beef.

Effectiveness of eight EOs as antimicrobial agents for fish preservation on 18 genera of bacteria, which included some important food pathogen and spoilage bacteria, was investigated. Clove EO showed the highest inhibitory effect, followed by rosemary and lavender [128]. It was found that citrus EO incorporated into different edible biopolymer film has the potential to preserve fish fillets [129]. Also, using EO in a coating for shrimps appears effective in inhibiting the respective natural spoilage flora [130].

The antimicrobial effect of nine EOs on *Photobacterium phosphoreum* on the shelf life of modified atmosphere-packed cod fillets was determined. The antimicrobial effect of EO was studied in a liquid medium and in product storage trials. Oils of oregano and cinnamon had strongest antimicrobial activity, followed by lemongrass, thyme, clove, bay, marjoram, sage, and basil oils, whereas oregano oil (0.05%, v/w) reduced the growth of *P. phosphoreum* in naturally contaminated cod fillets and extended shelf life from 11–12 days to 21–26 days at 2°C [131].

5.2. Fruit and vegetable

The shelf life of unpasteurized fruit juices is limited by microbial enzymatic spoilage; moreover, these products could be contaminated by some pathogens. Some EOs could be used to prevent this kind of problem. Lemongrass and geraniol have been found effective against *E. coli, Salmonella* sp., and *Listeria* spp. in apple, pear, and melon juices [47, 132].

Carvacrol and cinnamaldehyde were very effective at reducing the viable count of the natural flora on kiwifruit but less effective on honeydew melon. It is possible that this difference is due to difference in pH between the fruits: the lower the pH, the more effective EOs and their components generally are [20, 133].

The antimicrobial activity of basil, caraway, fennel, lemon balm, marjoram, nutmeg, oregano, parsley, rosemary, sage, and thyme EO against food-borne pathogens and key spoilage bacteria pertinent to ready-to-eat vegetables was evaluated [134]. On a carrot model product, basil, lemon balm, marjoram, oregano, and thyme EOs were deemed organoleptically acceptable, but only oregano and marjoram EOs were deemed acceptable for lettuce. It was found that selected EOs may be useful as natural and safe additives for promoting the safety and quality of ready-to-eat vegetables [135]

Listeria strains were more sensitive than spoilage bacteria, and oregano and thyme were the most active EOs using food model media based on lettuce, meat and milk [136]. This work shows that EOs might be more effective against food-borne pathogens and spoilage bacteria when applied to foods containing a high protein level at acidic pH, as well as moderate levels of simple sugars.

The antimicrobial potential of oregano EO on *S. typhimurium* ATCC 13311 on tomatoes was tested. Tomatoes treated with 100 ppm oregano oil resulted in 2.78 log reduction, after 20 min [137]. Oregano oil was effective at inhibiting *E. coli* O157:H7 and reducing final populations in eggplant salad compared to the untreated control. Also, it was found that, in vegetable dishes, the antimicrobial activity of oil decrease in storage temperature and/or a decrease in the pH increases [138].

5.3. Cereals and dairy products

Nielsen *et al.* [139] found that cinnamon, mustard, garlic and clove EO have strong effect in active packaging to prevent bread from fungal contamination. Although oregano oleoresin weakly prevent the growth of most important spoilage fungi of bread, vanilla EO had no preventative effect against these fungi [139].

It was found that sage oil was ineffective against *B. cereus* in rice was while carvacrol was very effective at extending the *B. cereus* lag phase, reducing the final population compared to a control [20, 140, 141].

Orange, lemon, grapefruit, madrine, terpeneless lime, orange, D-limonene, terpineol, and geraniol were tested against *Salmonella, E. coli, S. aureus,* and *Pseudomonas* spp. in different types of milk. Terpineol was the most effective oil in vitro, thus it was used in combination with orange oil for a validation in milk [142]. For mint oil, it was found that it is effective against *S. enteritidis* in cucumber salad and low-fat yoghurt [143].

6. Legal aspects of the herbal extracts use in foods and safety data

The new Regulation (EC) N°1334 of the European Parliament and of the Council on flavorings and certain food ingredients with flavoring properties for use in and on foodstuffs entered into force in 2009 [138].

The new Regulation stipulates new labeling requirements for both flavoring manufacturers and (final) food manufacturers. These include labeling as "natural flavoring substance(s)" may only be used for flavorings where the flavoring part contains exclusively natural flavoring substances.

The risk management of certain substances naturally present in certain food ingredients with flavoring properties (e.g. herbs, spices) and/or flavorings is based on the "major contributor approach": maximum levels are established for the presence of these undesirable substances in food, which contribute most to the human intake of these substances.

European Commission has been registered numerous herbal extracts' components for use as flavorings in foodstuffs. The flavorings registered are considered to present no risk to the health of the consumer and include, among other, carvacrol, carvone, cinnamaldehyde, p-cymene, eugenol, limonene, menthol, and thymol.

New flavorings may only be evaluated for registration after carrying out serious toxicological and metabolic studies, which could require a considerable financial expense. Also, the legislative is different in different countries, for example, estragole and methyl eugenol were deleted from the European Commission list in 2001 due to their being genotoxic. However, since today, estragole is on the EAFUS (Everything Added to Food in the United States) list. The EAFUS list is a list of substances that the United States Food and Drug Administration (FDA) has classified the as generally recognized as safe (GRAS) or as approved food additives.

Till today, many researchers found that a significant number of herbal extracts' components are GRAS and/or approved food flavorings. However, some research data indicate irritation and toxicity: cinnamaldehyde, carvacrol, carvone, and thymol appear to have no significant effects in vivo, whereas in vitro they exhibit mild-to-moderate toxic effects at the cellular level [20]. Also, it was found that eugenol, menthol, and thymol, when applied in root canal treatments, could cause irritation of mouth tissues [144]. Some oils used in the fields of medicine, paramedicine, and aromatherapy have been shown to exhibit spasmolytic or spasmogenic properties [145].

7. Conclusion

Due to consumer awareness and negative perception of artificial preservatives in food, in recent years attention is shifting toward alternatives that the consumers recognize as natural. Thus, herbal extracts, particularly EOs, are getting more space in food industry to prevent the propagation of bacteria that are responsible for the spoilage of food or for the spread of so-called food-borne diseases.

Herbal extracts have antimicrobial potential and thus can be used for preventing food deterioration and shelf life extension. However, if herbal extracts were to be more widely applied as antibacterials in foods, the cytotoxic property and organoleptic impact are extremely important issues to consider. They may vary according to extract composition where the method of extraction plays a crucial role.

Therefore, herbal extracts should be used very carefully and with considerable precautions about the concentrations and product application, target consumer, major constituents of the oil, and toxicology profile.

Author details

Biljana Damjanović-Vratnica*

Address all correspondence to: biljanad@ac.me

Faculty of Metallurgy and Technology University of Montenegro Podgorica, Montenegro

References

[1] Food safety fact sheet [Internet]. Geneva: World Health Organization; 2014. Available from: http://www.who.int/mediacentre/factsheets/fs399/en/ [Accessed: 2015-07-28]

[2] Smith-Palmer A, Stewart J, Fyfe L. Antimicrobial properties of plant essential oils and essences against five important food-borne pathogens. Letters in Applied Microbiology. 1998; 26: 118-122. DOI: 10.1046/j.1472-765X.1998.00303.x

[3] Kotzekidou P, Giannakidis P, Boulamatsis A. Antimicrobial activity of some plant extracts and essential oils against foodborne pathogens *in vitro* and on the fate of inoculated pathogens in chocolate. LWT–Food Science Technology. 2008; 41: 119-127. DOI: 10.1016/j.lwt.2007.01.016

[4] Commission regulation (EC) No 2073/2005 on microbiological criteria for foodstuffs. Official Journal of European Union, L388. 2005; 48: 1-25.

[5] Gupta S, Abu-Ghannam N. Recent advances in the application of non-thermal methods for the prevention of salmonella in foods, In: Mahmoud BSM, editor. *Salmonella* - A Dangerous Food-Borne Pathogen. InTech, Rijeka, Croatia; 2012. p. 287-304. Available from: http://www.intechopen.com/books/salmonella-a-dangerous-food-borne-pathogen/recent-advances-in-theapplication-of-non-thermal-methods-for-the-prevention-of-salmonella-in-foods

[6] Rizzo PV, Menten JFM Racanicci AMC Santarosa J. Foundation and perspectives of the use of plant extracts as performance enhancers in broilers. *Revista Brasileira de Ciência Avícola*. 2008; 10: 195-204. ISSN 1516-635X

[7] Mišić D. The applicability of supercritical extracts in clinical treatment of bacterial infections in humans and animals. In: Rój E, editor. Supercritical CO_2 Extraction and Its Applications. OIC Poland, Lublin; 2014. p. 57-98. ISBN: 978-83-86499-96-0

[8] Anitescu G, Doneanu C, Radulescu V. Isolation of coriander oil: comparison between steam distillation and supercritical CO_2 extraction. Flavour and Fragrance Journal. 1997; 12: 173-176. DOI: 10.1002/(SICI)1099-1026(199705)12:3<173::AID-FFJ630>3.0.CO;2-1

[9] Pourmortazavi SM, Hajimirsadeghi SS. Supercritical fluid extraction in plant essential and volatile oil analysis. Journal of Chromatography A. 20071163: 2-24. DOI: 10.1016/j.chroma.2007.06.021

[10] Vinatoru M. An overview of the ultrasonically assisted extraction of bioactive principles from herbs. Ultrasonics Sonochemistry. 2001; 8: 303-313. DOI:10.1016/j.phpro. 2010.01.024

[11] Paulsen BS. Highlights through the history of plant medicine. In: Proceedings from a Symposium Held at the Norwegian Academy of Science and Letters; 13–14 November 2008; The Norwegian Academy of Science and Letters, Oslo, Norway. 2010. p. 18-29

[12] Bernhoft A. A brief review on bioactive compounds in plants. In: Proceedings from a Symposium Held at The Norwegian Academy of Science and Letters; 13–14 November 2008; The Norwegian Academy of Science and Letters, Oslo, Norway. 2010. p. 11-19

[13] Croteau R, Kutchan TM, Lewis NG. Natural products (secondary metabolites). In: Buchanana B, Gruissem W, Jones R, editors. Biochemistry and Molecular Biology of Plants. American Society of Plant Biologists, Rockville, MD, USA; 2000. p. 1250-1268. ISBN: 0943088372 9780943088372

[14] Martin JF, Demain AL. The filamentous fungi. In: Smith JE, Berry DR, editors. Developmental Mycology. Edward Arnold, London; 1978. p. 358-375. ISBN: 0713126426 9780713126426

[15] Pichersky E, Sharkey TD, Gershenzon J. Plant volatiles: A lack of function or a lack of knowledge? Trends in Plant Science. 2006; 11: 421. DOI: 10.1016/j.tplants.2006.07.007

[16] McCaskill D, Croteau R. Some caveats for bioengineering terpenoid metabolism in plants. Trends in Biotechnology. 1998; 16: 349-355. DOI: 10.1016/ S0167-7799(98)01231-1

[17] Gershenzon J, Dudareva N. The function of terpene natural products in the natural world. Nature Chemical Biology. 2007; 3: 408-414. DOI: 10.1038/nchembio.2007.5

[18] Dudareva N, Pichersky E. Biochemical and molecular genetic aspects of floral scent. Plant Physiology. 2000; 122: 627-633. DOI: 10.1104/pp.122.3.627

[19] Azmir J, Zaidul ISM, Rahman MM, Sharif KM, Mohamed A, Sahena F, Jahurul MHA, Ghafoor K, Norulaini NAN, Omar AKM. Techniques for extraction of bioactive compounds from plant materials: A review. Journal of Food Engineering. 2013; 117: 426-436. DOI: 10.1016/j.jfoodeng.2013.01.014

[20] Burt S. Essential oils: their antibacterial properties and potential applications in foods - A review. International Journal of Food Microbiology. 2004; 94: 223-253. DOI: 10.1016/j.ijfoodmicro.2004.03.022

[21] Bakkali F, Averbeck S, Averbeck D, Idaomar M. Biological effects of essential oil: A review. Food and Chemical Toxicology. 2008; 46: 446-475. DOI:10.1016/j.fct. 2007.09.106

[22] Guenther E. The essential oils. D. Van Nostrand, New York; 1948. Reproduction by Nabu Press 2011. 454 p. ISBN-10: 1178571564

[23] Schmidt E. Production of essential oils. In: Baser KHC, Buchbauer G, editors. Handbook of Essential Oils. Science, Technology and Applications. CRC Press, Taylor & Francis Group, Boca Raton, London, New York; 2010. p. 83-118. ISBN: 978-1-4200-6315-8

[24] Bauer K, Garbe D, Surburg H. Common Fragrance and Flavor Materials: Preparation, Properties and Uses. Wiley-VCH, Weinheim; 2001. 29 p. ISBN: 3-527-30364-2

[25] Lawless J. The Encyclopedia of Essential Oils. 2nd ed. Thorsons, London; 2002. p.14. ISBN: 10-0-00-714518-7

[26] Carson CF, Riley TV. Antimicrobial activity of the major components of the essential oil of *Melaleuca alternifolia.* Journal of Applied Bacteriology. 1995; 78: 264-269. DOI: 10.1111/j.1365-2672.1995.tb05025.x

[27] Kubeczka K. History and sources of essential oil research. In: Baser KHC, Buchbauer G, editors. Handbook of Essential oils. Science, Technology and Applications. Taylor & Francis Group, Boca Raton, London, New York;. 2010; p. 3-38. ISBN: 978-1-4200-6315-8

[28] Lawless J. The Encyclopedia of Essential Oils. 2nd ed. Thorsons, London; 2002. p. 15. ISBN: 10-0-00-714518-7.

[29] Bergougnoux V, Caissard JC, Jullien F, Magnard JL, Scalliet G, Cock JM, Hugueney P, Baudino S. Both the adaxial and abaxial epidermal layers of the rose petal emit volatile scent compounds. Planta, 2007; 226: 853-866. DOI: 10.1007/s00425-007-0531-1

[30] Vigan M. Essential oils: renewal of interest and toxicity. European Journal of Dermatology. 2010; 20: 685-692. DOI: 10.1684/ejd.2010.1066.

[31] Fahn A. Secretory tissues in vascular plants. New Phytologist. 1988; 108: 229-257. DOI: 10.1111/j.1469-8137.1988.tb04159.x

[32] Svoboda KP, Svoboda TG, Syred AD. Secretory Structures of Aromatic and Medicinal Plants. Annals of Botany.2000; 86: 1199-1200. DOI: 10.1006/anbo.2000.1279

[33] Obst JR. Special (secondary) metabolites from wood. In: Bruce A, Palfreyman JW, editors. Forest Products Biotechnology. Taylor & Francis, London; 1998. p. 151-165, ISBN: 0-7484-0415-5

[34] Hammer KA, Carson CF. Antibacterial and antifungal activities of essential oils. In: Thormar H, editor. Lipids and Essential Oils as Antimicrobial Agents. John Wiley & Sons, Chichester, England; 2011. p. 255-306. ISBN: 978-0-470-74178-8.

[35] Franz C, Novak J. Sources of essential oils, In: Baser KHC, Buchbauer G, editors. Handbook of Essential Oils. Science, Technology and Applications. Taylor & Francis Group, Boca Raton, London, New York; 2010. p. 39-82. ISBN: 978-1-4200-6315-8

[36] Hammer KA, Carson CF, Riley TV, Nielsen JB. A review of the toxicity of *Melaleuca alternifolia* (tea tree) oil. Food and Chemical Toxicology. 2006; 44: 616-625. DOI: 10.1016/j.fct.2005.09.001

[37] Hussain AI, Anwar F, Hussain Sherazi ST, Przybylski R. Chemical composition, antioxidant and antimicrobial activities of basil (*Ocimum basilicum*) essential oils depends on seasonal variations. Food Chemistry. 2008; 108: 986-995. DOI: 10.1016/j.foodchem.2007.12.010

[38] Langengeim JH. Higher plant terpenoids: a phytocentric overview of their ecological roles. Journal of Chemical Ecology. 1994; 20: 12231280. DOI: 10.1007/BF02059809.

[39] Pengelly A. The Constituents of Medicinal Plants. Allen & Unwin, Sydney; 2004. 184 p. ISBN: 1-74114-052-8.

[40] Russo M, Galletti GC, Bocchini P, Carnacini A. Essential oils chemical composition of wild populations of Italian oregano spice (*Origanum vulgare* ssp. hirtum (Link) Ietswaart): A preliminary evaluation of their use in chemotaxonomy by cluster analysis: 1. Inflorescences. Journal of Agricultural and Food Chemistry. 1998; 46: 3741-3746. DOI: 10.1021/jf980087w

[41] Senatore F. Influence of harvesting time on yield and composition of the essential oils of a thyme (*Thymus pulegioides* L.) growing wild in Campania (Southern Italy). Journal of Agricultural and Food Chemistry. 1996; 44: 1327-1332. DOI:10.1021/jf950508z

[42] Delaquis PJ, Stanich K, Girard B, Mazza G. Antimicrobial activity of individual and mixed fractions of dill, cilantro, coriander and eucalyptus essential oils. International Journal of Food Microbiology. 2002; 74: 101-109. DOI: 10.1016/S0168-1605(01)00734-6

[43] Edris AE. Pharmaceutical and therapeutic potentials of essential oils and their individual volatile constituents: A Review. Phytotherapy Research. 2007; 21: 308-323. DOI: 10.1002/ptr.2072

[44] Miguel MG. Antioxidant and anti-inflammatory activities of essential oils: A short review molecules. 2010; 15: 9252-9287. DOI: 0.3390/molecules15129252

[45] Djilani A, Dicko A. The therapeutic benefits of essential oils. In: Bouayed J, editor. Nutrition, Well Being and Health. InTech, Rijeka, Croatia; 2012. ISBN: 978-953-51-0125-3. Available from: http://www.intechopen.com/books/nutrition-well-being-and-health/the-therapeutic-benefits-of-essential-oils

[46] Carson CF, Hammer KA. Chemistry and bioactivity of essential oils. In: Thormar H, editor. Lipids and Essential Oils as Antimicrobial Agents. John Wiley & Sons, Chichester, England; 2011. p. 255-306. ISBN: 978-0-470-74178-8.

[47] Perricone M, Arace E, Corbo MR, Sinigaglia M, Bevilacqua A. Bioactivity of essential oils: a review on their interaction with food components. Frontiers in Microbiology. 2015; 6: 76. DOI: 10.3389/fmicb.2015.00076

[48] Aleksic V, Knezevic P. Antimicrobial and antioxidative activity of extracts and essential oils of *Myrtus communis* L. Microbiological Research. 2014; 169: 240-254. DOI: 10.1016/j.micres.2013.10.003

[49] Sell C. Chemistry of essential oils. In: Baser KHC, Buchbauer G, editors. Handbook of Essential Oils. Science, Technology and Applications. Taylor & Francis Group, Weinheim, Germany; 2010. p. 121-149. ISBN: 978-1-4200-6315-8

[50] Baser KHC, Demirci F. Chemistry of essential oils. In: Berger RG, editor. Flavour and Fragrances Chemistry, Bioprocessing and Sustainability. Springer, Berlin, Heidelberg, New York; 2007. p. 43-86. DOI: 10.1007/978-3-540-49339-6_4

[51] Faleiro ML, Miguel MG, Ladeiro F, Venancio F, Tavares R, Brito JC, Figueiredo AC, Barroso JG, Pedro LG. Antimicrobial activity of essential oils isolated from Portuguese endemic species of *Thymus*. Letters in Applied Microbiology. 2002; 36: 35-40. DOI: 10.1046/j.1472-765X.2003.01259.x

[52] Raut JS, Karuppayil SM, A status review on the medicinal properties of essential oils. Industrial Crops and Products. 2014; 62: 250-264. DOI: 10.1016/j.indcrop.2014.05.055

[53] Damjanović-Vratnica B, Perović A, Šuković D, Perović S. Effect of vegetation cycle on chemical content and antibacterial activity of *Satureja montana* L. Archive of Biological Science Belgrade. 2011; 63: 1173-1179. DOI: 10.2298/ABS1104173D

[54] Marino M, Bersani C, Comi G.. Impedance measurements to study the antimicrobial activity of essential oils from *Lamiacea* and *Compositae*. International Journal of Food Microbiology. 2001; 67: 187-195. DOI: 10.1016/S0168-1605(01)00447-0

[55] Couladis M, Tzakou O, Mimica-Dukić N, Jančić R, Stojanović D. Esssential oil of *Salvia officinalis* L. from Serbia and Montenegro. Flavour and Fragrance Journal. 2002; 17: 119-126. DOI: 10.1002/ffj.1065

[56] Outaleb T, Hazzit M, Ferhat Z, Baaliouamer A, Yekkour A, Zitouni A, Sabaou N. Composition, antioxidant and antimicrobial activities of Algerian *Rosmarinus officinalis* L. extracts. Journal of Essential oil Bearing Plants. 2015; 18: 654-655. DOI: 10.1080/0972060X.2014.960276

[57] Daferera DJ, Ziogas BN, Polissiou MG. The effectiveness of plant essential oils in the growth of *Botrytis cinerea, Fusarium* sp. and *Clavibacter michiganensis* subsp. michiganensis. Crop Protection. 2003; 22: 39-44. DOI: 10.1016/S0261-2194(02)00095-9

[58] Pintore G, Usai M, Bradesi P, Juliano C, Boatto G, Tomi F, Chessa M, Cerri R, Casanova J. Chemical composition and antimicrobial activity of *Rosmarinus officinalis* L. oils from Sardinia and Corsica. Flavour and Fragrance Journal. 2002; 17: 15-19. DOI: 10.1002/ffj.1022

[59] Daferera DJ, Ziogas BN, Polissiou MG. GC-MS analysis of essential oils from some Greek aromatic plants and their fungitoxicity on *Penicillium digitatum*. Journal of Agricultural and Food Chemistry, 2000; 48: 2576-2581. DOI: 10.1021/jf990835x

[60] Teixeira B, Marques A, Ramos C, Serrano C, Matos O, Neng N, Nogueira J, Saraiva. JA, Nunes ML. Chemical composition and bioactivity of different oregano (*Origanum vulgare*) extracts and essential oil. Journal of the Science of Food and Agriculture. 2013; 93: 2707-2714. DOI: 10.1002/jsfa.6089

[61] Lukas B, Schmiderer C, Novak J. Phytochemical diversity of *Origanum vulgare* L. subsp. *vulgare* (*Lamiaceae*) from Austria. Biochemical Systematics and Ecology. 2013; 50: 106-113. DOI: 10.1016/j.bse.2013.03.037

[62] Bezić N, Skočibusić M, Dunkic V. Phytochemical composition and antimicrobial activity of *Satureja Montana* L. and *Satureja cuneifolia* Ten. essential oils. Acta Botanica Croatica. 2005; 64: 313-322. ISSN: 0365-0588

[63] Skočibušic M, Bezić N. Chemical composition and antimicrobial variability of Satureja montana L. essential oils in the ontogenesis. Journal of Essential Oil Research. 2004; 16: 387-391. DOI: 10.1080/10412905.2004.9698751

[64] Brud WS, Industrial uses of essential oils. In: Baser KHC, Buchbauer G, editors. Handbook of Essential Oils. Science, Technology and Applications. Taylor & Francis Group, Boca Raton, London, New York; 2010. p. 843-853. ISBN: 978-1-4200-6315-8

[65] Van de Braak SAAJ, Leijten GCJJ. CBI, Centre for the Promotion of Imports from Developing Countries, Rotterdam; 1999. 116 p. (Monograph)

[66] Smith RM. Before the injection—modern methods of sample preparation for separation techniques. Journal of Chromatography A. 2003; 1000: 3-27. DOI: 10.1016/S0021-9673(03)00511-9

[67] Clarke S. Essential Chemistry for Aromatherapy, 2nd edition. Churchill Livingstone, Elsevier; 2008. 302 p. ISBN: 978-0-4431-0403-9

[68] Vankar PS. Essential oils and fragrances from natural sources. Resonance 2004; 9: 30-41. DOI: http://dx. doi.org/10.2307/10.1007/BF02834854

[69] Silva LV, Nelson DL, Drummond MFB, Dufosse L, Gloria MBA. Comparison of hydrodistillation methods for the deodorization of turmeric. Food Research International. 2005; 38: 1087-1096. DOI: 10.1016/j.foodres.2005.02.025

[70] Luque de Castro, MD, Garcia-Ayuso, LE. Soxhlet extraction of solid materials: an outdated technique with a promising innovative future. Analytica Chimica Acta. 1998; 369: 1-10. DOI: 10.1016/S0003-2670(98)00233-5

[71] Chemat F, Tomao V, Virot M. Ultrasound-assisted extraction in food analysis. In: Otles S, editor. Handbook of Food Analysis Instruments. CRC Press - Taylor & Francis Group, Boca Raton, London, New York; 2008. p. 85-94. ISBN: 978-1-4200-4566-6

[72] Delsart C, Ghidossi R, Poupot C, Cholet C, Grimi N, Vorobiev E, Milisic V, Peuchot MM. Enhanced extraction of phenolic compounds from merlot grapes by pulsed electric field treatment. American Journal of Enology and Viticulture. 2012; 63: 205-211. DOI: 10.5344/ajev.2012.11088

[73] Bhattacharjee P, Singhal RS, Tiwari SR. Supercritical carbon dioxide extraction of cottonseed oil. Journal of Food Engineering. 2006; 79: 892- 989. DOI: 10.1016/j.jfoodeng.2006.03.009

[74] Puri M, Sharma D, Barrow CJ. Enzyme-assisted extraction of bioactives from plants. Trends in Biotechnology. 2012; 30: 37-44. DOI: 10.1016/j.tibtech.2011.06.014

[75] Fornari T, Vicente G, Vázquez E,. García-Risco MR, Reglero G. Isolation of essential oil from different plants and herbs by supercritical fluid extraction. Journal of Chromatography A. 2012; 1250: 34-48. DOI: http://dx.doi.org/10.1016/j.chroma.2012.04.051

[76] Ghafoor K, AL-Juhaimi FY, Choi YH. Supercritical fluid extraction of phenolic compounds and antioxidants from grape (*Vitis labrusca* B.) seeds. Plant Foods for Human Nutrition. 2012; 67: 407-414. DOI: 10.1007/s11130-012-0313-1

[77] Reverchon E, De Marco I. Supercritical fluid extraction and fractionation of natural matter. The Journal of Supercritical Fluids. 2006; 38: 146-166. DOI: 10.1016/j.supflu.2006.03.020

[78] Ibanez E, Herrero M, Mendiola JA, Castro-Puyana M. Extraction and characterization of bioactive compounds with health benefits from marine resources: macro and microalgae, cyanobacteria, and invertebrates. In: Hayes M. editor. Marine Bioactive Compounds: Sources, Characterization and Applications. Springer, Berlin, Heidelberg, New York; 2012. p. 55-98.

[79] Žižović I, Ivanović J, Milovanović S, Stamenić M. Impregnations using supercritical carbon dioxide. In: Rój E, editor. Supercritical CO_2 Extraction and Its Applications. OIC Poland, Lublin, Poland; 2014. p. 23-34. ISBN: 978-83-86499-96-0

[80] Nakatsu T, Lupo AT, Chinn JW, Kang RKL. Biological activity of essential oils and their constituents. Studies in Natural Product Chemistry. 2000; 21: 571-631. ISBN: 0-444-50469-9

[81] Hall-Stoodley L, Costerton JW, Stoodley P. Bacterial biofilms: from the natural environment to infectious diseases. Nature Review Microbiology. 2004; 2: 95-108. DOI: 10.1038/nrmicro821

[82] Božin B, Mimica-Dukić N, Simin N, Anackov G. Characterization of the volatile composition of essential oils of some *Lamiaceae* spices and the antimicrobial and antioxidant activities of the entire oils. Journal of Agriculture and Food Chemistry. 2006; 54: 1822-1828. DOI: 10.1021/jf051922u

[83] Holley RA, Patel D. Improvement in shelf-life and safety and safety of perishable foods by plant essential oils and smoke antimicrobials. Food Microbiology. 2005; 22: 273-292. DOI: 10.1016/j.fm.2004.08.006

[84] Pauli A, Schilcher H. *In Vitro* antimicrobial activities of essential oils monographed in the European Pharmacopoeia. In: Baser KHC, Buchbauer G, editors. Handbook of Essential Oils. Science, Technology and Applications. Taylor & Francis Group, Boca Raton, London, New York; 2010. p. 353-548. ISBN: 978-1-4200-6315-8

[85] Pandey AK, Kumar S. Perspective on plant products as antimicrobials agents: A review. Pharmacologia. 2013; 4: 469-480. DOI: 10.5567/pharmacologia.2013.469.480

[86] Tsuchiya H. Membrane interactions of phytochemicals as their molecular mechanism applicable to the discovery of drug leads from plants – Review. Molecules. 2015; 20: 18923-18966. DOI: 10.3390/molecules201018923

[87] Ultee A, Smid EJ. Influence of carvacrol on growth and toxin production by *Bacillus cereus*. International Journal of Food Microbiology. 2001; 64: 373-378. DOI: 10.1016/S0168-1605(00)00480-3.

[88] Kotzekidou P, Giannakidis P, Boulamatsis A. Antimicrobial activity of some plant extracts and essential oils against foodborne pathogens in vitro and on the fate of inoculated pathogens in chocolate. LWT – Food Science and Technology. 2008. 41: 119-127

[89] Mendonça-Filho RR. Bioactive phytocompounds: New approaches in the phytosciences, in modern phytomedicine. In: Ahmad I, Aqil F, Owais M, editors. Turning Medicinal Plants into Drugs. Wiley-VCH, Weinheim, Germany; 2006. p. 1-24. DOI: 10.1002/9783527609987.ch1

[90] Doughari JH, Human IS, Bennade S, Ndakidemi PA. Phytochemicals as chemotherapeutic agents and antioxidants: Possible solution to the control of antibiotic resistant verocytotoxin producing bacteria. Journal of Medicinal Plants Research. 2009; 3: 839-848. ISSN 1996-0875

[91] Xu J, Zhou F, Ji B.P, Pei RS, Xu N. The antibacterial mechanism of carvacrol and thymol against *Escherichia coli*. Letters in Applied Microbiology. 2008; 47: 174-179. DOI: 10.1111/j.1472-765X.2008.02407.x.

[92] Helander IM, Alakomi HL, Latva K, Mattila-Sandholm T, Pol I, Smid EJ, Gorris LGM, von Wright A. Characterization of the action of selected essential oil components on Gram-negative bacteria. Journal of Agriculture and Food Chemistry. 1998; 46: 3590-3595. DOI: 10.1021/jf980154m

[93] Zengin H, Baysal AH. Antibacterial and antioxidant activity of essential oil terpenes against pathogenic and spoilage-forming bacteria and cell structure-activity relationships evaluated by SEM microscopy. Molecules. 2014; 19: 17773-17798. DOI: 10.3390/molecules191117773

[94] Bard M, Albrecht MR, Gupta N, Guynn, CJ, Stillwell W. Geraniol interferes with membrane functions in strains of *Candida* and *Saccharomyces*. Lipids. 1988; 23: 534-538. ISSN: 0024-4201

[95] Mendanha SA, Moura SS, Anjos JL, Valadares MC, Alonso A. Toxicity of terpenes on fibroblast cells compared to their hemolytic potential and increase in erythrocyte membrane fluidity. Toxicology in Vitro. 2013; 27: 323-329. DOI: 10.1016/j.tiv.2012.08.022

[96] Yin Q, Shi X, Ding H, Dai X, Wan G, Qiao Y. Interactions of borneol with DPPC phospholipid membranes: A molecular dynamics simulation study. International Journal of Molecular Science, 2014; 15: 20365-20381. DOI: 10.3390/ijms151120365

[97] Cowan MM. Plant products as antimicrobial agents. Clinical Microbiology Reviews. 1999; 12: 564-582. ISSN: 0893-8512

[98] Oosterhaven K, Poolman B, Smid EJ. S-carvone as a natural potato sprout inhibiting, fungistatic and bacteristatic compound. Industrial Crops and Products. 1995; 4: 23-31. DOI: 10.1016/0926-6690(95)00007-Y

[99] Thoroski J. Eugenol induced inhibition of extracellular enzyme production by *Bacillus cereus*. Journal of Food Protection. 1989; 52: 399-403. ISSN 0362-028X

[100] Nazzaro F, Fratianni F, De Martino L, Coppola R, De Feo V. Effect of essential oils on pathogenic bacteria. Pharmaceuticals (Basel). 2013; 6: 1451-1474. DOI: 10.3390/ph6121451

[101] Wendakoon CN, Sakaguchi M. Inhibition of amino acid decarboxylase activity of *Enterobacter aerogenes* by active components in spices. Journal of Food Protection. 1995; 58: 280-283. ISSN: 0362-028X

[102] Gill AO, Holley RA. Mechanisms of bactericidal action of cinnamaldehyde against *Listeria monocytogenes* and of eugenol against *L. monocytogenes* and *Lactobacillus sakei*. Applied Environmental Microbiology. 2004; 70: 5750-5755. DOI: 10.1128/AEM.70.10.5750-5755.2004

[103] Hattori M, Kusumoto IT, Namba T, Ishigami T, Hara Y. Effect of tea polyphenols on glucan synthesis by glucosyltransferase from *Streptococcus mutans*. Chemical and Pharmaceutical Bulletin. 1990; 38: 717-720. ISSN: 0009-2363

[104] Huber B, Eberl L, Feucht W, Polster J. Influence of polyphenols on bacterial biofilm formation and quorum-sensing. Zeitschrift für Naturforschung B. 2003; 58: 879-884. DOI: 10.1515/znc-2003-11-1224

[105] Ferrazzano GF, I. Amato A, Ingenito A, Pinto Z, Pollio A. Plant polyphenols and their anti-cariogenic properties: A review. Molecules. 2011; 16: 1486-1507. DOI: 10.3390/molecules16021486

[106] Kumar S, Pandey AK. 2013. Phytochemicals analysis, cytotoxic and radical scavenging activities of *Parthenium Hysterophorus*. In: Sharma S, editor. Proceedings of Inter-

national Conference on Health, Environment and Industrial Biotechnology, McGraw Hill Education (India). 2013; p. 271-274. ISBN-13: 9789332901377

[107] Cyboran S, Strugała P, Włoch A, Oszmiański J, Kleszczyńska H. Concentrated green tea supplement: Biological activity and molecular mechanisms. Life Sciences. 2015; 126: 1-9. DOI: 10.1016/j.lfs.2014.12.025

[108] Tsuchiya H. Structure-dependent membrane interaction of flavonoids associated with their bioactivity. Food Chemistry. 2010; 120: 1089-1096. DOI:10.1016/j.foodchem. 2009.11.057

[109] Arora A, Byrem TM, Nair MG, Strasburg GM. Modulation of liposomal membrane fluidity by flavonoids and isoflavonoids. Archives of Biochemistry and Biophysics. 2000; 373: 102-109. DOI:10.1006/abbi.1999.1525

[110] Selvaraj S, Krishnaswamy S, Devashya V, Sethuraman S, Krishnan UM. Influence of membrane lipid composition on flavonoid-membrane interactions: Implications on their biological activity. Progress in Lipid Research. 2015; 58: 1-13. DOI: 10.1016/ j.plipres.2014.11.002

[111] Critchfield JW, Butera ST, Folks TM. Inhibition of HIV activation in latently infected cells by flavonoid compounds. AIDS Research and Human Retroviruses, 1996; 12: 39-46. ISSN: 0889-2229

[112] Wu T, He M, Zang X, Zhou Y, Qiu T, Pan S, Xu X. A structure-activity relationship study of flavonoids as inhibitors of *E. coli* by membrane interaction effect. Biochimica et Biophysica Acta. 2013; 1828: 2751-2756. DOI: 10.1016/j.bbamem.2013.07.029

[113] Rosenkranz V, Wink M. Alkaloids induce programmed cell death in bloodstream forms of trypanosomes (*Trypanosoma b. brucei*). Molecules. 2008; 13: 2462-2473. DOI: 10.3390/molecules13102462

[114] Gąsiorowska J, Wesołowska O, Michalak K. Interaction of plant alkaloid, berberine, with zwitterionic and negatively charged phospholipid bilayers. Current Topics in Biophysics. 2011; 34: 45-51. DOI: 10.2478/v10214-011-0007-0

[115] Dzink JL, Socransky SS. Comparative in vitro activity of sanguinarine against oral microbial isolates. Antimicrobial Agents and Chemotherapy. 1985; 27: 663-665. DOI: 10.1128/AAC.27.4.663

[116] Stermitz FR, Lorenz P, Tawara JN, Zenewicz LA, Lewis K. Synergy in a medicinal plant: Antimicrobial action of berberine potentiated by 5′-methoxyhydnocarpin, a multidrug pump inhibitor. Proceedings of the National Academy of Sciences of the United States of America. 2000; 97: 1433-1437. DOI: 10.1073/pnas.030540597

[117] Damjanović B, Lepojević Ž, Živković V, Tolić A. Extraction of fennel (*Foeniculum vulgare* Mill.) seeds with supercritical CO2: Comparison with hydrodistillation. Food Chemistry. 2005; 92: 143-149. DOI:10.1016/j.foodchem.2004.07.019

[118] Sokmen A, Gulluce M, Akpulat HA Daferera D, Tepe B, Polissiou M,... Sahin F. The in vitro antimicrobial and antioxidant activities of the essential oils and methanol extracts of endemic *Thymus spathulifolius*. Food Control. 2004; 15: 627-634. DOI: 10.1016/j.foodcont.2003.10.005

[119] Witkowska AM, Hickey DK, Alonso-Gomez M, Wilkinson M. Evaluation of antimicrobial activities of commercial herb and spice extracts against selected food-borne bacteria. Journal of Food Research. 2013; 2: 37-54. DOI: 10.5539/jfr.v2n4p37

[120] Smid EJ, Gorris LGM. Natural antimicrobials for food preservation. In: Rahman MS. editor. Handbook of Food Preservation. Marcel Dekker, NY; 1999. p. 285-308. ISBN-10: 0824702093

[121] Gill AO, Delaquis P, Russo P, Holley RA. Evaluation of antilisterial action of cilantro oil on vacuum packed ham. International Journal of Food Microbiology. 2002; 73: 83-92. DOI: 10.1016/S0168-1605(01)00712-7

[122] Barbosa LN, Rall VLM, Fernandes AAH, Ushimaru PI, Probst I, Fernandes A. Essential oils against foodborne pathogens and spoilage bacteria in minced meat. Foodborne Pathogene Disease. 2009; 6: 725-728. DOI: 10.1089/fpd.2009.0282

[123] Tassou C, Drosinos EH, Nychas GJE. Inhibition of resident microbial flora and pathogen inocula on cold fresh fish fillets in olive oil, oregano, and lemon juice under modified atmosphere or air. Journal of Food Protection. 1996; 59: 31-34. ISSN0362-028X

[124] Skandamis PN, Nychas GJE. Effect of oregano essential oil on microbiological and physico-chemical attributes of minced meat stored in air and modified atmospheres. Journal of Applied Microbiology. 2001; 91: 1011-1022. DOI: 10.1046/j.1365-2672.2001.01467.x

[125] Burt SA, Fledderman MJ, Haagsman HP, van Knapen F, Veldhuizen EJA. Inhibition of *Salmonella enterica* serotype *Enteritidis* on agar and raw chicken by carvacrol vapour. International Journal of Food Microbiology. 2007; 119: 346-350. DOI:10.1016/j.ijfoodmicro.2007.05.002

[126] Fisher K, Phillips CA. The effect of lemon, orange and bergamot essential oils and their components on the survival of *Campylobacter jejuni, Escherichia coli* O157, *Listeria monocytogenes, Bacillus* cereus and *Staphylococcus aureus* in vitro and in food systems. Journal of Applied Microbiology. 2006; 10: 1232-40. DOI: 10.1111/j.1365-2672.2006.03035.x

[127] Hayouni EA, Chraief I, Abedrabba M, Bouix M, Leveau JY, Mohammed M, Hamdi M. Tunisian *Salvia officinalis* L. and *Schinus molle* L. essential oils: Their chemical compositions and their preservative effects against *Salmonella* inoculated in minced beef meat. International Journal of Food Microbiology. 2008; 125: 242-251. DOI: 0.1016/j.ijfoodmicro.2008.04.005

[128] Gómez-Estaca J, López de Lacey A, López-Caballero ME, Gómez-Guillén MC, Montero P. Biodegradable gelatin-chitosan films incorporated with essential oils as antimicrobial agents for fish preservation. Food Microbiology. 2010; 27: 889-896. DOI: 0.1016/j.fm.2010.05.012

[129] Iturriaga L, Olabarrieta I, de Marañón IM., Antimicrobial assays of natural extracts and their inhibitory effect against *Listeria innocua* and fish spoilage bacteria, after incorporation into biopolymer edible films. International Journal of Food Microbiology. 2012; 158: 58-64. DOI:10.1016/j.ijfoodmicro.2012.07.001

[130] Ouattara B, Sabato SF, Lacroix M. Combined effect of antimicrobial coating and gamma irradiation on shelf life extension of pre-cooked shrimp (*Penaeus* spp.). International Journal of Food Microbiology. 2001; 68: 1-9. DOI: 10.1016/S0168-1605(01)00436-6

[131] Mejlholm O, Dalgaard P. Antimicrobial effect of essential oils on the seafood spoilage micro-organism Photobacterium phosphoreum in liquid media and fish products. Letters in Applied Microbiology. 2002; 34: 27-31. DOI: 10.1046/j.1472-765x.2002.01033.x

[132] Raybaudi-Massilia RM, Mosqueda-Melgar J, Martin-Belloso O. Antimicrobial activity of essential oils on *Salmonella enteritidis*, *Escherichia coli* and *Listeriainnocua* in fruit juices. Journal of Food Protection 2006; 69: 1579-1586. ISSN: 0362-028X

[133] Roller S, Seedhar P. Carvacrol and cinnamic acid inhibit microbial growth in fresh-cut melon and kiwifruit at 4°C and 8°C. Letters in Applied Microbiology. 2002; 35: 390-394. DOI: 10.1046/j.1472-765X.2002.01209.x

[134] Gutierrez J, Rodriguez G, Barry-Ryan C, Bourke P. Efficacy of plant essential oils against foodborne pathogens and spoilage bacteria associated with ready-to-eat vegetables: antimicrobial and sensory screening. Journal of Food Protection. 2008; 71: 1846-1854. ISSN: 0362-028X

[135] Gutierrez J, Barry-Ryan C, Bourke P. Antimicrobial activity of plant essential oils using food model media: efficacy, synergistic potential and interactions with food components. Food Microbiology. 2009; 26: 142-150. DOI: 10.1016/j.fm.2008.10.008

[136] Gündüz GT, Gönül SA, Karapinar M. Efficacy of sumac and oregano in the inactivation of *Salmonella typhimurium* on tomatoes. International Journal of Food Microbiology. 2010; 141: 39-44. DOI: 10.1016/j.ijfoodmicro.2010.04.021

[137] Skandamis PN, Nychas GJE. Development and evaluation of a model predicting the survival of Escherichia coli O157:H7 NCTC 12900 in homemade eggplant salad at various temperatures, pHs and oregano essential oil concentrations. Applied and Environmental Microbiology. 2000; 66: 1646-1653. DOI: 10.1128/AEM.66.4.1646-1653.2000

[138] EUR Lex, Access to European Union Law – Food flavorings [Internet]. Available from: http://eur-lex.europa.eu/legal-content/EN/TXT/?uri=uriserv:sa0006 [Accessed: 2015-07-28]

[139] Nielsen PV, Rios R. Inhibition of fungal growth on bread by volatile compounds from spices and herbs and the possible application in active packaging with special emphasis on mustard essential oil. Journal of Food Microbiology. 2000; 60: 219-229. DOI: 10.1016/S0168-1605(00)00343-3

[140] Shelef LA. Antimicrobial effects of spices. Journal of Food Safety. 1983; 6: 29-44. DOI: 10.1111/j.1745-4565.1984.tb00477.x

[141] Ultee A, Slump RA, Steging G, Smid EJ. Antimicrobial activity of carvacrol toward *Bacillus cereus* on rice. Journal of Food Protection. 2000; 63: 620-624. ISSN: 0362-028X

[142] Fisher K, Phillips C. Potential antimicrobial uses of essential oils in food: is citrus the answer? Trends in Food Science and Technology. 2008; 19: 156-164. DOI: 10.1016/j.tifs.2007.11.006

[143] Tassou C, Drosinos EH, Nychas GJE. Effects of essential oil from mint (*Mentha piperita*) on *Salmonella enteritidis* and *Listeria monocytogenes* in model food systems at 4°C and 10°C. Journal of Applied Bacteriology. 1995; 78: 593-600. DOI: 10.1111/j.1365-2672.1995.tb03104.x

[144] Stammati A, Bonsi P, Zucco F, Moezelaar R, Alakomi H, Von Wright A. Toxicity of selected plant volatiles in microbial and mammalian short-term assays. Food and Chemical Toxicology. 1999; 37: 813-823. DOI: 10.1016/S0278-6915(99)00075-7

[145] Madeira SVF, Matos FJA, Leal-Cardoso JH, Criddle DN. Relaxant effects of the essential oil of *Ocimum gratissimum* on isolated ileum of the guinea pig. Journal of Ethnopharmacology. 2002; 8: 1-4. DOI: 10.1016/S0378-8741(02)00049-1

5

Food Safety – Problems and Solutions

Aslı Uçar, Mustafa Volkan Yilmaz and Funda Pınar Çakıroğlu

Abstract

When certain disease-causing bacteria, viruses or parasite contaminate food, they can cause food-related diseases. Another word for such a bacterium, virus, or parasite is "pathogen". Since food-related diseases can be serious, or even fatal, it is important to know and practice safe food-handling behaviors to help reduce the risk of getting sick from contaminated food. According to the Codex Alimentarius Commission (CAC), "food safety is the assurance that food will not cause harm to the consumer when it is prepared and/or eaten according to its intended use". Foodborne diseases are widespread throughout the world. The process by which a foodborne disease spreads begins with the features of the disease, contaminating the food, which in turn threatens both individual and public health by means of the foods. Healthy, or what can be termed as safe food, is food that has not lost its nutritional value, that is clean, in physical, chemical and microbiological terms and that is not stale. The factors causing the contamination of the food may threaten the safe consumption of it and thereby make the foods harmful to human health. For this reason, it is necessary to utilize various resources to prevent the food from being contaminated in all stages of the food chain, from harvest to consumption. The aim of this chapter is to determine the factors affecting food safety and proffer effective intervention strategies against food-related diseases.

Keywords: Food safety, food hygiene, food handlers hygiene, kitchen and equipment hygiene, nutrition

1. Introduction

The diseases caused by food, or the foodborne diseases, are described as the illnesses with which people are infected by the foods they eat [1]. These diseases are a widespread public health issue and are expensive to treat [2]. Foodborne diseases result from the consumption of contaminated foods and products. Contamination of the food at any stage, from production to consumption, produces bacteria, viruses, parasites, chemical agents and toxins, which eventually cause the foodborne diseases [1].

These diseases are seen as a pervasive, permanent problem that can lead to morbidity and, occasionally, to mortality. Foodborne diseases are increasing worldwide, particularly in the developing countries, due to neglect of personal hygiene and food hygiene [3].

Foodborne illnesses pose a threat to international public health safety and economic development. With the increasing amount of trade, travel and immigration, the rate at which dangerous contaminants and pathogens pass through the borders has also risen. Every year, approximately 2.2 million people, a majority of whom are children living in developing countries, die as a result of food and water contamination [1]. Typhoid fever occurs in 16.6 million people and causes 600,000 deaths every year around the world. In the United States, contaminated foods are seen as being responsible for nearly 76 million infections, 325,000 hospital cases and 5000 deaths every year [4]. According to 2011 data from the Centers for Disease Control and Prevention (CDC), it was estimated that in the United States, one out of every six persons was infected with foodborne illness (48 million people) and that foodborne illnesses resulted in 128,000 hospital cases and 3000 deaths [5].

In 2013, FoodNet, a CDC-established program that tracks foodborne illnesses in the United States, found that foodborne illnesses were responsible for 19,056 infections, 4,200 hospital cases and 80 deaths. The incidence of bacteria responsible for causing diseases in every 100.000 people was determined to be 15.19 for *Salmonella*, 13.82 for *Campylobacter*, 4.82 for *Shigella*, 2.48 for *Cryptosporidium*, 1.15 for *STEC non-O157*, 0.51 for *Vibrio*, 0.36 for *Yersinia*, 0.26 for *Listeria* and 0.03 for *Cyclospora*. Incidences of *Cyclospora*, *Listeria* and *Vibrio* were found to be the highest among the elderly, aged 65 years and older, whereas for all of the other pathogens, the highest incident rates were found among children younger than 5 years of age [6].

The diseases caused by *Salmonella* and *Campylobacter*, the main agents responsible for foodborne infections, are dramatically increasing in some countries, including Denmark, Finland, Iceland, the Netherlands, Norway, Sweden, Switzerland and the United Kingdom. In 1999 and 2000, the number of reported foodborne disease cases was 84,340 and 77,515, respectively in Turkey. While *Salmonella* is the most frequently encountered disease agent, the actual figures on foodborne infections and toxins are not reflected, as it is not mandatory to report these diseases [7].

The world's growing population and the consumers' desire to be provided with a wider range of foods have resulted in a longer and more complex food chain. Today, foods reach consumers after being collected from fields, farms and factories and then pass onto many countries, traveling distances of thousands of kilometers. With this global food distribution, an infection that occurs at any point within the food chain has the potential of affecting any given population in the world. It is therefore essential, given the number of interactions taking place between the actors involved in the food chain and the long distances between them, that multisectorial and international collaboration take place. As no country can provide food safety on its own, safety measures need to be enhanced in many countries [8].

While experts on food safety and health have determined that millions of foodborne disease cases are reported every year, the actual numbers are clouded by uncertainty, as most cases go unreported. Furthermore, foodborne diseases are difficult to diagnose, since they have various symptoms, including fatigue, chills, mild fever, vertigo, upset stomach, dehydration caused by diarrhea, severe cramps and, in some cases, even death. In many of the reported cases, foods prepared outside of the home are the primary cause of foodborne diseases, though it is not uncommon for home-made foods to also cause diseases [9]. Studies conducted on the distribution of foodborne diseases across the world have demonstrated that a majority of these diseases occur during the processing of the food in the preparation stage at home or at food production sites [10]. In fact, most foodborne diseases can be prevented if the regulations governing food safety were complied with, from production stages to consumption [11].

Improper heating of the food, such as undercooking, re-heating and waiting in the heat, or improper cooling of the food account for 44% of the foodborne illnesses. Inadequate preparation and improper cooking practices, such as those involving cross-contamination, insufficient processing, poor hygiene and the re-use of leftovers, are responsible for causing 14% of these diseases [7].

As indicated by these figures, foodborne diseases are widespread throughout the world. The process by which a foodborne disease spreads begins with the features of the disease contaminating the food, which in turn threaten both individual and public health by means of the foods. Healthy, or what can be termed as safe, food is food that has not lost its nutritional value, that is clean, in physical, chemical and microbiological terms and that is not stale. The factors causing the contamination of the food may threaten the safe consumption of it and thereby make the foods harmful to human health. For this reason, it is necessary to utilize various resources to prevent the food from being contaminated at all stages of the food chain, from harvest to consumption [12].

This study conducts an analysis of the factors responsible for jeopardizing food safety and food safety policies throughout the world.

2. The factors that affect food safety

Foods are the basic building blocks of living things, yet they may pose a threat and become harmful to human health in some situations [13]. Many people throughout the world become ill because of the food they eat. These diseases associated with food consumption are referred to as foodborne diseases, and they may result from dangerous microorganisms [14]. Foods can become harmful to human health or even fatal when combined with bacteria, mold, viruses, parasites and chemical toxins [13]. Therefore, it is absolutely necessary that consumers be provided with a safe food supply. The factors involved in the potential threat caused by foods are inappropriate agricultural practices, poor hygiene at any stage of the food chain, lack of preventive controls during processing and preparation of the food, incorrect use of the chemical materials, contaminated raw materials, food and water and inappropriate storage [15].

These issues were classified into three categories: **food hygiene**, **personal hygiene of food handlers** and **kitchen sanitation**.

2.1. Food hygiene

Many factors serve to undermine food hygiene. The hygienic quality of the foods is negatively influenced by purchasing low-quality or stale foods, storing food in inappropriate conditions, cooking large amounts of food, more than is necessary, and letting it sit in inappropriate environments, storing raw and cooked foods together and preparing, cooking and storing food using incorrect methods [13]. If foods are contaminated at any stage, from production to consumption, the hygiene of the food is compromised, depending on the temperature, humidity and pH values of the environment it is stored in, and the food then becomes potentially harmful to human health. An infection or intoxication caused by the consumption of a contaminated food or drink is called food poisoning [16]. The causes of food poisoning are classified as microorganisms, parasites, chemicals, naturally created food toxins, naturally created fish toxins, metabolic disorders, allergic reactions and radioactive substances [17].

Salmonella, Campylobacter and *Enterohemorrhagic Escherichia coli* (EHEC) are foodborne pathogens that affect millions of people every year. Symptoms of food poisoning caused by these pathogens include fever, headache, upset stomach, vomiting, abdominal pain and diarrhea. Although food poisoning is mostly caused by bacteria, some parasites and viruses can also be factors. Parasites such as *Trichinella spiralis* and *Toxoplasma gondii* can remain alive by using the nutritional elements in the carrier. Viruses such as *Hepatitis A* can behave like parasites and infect people as well as the entire food chain [9,18].

Staphylococcus aureus, Clostridium perfringens, Salmonella, Streptococcus, Shigella, Clostridium botulinum, E. coli 0157:H7, Campylobacter and *Bacillus cereus* are the microorganisms that most frequently cause food poisoning [9,13,18,19].

S. aureus is a gram-positive coccal bacterium about 0.5–1.0 μm in diameter. The optimum growth temperature is 37° C. The normal ecological habitat of *S. aureus* is human body [16]. *S. aureus* can be cultured from multiple sites of the skin and mucosal surfaces of carriers; the primary reservoir of staphylococci is thought to be the nostrils of the nose. Spread of *S aureus* generally is through human-to-human contact. Carriage of *S. aureus* in the nose appears to play a key role in the epidemiology and pathogenesis of infection. In the general population, a mean carriage rate of 37.2% was found [20]. *S. aureus* is conveyed to the food by the person handling it. Persons with skin, nose or throat infections or inflammatory wounds pass this microorganism onto the food. The foods posing a particular risk for containing *Staphylococcus* include cooked meat, potato salad, desserts with milk, such as custard, and chicken, fish and other meat salads [9,13,21]. It causes food poisoning by releasing enterotoxins into food. After 3–6 hours of consuming contaminated food, symptoms begin. The most marked and severe symptoms are nausea and vomiting. The others are stomachache and diarrhea [16]. This bacterium cannot be eliminated by cooking. Therefore, it is necessary to keep foods refrigerated; the use of aprons and gloves by staff reduces skin-to-skin contact and, therefore, the staff should further follow the rules of hygiene and minimize physical contact with food [9,13,21]. For staphylococcal food poisoning, phage typing can be performed to determine whether the staphylococci recovered from the food were the source of infection. Depending on the type of infection present an appropriate specimen is obtained accordingly and sent to the laboratory for definitive identification using biochemical or enzyme-based tests. Furthermore, for differentiation on the species level, catalase, coagulase, DNAse, lipase and phosphatase tests are all done [19].

C. perfringens is a gram-positive, rod-shaped anaerobic, spore-forming pathogenic bacterium [19]. *C. perfringens* is very common in nature. Especially, it is found in the digestive system of humans as well as of animals and in soil [13]. This bacterium is the third most common cause of foodborne illnesses in the United Kingdom and United States. According to some estimates, this type of bacteria causes nearly a million illnesses each year. Poisoning occurs after contamination of foods. Cooking kills the growing *C. perfringens* cells that cause food poisoning, but not necessarily the spores that can grow into new cells. If cooked food is not promptly served or refrigerated, the spores can grow and produce new cells [22]. Because the spores of some strains are resistant to temperatures as high as 100°C for more than 1 h, their presence in foods may be unavoidable. Furthermore, the oxygen level may be sufficiently reduced during cooking to permit growth of the clostridia. Spores that survive cooking may germinate and grow rapidly in foods that are inadequately refrigerated after cooking [23]. It easily reproduces in meat that has been sitting at room temperature for a long time after being cooked, in undercooked and repeatedly re-heated meat and in meat served cold [13,19]. *C. perfringens* cells lose their viability when foods are frozen or held under prolonged refrigeration unless special precautions are taken. Such losses may make it difficult to establish *C. perfringens* as the specific cause of a food poisoning outbreak. It is recommended that samples that cannot be examined immediately be treated with buffered glycerin salt solution and stored or shipped frozen to the laboratory [23]. It can be detected in contaminated food [if not heated properly) and feces. Incubation time is between 6 and 24 hours after consuming of contaminated food [24]. Usual symptoms typically include abdominal cramping, diarrhea; vomiting and fever. Very rare, fatal cases of clostridial necrotizing enteritis (also known as pigbel) have been known to involve "Type C" strains of the organism, which produce a potently ulcerative β-toxin. Many cases of *C. perfringens* food poisoning likely remain subclinical, as antibodies to the toxin are common among the population. This has led to the conclusion that most of the population has experienced food poisoning due to *C. perfringens*. Despite its potential dangers, *C. perfringens* is used as the leavening agent in salt rising bread. The baking process is thought to reduce the bacterial contamination, precluding negative effects [25]. This microorganism can be prevented from harming the food by cooling the cooked food rapidly, saving leftover food in shallow containers and storing food in appropriate conditions [13,19].

Salmonella is a rod shaped bacteria and can live in a variety of habitats. Some strains of *Salmonella* live in water, soil, food plants and feces of contaminated humans. Generally the bacterium is able to access those places through cross-contamination of already-infected organisms or feces. When present in water, *Salmonella* can live for several weeks; in soil the bacteria can live up to several years, while in feces the bacteria may only survive a few days. The bacteria can survive in salt water and cooler environments, but a too hot of an environment may kill the bacteria [19,26]. Salmonellosis in humans is generally contracted through the consumption of contaminated food of animal origin (mainly eggs, meat, poultry and milk), although other foods, including green vegetables contaminated by manure, have been implicated in its transmission. Person-to-person transmission through the fecal–oral route can also occur. Human cases also occur where individuals have contact with infected animals, including pets [27]. Usually, symptoms start 12–72 hours after ingestion of bacteria. It is usually characterized by acute onset of diarrhea, fever, abdominal cramps, nausea and vomiting. In most cases, the illness lasts for 4–7 days, and most people recover without treatment. But, *Salmonella* can cause more serious illness in older adults, infants and persons with chronic diseases [26]. *Salmonella* food poisoning has long been, and continues to be, an important global public health problem. In much of Europe and North America, *Salmonella* is mostly found in raw or undercooked chicken, meat, eggs and fish and in unpasteurized milk. It is very easy to control and can be killed by cooking foods at sufficiently high temperatures. To prevent *Salmonella* contamination, the food should be stored at appropriate temperatures, sanitation and hygiene rules should be followed and rodents and flies should be removed from the work environment [9,13,27].

Streptococcus is a gram-positive and nonmotile bacterium and the name refers to the bacterium's characteristics of grouping in chains that resemble a string of beads. The natural habitat of the bacteria are pharynx, rectum and skin [19]. Certain *Streptococcus* species are responsible for many cases of pink eye, meningitis, bacterial pneumonia, endocarditis, erysipelas and necrotizing fasciitis [18]. *Streptococcus* is found inside the human mouth, on hands and in nose secretions and can be conveyed to foods through contact with these secretions. For this reason, infected persons should not prepare food, and the food should be kept in appropriate conditions after being cooled rapidly [19,21].

Shigella is a nonmotile, gram-negative, facultative anaerobic, non–spore-forming, rod shaped bacterium. It is one of the leading bacterial causes of diarrhea worldwide [28]. *The primary host and natural reservoir known at this point for Shigella* is the human gastrointestinal tract. *Shigella* can survive in fecal-contaminated material but has a low survival rate without the optimal acidic environment in the intestinal tract [29,30]. They can easily multiply between 10 and 48 °C [16]. The optimum growth temperature for this bacterium is 37 °C [30]. There are two different mechanisms for pathogenicity. Firstly, bacterial number increases very fast in intestine and then settles into mucosal entry and colon. Because they may cause leakage of blood into the lumen, bloody diarrhea occurs. Secondly, the production of endogenous toxin, which is known as *Shiga toxin*, results in diarrhea [16]. *Shigella* infection typically occurs by ingestion (fecal–oral contamination); depending on age and condition of the host, fewer than 100 bacterial cells can be enough to cause an infection. Food prepared by the contaminated person may easily become contaminated with *Shigella* bacteria [30]. Symptoms of shigellosis include mild to severe diarrhea, bloody diarrhea, fever, dehydration, nausea, vomiting and stomach cramps. They usually appear between 1 and 7 days after contracting the bacteria [31]. The diagnosis of shigellosis is made by isolating the organism from diarrheal fecal sample cultures. It can infect the food in any environment where hygiene rules are not followed. The most important protective factor against *Shigella* is to follow proper personal hygiene rules [13,18,21]. Hand washing before handling food and thoroughly cooking all food before eating decrease the risk of getting shigellosis [32].

C botulinum is a gram-positive, rod-shaped, anaerobic, spore-forming, motile bacterium with the ability to produce the neurotoxin botulinum [33,34]. The natural habitats of the *C. botuli-*

num are soils and marine sediments throughout of the world. Since it is found in the soil, it may contaminate vegetables cultivated in or on the soil. It also colonizes the gastrointestinal tract of fishes, birds and mammals [35]. *C. botulinum* is not a very common bacterium, yet it is very dangerous when it does infect a person [9,18]. Foodborne botulism generally occurs 18 to 36 hours after exposure [range 6 hours to 8 days]. Initial symptoms can include nausea, vomiting, abdominal cramps or diarrhea. After the onset of neurological symptoms, constipation is typical. Dry mouth, blurred vision and diplopia are usually the earliest neurological symptoms. They are followed by dysphonia, dysarthria, dysphagia and peripheral muscle weakness. Symmetric descending paralysis is characteristic of botulism [35]. There are no fever and no loss of consciousness. The symptoms are not caused by the bacterium itself, but by the toxin produced by the bacterium. Incidence of botulism is low, but the mortality rate is high if prompt diagnosis and appropriate, immediate treatment (early administration of antitoxin and intensive respiratory care) are not given. The disease can be fatal in 5 to 10% of cases [36]. Because it is an anaerobic bacterium, it can only grow in the absence of oxygen. Foodborne botulism occurs when *C. botulinum* grows and produces toxins in food prior to consumption. The growth of the bacteria and the formation of toxin occur in products with low oxygen content and certain combinations of storage temperature and preservative parameters. Canned foods improperly prepared and heated and particularly low-acid foods, such as green beans, spinach, mushrooms, meat and vegetables, are very risky in regard to *C. botulinum* contamination [9,18]. Occasionally, commercially prepared foods are involved. Though spores of *C. botulinum* are heat resistant, the toxin produced by bacteria growing out of the spores under anaerobic conditions is destroyed by boiling (for example, at internal temperature >85°C for 5 minutes or longer). Therefore, ready-to-eat foods in low-oxygen packaging are more frequently involved in botulism [36]. Food inside cans that are deformed or warped should under no circumstances be consumed [9,18]. Prevention of foodborne botulism is based on good practice in food preparation, particularly preservation and hygiene. Botulism may be prevented by the inactivation of the bacterial spores in heat-sterilized or canned products or by inhibiting bacterial growth in other products. Commercial heat pasteurization may not be sufficient to kill all spores, and therefore the safety of these products must be based on preventing bacterial growth and toxin production. Refrigeration temperatures combined with salt content and/or acidic conditions will prevent the growth of the bacteria and formation of toxin [18,35,36].

E. coli is a gram-negative, facultative anaerobic, rod-shaped bacterium [18]. *E. Coli 0157:H7* is a very common bacterium found in the human intestines [9]. Provided resource availability and key abiotic conditions (availability of energy and nutrient sources, pH, moisture and temperature) are propitious, *E. coli* populations can survive and even grow in open environments such as soil, manure and water. There are also possibilities for migration between these habitats [37]. Whether food is prepared at home, in a restaurant or in a grocery store, unsafe handling and preparation can cause contamination. Common causes of food poisoning include failing to wash hands completely before preparing or eating food, using utensils, cutting boards or serving dishes that are not clean, causing cross-contamination, consuming dairy products or food containing mayonnaise that have been left out too long, consuming foods that have not been stored/cooked at the right temperature, especially meats and poultry, consuming raw seafood products, drinking unpasteurized milk and consuming raw produce that has not been properly washed *[38]*. Fecal contamination of water or foods is responsible for causing the infection [39]. It can be conveyed through raw or undercooked minced meat or unpasteurized milk. Infection by this bacterium can cause hemolytic–uremic syndrome, which can cause bloody colitis with severe abdominal pain, bloody diarrhea, nausea and vomiting, and the syndrome can lead to renal failure, brain damage, heart attack, paralysis and even death [9]. People with weakened immune systems, pregnant women, young children, and older adults are at increased risk for developing these complications [38]. Person-to-person contact is an important mode of transmission through the oral-fecal route. The duration of excretion of EHEC is about 1 week or less in adults, but can be longer in children. Visiting farms and other venues where the general public might come into direct contact with farm animals has also been identified as an important risk factor for EHEC infection [39]. To protect against contamination, the meat should be cooked very well and foods that include meat should be prepared in perfectly hygienic conditions [9]. The prevention of infection requires control measures at all stages of the food chain, from agricultural production on the farm to processing, manufacturing and preparation of foods in both commercial establishments and household kitchens [39].

Campylobacter is a gram-negative, microaerophilic, non-fermentative bacterium. It grows best in habitats with an oxygen level lower than 5%, and it is typically found in the intestinal tract of animals. They are able to move via flagella. The optimum growth temperature is 42–45 °C and they cannot proliferate in room temperature [16]. *Campylobacter* is one of the bacteria most frequently responsible for causing gastroenteritis. While its actual incidence is not known exactly, in high-income countries, its incidence ranges between 4.4 and 9.3 in every 1000 people each year *[27]*. The infection can pass through animal-sourced food, particularly those of flying bird species, domestic animals, such as cats and dogs, contaminated surface streams, unpasteurized milk and direct physical contact with infected animals. Infection from this bacterium leads to foodborne diseases usually through uncooked meat and other products and raw or unpasteurized milk. Contaminated water and ice are also sources of infection [39]. *Campylobacter jejuni* can be found in both fresh water and sea water and can live for 5 weeks in this environment. Wild birds, farm animals, farm areas and surfaces of still waters create an ideal ecological system for different types of *Campylobacter*. The presence of this microorganism in nature indicates that fecal contamination has occurred in that area. These microorganisms are unable to survive and reproduce anywhere except for in their hosts, and they die in sunlight. Contamination occurs in the environment and on other animals particularly through the droppings of wild birds, such as ducks, geese and seagulls [21]. The most frequent symptoms associated with this infection are Guillain-Barre syndrome (GBS), reactive arthritis (ReA) and irritable bowel syndrome (IBS) *[27]*. To prevent its infection, it is necessary to take protective measures in all stages of the food chain, from production to consumption, and to apply these protective measures in both industrial and domestic environments [39].

B. cereus is a gram-positive, rod-shaped, soil-dwelling, facultative anaerobic bacterium. It can grow between 10 and50 °C, but the optimum growth temperature is 28–37 °C. It can be destroyed at the boiling temperature in 5–30 minutes [16]. *B. cereus* is present in dust and soil and can cause contamination in cereals, particularly those made of rice, in food that sits out for a long time and in meat products [19,21]. The microbe is able to grow either in the presence or in the absence of oxygen. Its spores are sufficiently heat resistant to survive pasteurization treatment of milk and standard cooking temperatures reached in domestic kitchens. It cannot survive the high-temperature treatment used to process canned foods. The symptoms start after 30 minutes of ingestion. Firstly, nausea and vomiting can be seen and then diarrhetic syndrome generally starts [40]. In case of foodborne illness, the diagnosis of *B. cereus* can be confirmed by the isolation of more than 105 *B. cereus* organisms per gram from epidemiologically implicated food, but such testing is often not done because the illness is relatively harmless and usually self-limiting [41]. To prevent infection, foods should be washed thoroughly, not left at room temperature for a long time after being cooked and not left in an open container on the kitchen floor [19,21].

Giardia duodenalis, Cryptosporidium parvum, Cyclospora cayetanensis and *T. gondii* are the parasites that pose the biggest threat to food safety. *G. duodenalis,* previously called *Giardia lamblia,* and now commonly known as "intestinalis", is a microscopic parasite, which lives in human and animal intestines. It usually infects people through contaminated water and uncooked meat. The parasite attaches to the epithelium by a ventral adhesive disc and reproduces via binary fission. The most frequent symptoms associated with this parasite are diarrhea, abdominal cramps, gas and nausea. Giardia lamblia is difficult to detect, which often leads to a delay in diagnosis or misdiagnosis; several tests should be conducted over a 1-week period. Giardiasis is a global disease that infects nearly 33% of people in developing countries and 2% of adults and 6–8% of children in developed countries worldwide. It is especially important that proper hand hygiene be practiced for protection against this parasite [42].

C. parvum is a unicellular microscopic parasite protected by a shell. It is responsible for causing the disease cryptosporidiosis, also known as "Crypto", and it is the main cause of worldwide diseases originating from water and food. Generally, it infects people through soil, foods, water and infected animal parts that had contact with feces. Its most widespread symptoms are diarrhea, abdominal cramps, upset stomach and mild fever. Some cases, however, may not cause any symptoms. In order to protect against this parasite, proper personal hygiene should be performed, only clean water should be used for drinking and vegetables or fruits that have been fertilized with unprocessed fertilizer should not be consumed [42]. This parasite is transmitted by infected food handlers or processors' contact to the food or by foods produced with using animal feces as a fertilizer. When the parasite is involved, watery diarrhea is commonly seen as the main symptom. There is no known effective treatment method. [17].

C. cayetanensis is a unicellular microscopic parasite that causes cyclosporiasis. Although the incidence of cyclosporiasis is reported to be increasing in many countries, this organism is not very well known. In the incubation stage of the parasite, *C. cayetanensis oocysts* infect people through the consumption of water contaminated with this parasite or by eating anything that has come into contact with the feces of an animal with cyclosporiasis disease. The disease may cause symptoms such as diarrhea, loss of appetite, abdominal ventricosity, nausea and vomiting. To protect against infection from this parasite, it is necessary to perform proper personal hygiene and to wash or peel vegetables and fruits before eating them [43].

T. gondii is a unicellular microscopic parasite that causes toxoplasmosis disease, and it exists all over the world. It is among the top three causes of death from foodborne diseases worldwide. It can infect people through the consumption of uncooked meat or by drinking from fresh water sources, such as lakes or streams. This parasite can also infect the fetus through the mother. Moreover, as cats are major hosts for *T. gondii*, domestic cats should receive proper cleaning and hands must be washed after handling them. Symptoms associated with infection from this parasite include diarrhea, upset stomach, vomiting and abdominal pain [44]. In food-caused toxoplasmosis cases, *T. gondii* infects people through consumption of uncooked or undercooked meats. Especially, this parasite spreads faster by products that are contaminated with cat feces. With the higher consumption of raw meats, it shows increased prevalence in Europe and South Africa. *Toxoplasma* infections can be diagnosed by response of antibodies with serologic applications [17].

Infections caused by microorganisms are largely the result of the poor hygiene of the person responsible for preparing the food. These microorganisms can rapidly reproduce in temperatures outside the safe ranges specified by food safety regulations [1].

Cooked foods should not be left to sit in room temperature for longer than two hours. These foods should be cooled rapidly and kept refrigerated, preferably at a temperature under 5°C. Microorganisms can reproduce very rapidly at room temperature. Temperatures below 5°C and above 60°C cause the reproduction of microorganisms to slow down or stop. Table 1 presents the measures that should be followed to ensure food safety [1].

Step	Danger	Measure
Supply/Purchase	Contamination of raw foods	Purchase foods from reliable suppliers. Make sure that hygienic conditions are provided and maintained during supply and transportation.
	Contamination of ready-to-eat foods	Purchase foods from reliable suppliers. Purchase foods from companies that apply HACCP* system.
Storage	Contamination	Keep foods in wrapped or closed containers. Perform pest control.
	Reproduction of bacteria	Monitor the time and temperature of storage. Follow the FIFO** principle.
Preparation	Contamination resulting from personal hygiene	Wash hands before touching the food. Prevent cross-contamination by surfaces and containers. Separate cooked foods from raw foods. Use boiling water, especially if the food will not undergo additionally cooking.
	Reproduction of bacteria	Pay close attention to the amount of time foods remain at room temperature.

Step	Danger	Measure
Cooking	Survival of the pathogen	Make sure that the food is cooked well (the food in its entirety should have a temperature of 70°C)
Cooling and keeping at cool temperatures	Reproduction of the bacteria and spores which did not die in high temperature; toxin production	Make sure that the temperature of the food drops below 5°C as soon as possible when cooling it.
		Do not let foods remain at room temperature longer than two hours. Avoid storing too much food in the refrigerator or in the cool spaces in it. Beware of the thermal agitations in long-term cold storage.
	Contamination by various sources	Wrap the foods appropriately and prevent their direct or indirect contact with raw foods Make sure that the food containers are clean when storing the cooked foods.
Waiting in high temperature	Reproduction of the bacteria and spores which did not die in high temperature; toxin production	Keep temperature of the food above 60°C.
Re-heating	Survival of the bacteria	Re-heat the food properly.
Service	Reproduction of bacteria, production of spores, and toxins	Re-heat the food properly.
	Contamination	Do not touch the food with hands. Serve the food hot. Prevent contact between uncooked foods and unclean containers.

*HACCP: Hazard Analysis Critical Control Point
**FIFO: First In First Out Method

Table 1. Measures to Be Followed to Ensure the Provision of Food Safety [45]

2.2. Personal Hygiene of Food Handlers

The food processing stage is one of the most important stages in the food chain, and those responsible for performing the duties involved in this stage assume major responsibilities in the prevention of food poisoning cases [46,47]. The food processing staff should include healthy individuals who do not have any diseases, and they should undergo regular medical check-ups. In addition to being healthy, it is also important that the workers take particular care for their personal hygiene and execute proper food handling behavior. This is especially important because food handlers can cause cross-contamination between raw and cooked foods, and they may jeopardize food hygiene by improper preparation, cooking and storage of foods [47]. A study confirmed by the Food and Drug Administration (FDA) determined that

81 foodborne diseases were caused by foods contaminated via food processing workers [48]. It should be noted that food workers have the power to make a remarkable impact on public health. In reducing the foodborne diseases or food poisoning, the personal hygiene practices of workers at food production sites are a key factor [49]. It is well known that proper personal hygiene is the best way to mitigate the risks associated with contamination by most of the bacteria generally seen as being responsible for foodborne diseases (Table 2) [40].

Pathogen	Foods involved/sources	Prevention
Campylobacter	Unpasteurized dairy, poultry and meats, infected food handler	Cook all foods thoroughly. Use only pasteurized dairy products. Wash hands properly.
Hepatitis A	Water, ice, shellfish, salads, cold cuts, sandwiches, fruits, fruit juices, milk, milk products, vegetables, any food that will not undergo further heat treatment	Purchase shellfish from approved suppliers. Prevent cross–contamination through hands. Ensure food handlers practice proper hand washing, and prevent bare hand contact with food.
Norovirus	Raw fruit, raw vegetables, prepared salads, raw shellfish	Cook foods thoroughly. Wash hands. Use certified shellfish. No bare hand contact with food.
(Staph) *S.* aureus	Ready-to-eat foods, i.e., sandwiches, salads, ham and other meats, potato salads, custards, warmed-up foods; food handlers' infected cuts, throat, nose and acne	Practice proper hand washing and hygiene. Avoid cross-contamination. Keep bare hand contact with foods to a minimum. Prohibit workers who have cuts and lesions from handling the food. Cool foods rapidly.
Salmonella	Undercooked or raw meats, poultry & shell eggs, poultry and egg salads, egg custards and sauces, protein-containing foods, pets and infected food handlers	Avoid cross-contamination. Cool and refrigerate foods immediately. Cook meats / poultry thoroughly. Practice proper hand washing.
Shigella	Ready-to-eat foods associated with bare hand contact (salads, sandwiches, etc.), source: human feces and flies	Practice proper hand washing after using the bathroom. Use only approved water and foods Control flies. No bare hand Contact with food.

Table 2. Pathogens that threaten food hygiene, the foods they infect and preventive measures

In the Codex Alimentarius [50], the topics involving workers at food processing sites and food hygiene were classified under the following titles: **Health status, Illness and injuries, Personal cleanliness** and **Personal behavior**.

Persons who have any disease that may cause food infection or persons who are suspected to be hosts should not be allowed into the food processing site. Workers at food processing sites who have any symptoms associated with infections should be reported to the administration, and they must be examined by a physician [50]. Accordingly, the law that entered into force in Turkey in 1996 includes the provision: "The staff to be hired to work in the processing of foods and food additives cannot be employed without first providing a medical report. Those who are determined to be carriers should immediately be treated. Those who do not obtain a clean bill of health after completing their treatment should definitely not be employed. The owner or administrator of the workplace is responsible for ensuring that the workplace is in compliance with all of these practices" [51]. Similarly, it is reported that in certain regions in the United States, workers at food processing sites are required to show a health card to their employers [52]]. In the WHO consultation report, routine medical and microbiological examinations of food handlers are not generally recommended, but if food handlers are suffering from an illness that includes symptoms such as jaundice, diarrhea, vomiting, fever, sore throat, skin rash or skin lesions, like boils or cuts, they should report this to their supervisor before starting work [53]. If workers have a sore throat or fever, open cuts or infected wounds, Norovirus, diarrhea, vomiting or jaundice or if they have had contact with someone who has *Salmonella typhi, E. coli 0157:H7, Hepatitis A* or *Shigella,* they should immediately be removed from the food production sites. The employer must take the necessary precautions within 24 hours and report these workers to the relevant institutions. Many studies have recommended that food production sites include health service units and that it was important that workers report their diseases [54–57].

The hygiene practices that should be performed by food processing workers include precise adherence to personal hygiene regulations and the wearing of special, protective attire such as bonnets and gloves to help secure their hygiene. It is important that these clothes be regularly cleaned and cared for [50]. Reports have shown that the lack of personal hygiene among workers at food processing sites was among one of the practices that contributed to food borne diseases and that proper hand washing was the most commonly neglected practice. The practice of improper hand washing may be an important factor in the spreading of foodborne diseases by cross-contamination. It was reported that of the staff working in food production sites, 60% did not wash their hands correctly, and of the foodborne diseases, 25 to 40% were linked to staff working in food processing and food services industries [58,59]. It was also reported that food processing and food service workers were the asymptomatic carriers of the pathogens which caused food poisoning, due to their failure to wash their hands properly after using the restroom [60]. Another study determined that the foods became infected due to improper cleaning of contaminated hands after using the restroom [61]. Aycicek [62] took samples from bare hands and hands with gloves during the food preparation stage (180 in total) and found that the bacteria load on bare hands was significantly higher than that of the gloved hands ($p < 0.05$). The most common bacteria found were *S. aureus* (126/180), *Bacillus spp.* (19/180) and *E. coli* (14/180).

In addition, many studies have reported that workers in the food processing industry did not show enough care in washing their hands properly when necessary and in using protective attire (e.g., gloves, bonnet) [63–66].

The situations when workers should wash their hands are summarized below.

- Before starting to prepare food

- Before touching unpacked foods and clean equipment
- Before serving the foods and touching the clean utensils and equipment
- Before changing tasks from raw meat to ready-to-consume foods in order to prevent cross-contamination during food preparation practices
- After touching any part of the body
- After using the restroom
- After coughing, sneezing or using a handkerchief
- After smoking
- After eating or drinking something
- After touching unclean equipment and tools
- After leaving the kitchen and before entering again
- After washing the dishes
- After touching the handle of the refrigerator, door or any other place commonly used by people during food preparation practices
- After touching working clothes
- Before and after entering areas where foods are kept and stored
- After handling cash
- Before wearing gloves and after taking them off, hands should be properly washed according to hygiene rules [67]

The steps for proper hygienic hand washing are: wash hands and wrists with soap under clean running water, being sure to rub between the fingers; use a nail brush to clean nails; rub arms and wrists with soap and water; soap and rub hands together for 10 to 15 seconds; dry hands with hot air or paper towel and use a paper towel to turn off the tap [63,68]. It is now known that hand-drying methods are as important as hand-washing methods in the prevention of contamination by microorganisms [69]. Studies have demonstrated that among all forms of drying methods, hand dryers posed the greatest threat of contamination. It was found that hand drying machines are less than an ideal alternative for drying hands after they have been washed, as these machines allow the bacteria to be able to spread as far as one meter away; therefore, the use of paper towels should be encouraged [70,71].

In general, humans are the primary source of food contamination, posing a risk to food safety as carriers. Along with the many extraneous factors, such as hands, clothes, accessories, hair and mustache, internally derived factors, such as the breath, spit and wounds, can be sources of contamination. It was reported that food processing workers were capable of spreading 10,000 to 100,000 microorganisms every minute [72,73]. Thus, workers should avoid certain behaviors at the production site, such as smoking, coughing, sneezing, chewing and eating. They also should not wear accessories, watches or hairpins [50].

2.3. Kitchen sanitation

Another important issue in the provision of food safety is kitchen sanitation. A study conducted in child care centers in the states of North Carolina and South Carolina in the United States found that most kitchens were not in compliance with the FDA's 2009 Food Code [74]. In order to minimize the risk of foodborne illnesses in the production and processing of foods, it is necessary to minimize the risk of contamination in the placement, arrangement and decking of kitchen utensils, to set up the area so that it is equipped to perform maintenance, repair, cleaning and disinfection and to ensure that surfaces and materials in the kitchen are anti-toxic, that the kitchen has control mechanisms for temperature and humidity, if possible, and that effective measures are taken against pests [50].

2.3.1. Kitchen hygiene

Issues related to kitchen hygiene should be addressed prior to even completing the construction of the kitchen. The plan and interior design of the kitchen should be arranged in such a way as to facilitate proper hygiene practices (e.g., protection against cross-contamination) [50].

The kitchen should be constructed with durable materials that are easy to care for and clean. These materials should be free of any substances that can potentially render the food unsuitable for consumption, such as parasites, pathogenic microorganisms and toxins, or raw materials, food components and others substances used in the production of processed products that have been infected by foreign substances [75].

The surfaces should be designed in such a way as to not accumulate dirt, to prevent foreign substances from infecting foods and to not allow the creation of dense liquids or mold. Pests should also be prevented from entering the workplace. Drainages should be easy to clean and prevent pests such as rodents from entering and waste liquids from re-entering back into the kitchen environment [76].

There should be warnings written and hung on the walls of the workplace about the rules the staff should obey and the best hygiene practices to be performed. The staff should be provided with changing rooms that include a sufficient number of lockers to hold both work and civilian clothes. The staff should not keep any food in these lockers [75].

Ventilation systems should be capable of eliminating smoke, odors, soot and evaporation, keeping heat inside and preventing dust, dirt and pests from entering. Filters and other parts of the systems should be easily accessible for cleaning or changing. The kitchen should have natural or artificial lights that are equal to the natural light of the day, and the intensity and color of the lights should not impact the production or the quality of the foods in a negative way. There should be continuous control on humidity and temperature in the food storage sites [76].

To maintain a hygienic kitchen, the continuity of cleaning and disinfection procedures is as important as the layout plan of the kitchen. Therefore, a cleaning and disinfection plan should be developed for the kitchen, and all cleaning and disinfection practices should be done according to this plan and recorded. The staff should be trained on the sanitation and disinfection of the kitchen [75].

2.4. Equipment hygiene

Equipment that comes into regular contact with foods should be made of material able to be cleaned and disinfected, resistant to corrosion and non-toxic. The equipment should be arranged in a way as to enable it and the area around it to be cleaned sufficiently. When it is necessary that chemicals be used to clean the equipment, the instructions governing the use of those chemicals should be followed. Calibration checks of the equipment and tools should be made regularly, and these checks should be recorded [76].

3. Food safety systems

Effective food control systems are needed to improve the applicability and control of food safety [77]. Currently, the HACCP, ISO 22000 and PAS 220 are the most commonly used internationally approved food safety systems.

3.1. Hazard analysis and critical control points (HACCP)

HACCP was first used in the 1960s by the American Pillsbury company for the purpose of producing "zero defect" products for the US Army and NASA. Later, starting in the 1970s, it began to be used as a reference by the Food and Drug Administration (FDA) in official supervisions. It was adopted by the Codex Alimentarius Commission in 1992 and published as the HACCP international standard for the first time. Since then, the food industry and official authorities have been using it to protect against and control the risks of potential dangers that could threaten food safety [78].

Initially, HAACCP had three principles:

1. Identification and assessment of hazards associated with food products
2. Determination of critical control points to control identified hazards
3. Establishment of a system to monitor the critical control points

The HACCP, as it is applied today, has five starting steps and is governed by seven principles. The starting steps were created by Codex, and they should be completed prior to implementing the seven HACCP principles. The starting steps help to ensure that the HACCP system is implemented and managed in the most effective way possible [79].

The HAACP system is applicable for any company operating within the food chain, regardless of their size. In the implementation stage, the HACCP system should be supported by certain preliminary condition programs. A company interested in implementing this system should already be following the requirements of this preliminary condition program. Preliminary condition programs include national regulations, codes of practice or other food safety prerequisites. In general, preliminary condition programs involve factories and equipment, staff training, cleaning and sanitation, maintenance chemical control, waste management, storage and transportation [78].

Five starting steps	The seven principles of the HACCP system
1. Assemble HACCP team	1. Conduct a hazard analysis
2. Describe the product	2. Determine Critical Control Points
3. Identify intended use	3. Establish critical limits for each CCP
4. Construct flow diagram	4. Establish a monitoring system for each CCP
5. Conduct on-site confirmation of flow diagram	5. Establish corrective actions
	6. Establish verification procedures
	7. Establish documentation and record keeping

Table 3. HACCP Implementation in 12 Steps

HACCP is an internationally accepted system and in most countries, it is required that companies within the food industry implement this system.

3.2. ISO 22000

In 2005, The International Organization for Standardization (ISO) published a standard for the Food Safety Management System known as ISO 22000. The ISO 22000 system is a combination of preliminary condition programs, HACCP principles and implementation steps defined by the Codex Alimentarius Commission and ISO 9001:2000 standard components. After it was defined, it began to be used in more than 50 countries within 2 years [80].

The basic approach of the ISO 22000 standard is to implement a preventive system that serves to protect consumers from foodborne diseases. This standard controls all the processes in the food chain, including infrastructure, staff and equipment. In business establishments, the Food Safety Management System implementations include production control, product control, equipment control, maintenance, general hygiene practices, staff and visitor hygiene, transportation, storage, product information, training, the selection and evaluation of suppliers, communication and other similar issues [81].

The main goal of this standard is to have a system in place that determines the unacceptable risks that may result from process errors and to secure product safety and consumer health. Food safety supervision over product, design, production and quality control determines and eliminates the potential dangers. The fundamental role of ISO 22000 is not only to provide food safety but also to improve the sensory and nutritional quality of food, and it also plays a primary role in the quality assurance of service practices in industrial production. Lastly, this standard helps to reduce operational losses by instituting a more effective use of resources to increase productivity, and thereby, directs the establishment to a system of total quality [80].

3.3. PAS 220 (publicly available specification)

This standard was created by the major global food producers in cooperation with the Confederation of Food and Drink Industries (CIAA) with the purpose of eliminating the weaknesses of the ISO 22000 food safety system standard. Nestle, Unilever, Danone and Kraft, the sector leaders generally known as "G4", collectively published the PAS 220 standard, which refines the preliminary conditions programs. The PAS 220 standard is applicable for all types of companies and was made available in 2008. It was intended that the PAS 220 standard be used together with the internationally accepted ISO 22000 standard [82].

The content and topics of PAS 220 elaborate on the 10 sub-titles in the ISO 22000 standard and adds 5 of its own, resulting in the following 15 items [82]:

- Structure and placement of buildings
- Placement of work site, buildings and their wings
- Supporting plants (air, water, energy)
- Supporting services, including wastes and sewage
- Adequacy of the equipment, cleaning and preventive care
- Management of purchased materials
- Measures against cross-contamination
- Cleaning and sanitation
- Pest control
- Staff hygiene and workers' lodgings
- Re-processing
- Product recall procedures
- Storage
- Informing consumers about products
- Food defense, biodefense and bioterrorism

4. Good agricultural practices

Today, increasing attention is focused upon the impact farming practices are having on the environment, and there is an increasing emphasis on more sustainable methods of crop production. Systems need to be adopted that are more sensitive to environmental issues, genetic diversity, wildlife and their habitats and in some cases the social structures of rural communities. Furthermore, consumers around the world are more sophisticated and critical than in the past, demanding to know how and what has been used to produce their agriculturally derived products.

Good Agricultural Practices (GAP) are defined "practices that address environmental, economic and social sustainability for on-farm processes, and result in safe and quality food and non-food agricultural products" by the FAO. The aims of the GAP are as follows:

- Ensuring agricultural production harmless to environment, human and animal health
- Safety of natural resources,
- Ensuring traceability and sustainability in agriculture,
- Improving workers health and working conditions,
- Ensuring safety and quality of produce in the food chain.

General principles for GAP were first presented to the FAO Committee on Agriculture (COAG) in 2003 in the paper "Development of a Framework for Good Agricultural Practices" the annex of which broadly outlined farm-level GAP recommendations in 10 fields; which are "soil", "water", "crop and fodder production", "crop protection", "animal protection", "animal health and welfare", "harvest and on-farm processing and storage", "energy and waste management", "human welfare, health, and safety" and "wildlife and landscape".

Although there are some GAP used by different organizations to succeed different purposes and goals, GLOBALGAP (EUREPGAP) is the widespread certificate in agricultural produce worldwide. GLOBALGAP documents consists of; ISO 9001:2000 Quality Management System, ISO 14000 Environmental management system, OHSAS 18001 Work Health and Safety Management System and ISO 22000 Food Safety Management System principles [83].

5. Conclusion

Food safety ultimately deals with the consumption stage, where the existence and level of the dangers caused by foods are of chief concerns. The observance of rigorous control procedures throughout the course of the food chain is a fundamental necessity, given that risks to food safety can surface in any stage of the chain. Therefore, all parties involved in the food chain share the responsibility for ensuring food safety.

The design of a food safety system involves numerous factors. To begin with, minimum hygiene standards should be determined by laws and regulations, food producers must apply food safety measures and procedures and official bodies must supervise and inspect food industry companies to confirm that they are conducting their operations in a manner consistent with the regulations in force. Food poisoning cases that threaten public health globally occur as a result of the contamination of foods in any stage, from production to consumption. Although the factors jeopardizing food safety seem to be easy to control in theory, studies and current practices indicate that there is still a long way to go in practice.

Author details

Aslı Uçar*, Mustafa Volkan Yilmaz and Funda Pınar Çakıroğlu

*Address all correspondence to: aucar@ankara.edu.tr

Ankara University Faculty of Health Sciences, Department of Nutrition and Dietetics, Ankara, Turkey

References

[1] WHO, 2008. WHO initiative to estimate the global burden foodborne diseases. Geneva. Also Available at http://www.who.int/foodsafety/publications/foodborne_disease/burden_nov08/en Food safety fact sheet [Internet]. Geneva: World Health Organization; 2014.

[2] CDC, 2015. CDC and Food Safety. http://www.cdc.gov/foodsafety (accessed: 22.06.2015).

[3] SADAOC, 2002. Food Hygiene and the Problem of Street Food in West Africa. Six Monthly Bulletin on Food Security Policies and Strategies in West Africa. 6(1). Available at: http://www.sadaoc.bf/anglais/sadaocinfo6.htm.

[4] WHO, 2007. Food safety and food borne illness. Fact sheet No: 237, http://www.who.int/mediacentre/factsheets/fs237/en/index.html, Accessed: 10.07.2008.

[5] CDC, 2011. CDC Estimates of foodborne illness in the United States. http://www.cdc.gov/foodborneburden 01.07.2015.

[6] CDC, 2014. Incidence and Trends of Infection with Pathogens Transmitted Commonly Through Food – Foodborne Diseases Active Surveillance Network, 10 U.S. Sites, 2006–2013. Morbidity and Mortality Weekly Report. 63(15): 328–332.

[7] WHO, 2004. Food and Health in Europe: a new basis for action. WHO regional publications European Series, No.96

[8] Fukuda, K. 2015. Food safety in a globalized world. Bulletin of the World Health Organization. 93: 212.

[9] Duyff, R.L. 2002. American Dietetic Association complete food and nutrition guide (2nd Edition). John Wiley and Sons Inc., New Jersey.

[10] WHO, 2000. Food borne disease: A focus for health education, Geneva.

[11] Medeiros, L., V. Hillers, P. Kendall, and A. Mason. 2001. Evaluation of food safety education for consumers. Journal of Nutrition Education and Behavior. 33(S1): S27–S34.

[12] Bilici, S., F. Uyar, et al. 2006. "Food Safety (in Turkish)" Republic of Turkey, Ministry of Health, Basic Health Services Head Office, Nutrition and Physical Activities Department, Nutrition Education Series, Ankara, Turkey.

[13] Yiğit, V. & Duran, T. 1997. Institutional Nutrition Technology I (in Turkish). Ekin Publishing, Istanbul, Turkey.

[14] WHO, 2012. Five keys to growing safer fruits and vegetables: promoting health by decreasing microbial contamination, Geneva.

[15] WHO, 2003. Assuring food safety and quality: guidelines for strengthening national food control systems. Food and Agriculture Organization of the United Nations.

[16] Bas, M. 2004.Food Hygiene Safety and HACCP (in Turkish). 502 p., Sim Publishing., Ankara, Turkey.

[17] Tayfur, M. 2009. Food Hygiene, food related infections and poisonings (in Turkish). 337 p. Kuban Publishing, Ankara, Turkey.

[18] Jay, J.M. 1998. Modern Food Microbiology. Fifth Edition, Aspen Publishers, Maryland.

[19] Adams, M.R., Moss, M.O. 2008. Food Microbiology. Third Edition, pp 463, Royal Society of Chemistry Publishing, UK.

[20] Kluytmans J, van Belkum A, Verbrugh H; Van Belkum; Verbrugh (1997). "Nasal carriage of Staphylococcusaureus: epidemiology, underlying mechanisms, and associated risks". Clinical Microbiology Review. 10(3): 505–20.

[21] Hayes, P.R. 1995. Food Microbiology and Hygiene. Second Edition, Springer-Science +Business Media, UK.

[22] Anon, 2015. *Clostridium perfringens*. http://www.foodsafety.gov/poisoning/causes/bacteriaviruses/cperfringens/ (accessed:04.11.2015).

[23] Rhodehamel, E.J. Harmon and S.M. 1998. *Clostridium perfringens*, Bacteriological Analytical Manual, 8th Edition, Revision A, Chapter 16. http://www.fda.gov/Fodd/FoddScienceResearch/LaboratoryMethods/ucm070878.htm Accessed: 04.11.2015.

[24] Murray; et al. 2009. Medical Microbiology (6th ed.). Mosby Elsevier. ISBN 978-0-323-05470-6.

[25] Juckett, Gregory; et al. November 2008. "The Microbiology of Salt Rising Bread" (PDF). West Virginia Medical Journal (22). Retrieved 22 July 2015.

[26] Anon, 2015b. Salmonella. http://www.foodsafety.gov/poisoning/causes/bacteriaviruses/cperfringens/ (accessed: 04.11.2015).

[27] WHO, 2013. *Salmonella* (non-typhoidal). Fact sheet No: 139. http://www.who.int/mediacentre/factsheets/fs139/en/ Accessed: 17.06.2015.

[28] Mims, Cedric; Dockrell, Hazel; Goering, Richard; Roitt, Ivan; Wakelin, Derek; Zuckerman, Mark, eds. (2004). Medical Microbiology (3rd ed.). Mosby. p. 287. ISBN 978-0-7234-3259-3.

[29] Niyogi, SK. 2005. Shigellosis. Journal of Microbiology. 43(2): 133–143.

[30] Warren, BR., H. G. Yuk, and K. R. Schneider. (2007) Survival of *Shigella sonnei* on smooth tomato surfaces, in potato salad and in raw ground beef. International Journal of Food Microbiology. 116, 400–404.

[31] Anon, 2011 "Shigellosis." http://www.health.ny.gov/diseases/communicable/shigellosis/fact_sheet.htm accessed on 04/11/2015.

[32] Ram, P. K.; Crump, J. A.; Gupta, S. K.; Miller, M. A.; Mintz, E. D. (2008). "Part II. Analysis of data gaps pertaining to Shigella infections in low and medium human development index countries, 1984–2005". Epidemiology and Infection 136 (5): 577–603.doi:10.1017/S0950268807009351

[33] Lindström, M; Korkeala, H (Apr 2006). "Laboratory diagnostics of botulism.".Clinical Microbiology Reviews19 (2): 298–314.doi:10.1128/cmr.19.2.298-314.2006.

[34] Peck, MW. 2009. Biology and genomic analysis of *Clostridium botulinum*. Advances in microbial physiology. Advances in Microbial Physiology. 55: 183–265, 320. doi: 10.1016/s0065-2911(09)05503-9.ISBN 9780123747907.PMID 19573697.

[35] Nantel, A.J. 1999. *Clostridium botulinum*. International Programme on Chemical Safety Poisons information monograph 858 bacteria, WHO, Geneva.

[36] Anon, 2013. Botulism. http://www.who.int/mediacentre/factsheets/fs270/en/ accessed: 17.06.2015.

[37] Elsas J.D. Semenov, A., R. Costa, J.T. Trevors. 2010. Survival of *Escherichia coli* in the environment: fundamental and public health aspects. ISME Journal, 2011 Feb. 5(2): 173–183.

[38] Pietrangelo, A. 2015. *E. coli* infection. http://www.healthline.com/health/e-coli-infection#Overview1 Accessed: 04.11.2015.

[39] WHO, 2011. Enterohaemorrhagic *Escherichia coli* (EHEC). Fact sheet No: 125. http://www.who.int/mediacentre/factsheets/fs125/en/ Accessed: 17.06.2015.

[40] Anon, 2015. Food Safety & Sanitation Program. http://dec.alaska.gov/eh/fss/ (accessed: 30.07.2015).

[41] Anon, 1994. Bacillus cereus Food Poisoning Associated with Fried Rice at Two Child Day Care Centersfrom Morbidity and Mortality Weekly Report from Centers for Disease Control and Prevention. 18 March 1994 / Vol. 43 / No. 10 U.S.

[42] Centers for Disease Control and Prevention, 2015. http://www.cdc.gov/parasites/crypto/ (accessed 10.06.2015).

[43] Strausbaugh, L. J. and B. L. Herwaldt. 2000. *Cyclospora cayetanensis*: a review, focusing on the outbreaks of cyclosporiasis in the 1990s. Clinical Infectious Diseases. 31(4): 1040–1057.

[44] Tenter, A. M., A. R. Heckeroth and L. M. Weiss. 2000. "*Toxoplasma gondii*: from animals to humans." International Journal for Parasitology. 30(12): 1217–1258.

[45] Wisner, B. and J. Adams. 2002. Environmental health in emergencies and disasters: a practical guide, World Health Organization.

[46] Cruickshank, J. G. 1990. Food handlers and food poisoning: training programmes are best. British Medical Journal. 300: 207–208.

[47] Walker E., C. Pritchard and S. Forsythe. 2003. Food handlers_ hygiene knowledge in small food businesses. Food Control. 14: 339–343.

[48] Guzewich, J. and M. Ross. 1999. Evaluation of risks related to microbiological contamination of ready-to-eat food by food preparation workers and the effectiveness of interventions to minimize those risks. www.cfsan.fda.gov (accessed 09.01.2008).

[49] Green, L., C. Selman, V. Radke, D. Ripley, J. Mack, D. Reimann, T. Stigger, M. Motsinger and L. Bushnell. 2006. Food Worker Hand Washing Practices: An Observation Study. Journal of Food Protection. 69(10): 2417–2423.

[50] Codex Alimentarius. 2009. Food Hygiene. Codex Alimentarius Commission, Joint FAO/WHO Food Standards Programme, Rome.

[51] Official Journal of the Republic of Turkey 1996. Legislations About Food Production and Marketing.

[52] Southern Nevada Health District. 2005. Health Cards, available at: www.cchd.org/health_cards/foodhandler_childcare_requirements.htm (accessed September 9, 2006).

[53] Adams M. and Y. Motarjemi 1999. Basic Food Safety for Health Workers WHO/SDE/PHE/FOS/99.1 Distr.: General Original: English World Health Organization Geneva.

[54] Hobbs, B. C. and D. Roberts. 1993. Food poisoning and food hygiene (6th ed.). Cornwall: Edward Arnold.58. Cogan, T. A., Slader, J., BloomWeld, S. F., & Humphrey, T. J. 2002. Achieving hygiene in the domestic kitchen: The effectiveness of commonly used cleaning procedures. Journal of Applied Microbiology. 92: 885–892.

[55] Trickett, J. 1997. Food hygiene for food handlers. London: Macmillan Press.

[56] Ziady, L. E., N. Small and A. M. J. Louis. 1997. Rapid reference: infection control. Pp: 151. Pretoria: Kagiso Tertiary.

[57] Nel S., J.F.R. Lues, E.M. Buys, P. Venter. 2004. The personal and general hygiene practices in the deboning room of a high throughput red meat abattoir. Food Control. 15: 571–578.

[58] Cogan T. A., J. Slader, S.F. BloomWeld and T.J. Humphrey. 2002. Achieving hygiene in domestic kitchen: The effectiveness of commonly used cleaning procedures. Journal of Applied Microbiology. 92: 885–892.

[59] Collins, J. E. 2001. Impact of changing consumer lifestyles on the emergence/re-emergence of foodborne pathogens. Emerging Infectious Diseases. 3(4): 1–13.

[60] Temelli, S. and C.M.K. Şen. ve Anar, Ş. 2007. Determination of Hygiene Status of Meat Seperation and Cheese Production Workers' Hands (in Turkish). The Journal of the Faculty of Veterinary Medicine, Kafkas University.24 (1-2-3-4): 75–80.

[61] Taylor, J. H., K.L. Brown, J. Toivenen and J.T. Holah. 2000. A microbiological evaluation of warm air hand driers with respect to hand hygiene and the washroom environment. Journal of Applied Microbiology. 89: 910–919.

[62] Ayçiçek, H. 2004. Determination of Correct Hand Cleaning and Glove Using Principles in Food Industry (in Turkish). Gülhane Medicine Journal. 44 (3): 344–350.

[63] Clayton, D. and C., Griffith. 2004. Observation of food safety practices in catering using notational analysis. British Food Journal. 106(3): 211–227.

[64] Green, L., V. Radke, R. Mason, L. Bushnell, D. Reimann, J. Mack, M. Motsinger, T. Stigger and C. Selman2007. Factors Related to Food Worker Hand Hygiene Practices. Journal of Food Protection. 70 (3): 661–666.

[65] Pragle, A., A. Harding. and C. Mack. 2007. Food workers perspectives on handwashing behaviors and barriers in the restaurant environment. Journal of Environmental Health. 69: 27–32.

[66] Ural, D. 2007. A Research On Knowledge And Applications Of Personnel Working In Accomodation Businesses Related To Personal Hygiene (in Turkish). Gazi University, MsC Thessis: p. 96, Ankara.

[67] Anonymous. 2005. The handwashing handbook, www.globalhandwashing.org. (accessed 19.01.2008).

[68] Strohbehn, C., S. Beattie, J. Sneed, J. Meyer and P. Paez. 2007. Impact of employee training on mitigating contamination in retail food service operations. www.extension.iastate.edu. (accessed 21.01.2008).

[69] Patrick, D.R., G. Findon and T.E. Miller. 1997. Residual moisture determines the level of touch-contact-associated bacterial transfer following hand washing. Epidemiology and Infection. 119(3): 319–325.

[70] Gustafson, D.I., M.J. Horak, C.B. Rempel, S.G. Metz, D.R. Gigax and P. Hucl.2005. An empirical model for pollen-mediated gene flow in wheat. Crop Science. 45: 1286–1294.

[71] Oğur, R., Ö. Tekbaş, O. Hancı and U. Özcan. 2005. Microbiological Properties of air flow from Hand Drying Machines (in Turkish). Turkish Armed Forces. Jorunal of Preventive Medicine. 4 (1); 1–7.

[72] Hansen, S. 2001. Personal hygiene and food safety. www.ianr.unl.edu. (accessed 10.02.2008).

[73] Tayar, M. 2004.Hygiene and Sanitation in Food Industry (in Turkish). www.uludag.edu.tr. (accessed 03.02.2008).

[74] Wohlgenent, K.C., C.S. Cates, A. Fraser, B. Chapman, L.A. Jaykus and X. Chen. 2014. Sanitation in classroom and food preparation areas in child-care facilities in North Carolina and South Carolina. Journal of Environmental Health. 77(4): 20–27.

[75] Kaferstein, F. and M. Abdussalam. 1999. Food Safety in the 21st Century. Bulletin of the World Health Organization. 77(4): 347.

[76] Anon, 2011. Food hygiene regulations. Official Journal of the Republic of Turkey http://www.resmigazete.gov.tr/eskiler/2011/12/20111217-5.htm (accessed 03.07.2015).

[77] McMeekin, T., J. Baranyi, J. Bowman, P. Dalgaard, M. Kirk, T. Ross, S. Schmid and M. Zwietering. 2006. Information systems in food safety management. International Journal of Food Microbiology. 112(3): 181–194.

[78] Pierson, M. D. 2012. HACCP: Principles and Applications, Springer Science & Business Media.

[79] Mortimore, S. and C. Wallace. 2013. HACCP: A Practical Approach, Springer Science & Business Media.

[80] Surak, J. G. 2007. A recipe for safe food: ISO 22000 and HACCP. Quality Progress. 40(10): 21.

[81] Arvanitoyannis, I. S. and T. H. Varzakas. 2009. HACCP and ISO 22000 – A Comparison of the Two Systems. HACCP and ISO 22000: Application to Foods of Animal Origin: 3–45.

[82] Boisrobert, C., A. Stjepanovic, S. Oh and H. Leliveld. 2009. Ensuring Global Food Safety: Exploring Global Harmonization, Academic Press.

[83] FAO, 2003. Development of a Framework for Good Agricultural Practices. Committee on Agriculture. Seventeenth Session. Rome, 31 March–4 April 2003.

6

Potential Exposure and Risk Associated with Metal Contamination in Foods

Luciana M. Coelho, Diego R. Pessoa, Kênia M. Oliveira, Priscila A.R. de Sousa, Lícia A. da Silva and Nívia M.M. Coelho

Abstract

Humans require several trace elements as components of the diet. Some of these elements are required in extremely small quantities (only micrograms per day). On the other hand, in higher concentrations, some elements may also have deleterious, even lethal, effects. Metals such as arsenic, chromium (Cr), lead (Pb), and mercury (Hg) are naturally occurring chemical compounds. The contamination of food with these metals occurs mainly through human activities, such as farming and industry, or from contamination during food processing and storage. People can be exposed to these metals by ingesting contaminated food or water, and their accumulation in the body can lead to harmful effects over time. The main objective of this chapter is to provide a literature review on the various types of foodborne poisoning caused by the contamination of food with arsenic, Cr, Pb, and Hg and on food safety issues associated with the presence of these metals in food. Research findings from various studies carried out to examine the relationship between metal exposure and the adverse health effects of metals are addressed.

Keywords: Diseases, chemical contamination, metals, food

1. Introduction

Foods can be contaminated with harmful chemicals and microorganisms, which can cause illness in humans. Chemical contaminants can be classified according to the source of contamination and the mechanism by which they enter the food product. In the case of metal residues in food, contamination mostly occurs during food processing and storage [1].

In human nutrition, metals are well recognized by public health agencies, nutritionists, and researchers from various areas of knowledge. Humans require several metals as components of their diet. Some of the metals are required in extremely small quantities, while some, such as arsenic, chromium (Cr), lead (Pb), and mercury (Hg), in certain amounts can adversely affect the nervous system, kidneys, and other vital organs of the body, which can be life threatening in extreme cases.

In general, sources of contamination are contaminated food and beverages and packaging [1, 2].

Metals can often be inadvertently and unintentionally introduced into food products. If these contaminants are not detected, they can become a major safety hazard for consumers. Metals such as arsenic, Cr, Pb, and Hg exist as naturally occurring chemical compounds. These metals are of particular concern in food because of their toxicity, especially in the case of long-term (chronic) intake, because they can accumulate in the body and cause organ damage particularly in susceptible groups, such as children [2].

Arsenic may be present as a contaminant in many foods, such as grains, fruits, and vegetables, where the metal is present because it is absorbed in the plant through the soil and water, and also trace amounts of arsenic can enter the food chain through the application of agricultural chemicals like fertilizers, which may contain arsenic. While most crops do not readily take up much arsenic from the ground, rice is different because it takes up arsenic from soil and water more readily than other grains. Also, arsenic exposure occurs through the consumption of aquatic food, especially shellfish and animals that feed from the bottom of the sea [3].

Cr exposure occurs mainly through the diet. Food crops that are polluted through contaminated soil or water may contain high concentrations of this metal.

Pb is a toxic substance present in the environment in small amounts, and everyone is exposed to some Pb from daily actions such as inhaling dust, eating food, or drinking water. Tobacco smoking and the use of leaded petrol in vehicles are reported to be major sources of Pb exposure, although the Pb content in petrol has dramatically declined over recent decades, thereby reducing environmental exposure [3].

Hg exposure can occur through dental fillings that contain Hg compounds, occupational exposure, and herbal medicines. However, to date no studies have been able to show an association between amalgam fillings and ill health. Most dietary exposure is in the form of inorganic Hg. However, some fish may bioaccumulate the more toxic organic form, methylmercury, in significant quantities. Thus, diets rich in fish can be a cause of organic Hg exposure [3].

Thus, these metallic contaminants have been evaluated by international authorities, and safe reference values have been established. The maximum concentrations of these contaminants allowed by legislation are often well below toxicological tolerance levels, because such levels can often be reasonably achieved by using good food manufacturing practices. Even so, food contaminant testing is needed to assure the safety and quality of food products. Chemical analysis can be very useful in the food industry, with the development of new techniques to accurately and precisely quantify metals present in low concentrations in foods. These data can be applied in the area of toxicology to prevent diseases through prior diagnosis.

2. Arsenic

2.1. Chemistry of arsenic

Arsenic is a chemical element found in several oxidation states (+III, +V, 0, and -III) and various inorganic and organic forms. Arsenic rarely occurs as a pure element. The most common ores of arsenic are arsenopyrite (FeAsS) and arsenic sulfur compounds [orpiment (As_2S_3) and realgar (As_4S_4)]. The inorganic forms are considered to be of greater toxicity, and the ascending order of toxicity is elemental arsenic < arsenobetaine < methylated forms < arsenate < arsenite < arsine [4]. As(III) and As(V) are apparently of comparable bioavailability but differ in terms of their biochemistry. Preferentially, As(III) binds thiol groups, whereas As(V) does not. So, the oxidation state of arsenic can affect its toxicity [5, 6].

2.2. Occurrence in the environment

Arsenic compounds are used in glass and semiconductor production, as a preservative for wood, and as a feed additive to increase weight gain for poultry and swine [7]. Historically, arsenic compounds were used in agriculture as insecticides or herbicides. Due to its widespread use, it can contaminate the environment. Environmental pollution by arsenic occurs as a result of anthropogenic activities and natural phenomena such as volcanic eruptions and soil erosion [8].

Arsenic can be present as a contaminant in environmental compartments, such as water, soil, and plants, and ultimately can seriously affect the human health through exposures to these contaminated compartments. Inorganic arsenic species are the most important chemical forms of arsenic in natural waters [9].

2.3. Dietary sources of arsenic

Arsenic can be found in fish, shellfish, meat, poultry, dairy products, and cereals. However, in fish and shellfish, organic chemical species are found, and thus the arsenic is less toxic. Marine organisms tend to accumulate more arsenic than those living in freshwater or terrestrial environments, which typically have lower arsenic concentrations of around 0.25 mg kg^{-1} [10–12].

China has established the acceptable level of arsenic in rice as 200 ng g^{-1}, while the Codex Alimentarius Committee on Contaminants in Food considers levels of 200 and 300 ng g^{-1} in polished and raw rice, respectively, to be safe [13]. The Joint FAO/WHO Expert Committee on Food Additives (JECFA) reports that the provisional tolerable daily intake (PTDI) of inorganic arsenic is 0.002 mg kg^{-1} body weight, equivalent to 0.12 mg day^{-1} for an adult of 60 kg [14]. However, if contamination with arsenic trioxide occurs, it should be noted that the minimum lethal dose of this compound is 70–180 mg in humans [15]. In general, high levels of arsenic are found in rice and the concentration can vary from 10 to 510 μg kg^{-1}, when rice is irrigated with contaminated water, contributing to the daily intake of arsenic [16].

Due to the toxicity and the many diseases resulting from the ingestion of arsenic, the concentrations of this metalloid and its species in different types of food need to be investigated. Numerous studies on arsenic levels in food have been conducted, and the results have been published in journals, newspapers, and other media. Figure 1 shows the number of publications per year in the Web of Science on the contamination of food with arsenic, demonstrating the interest of the scientific community.

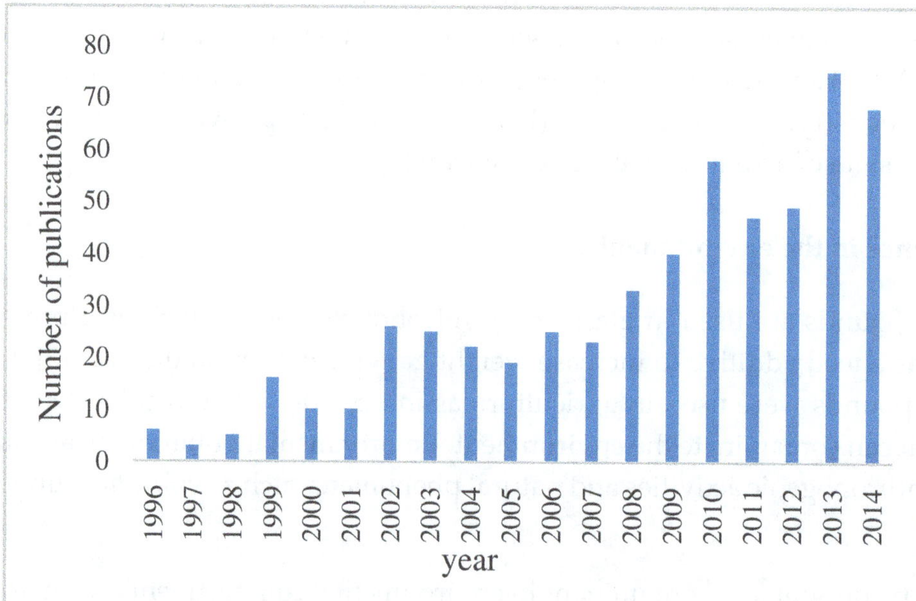

Figure 1. Publications on food contamination with arsenic.

The concentration of arsenic that is safe to human health is under discussion by organizations such as the World Health Organization (WHO), the Food and Agriculture Organization (FAO), the US Food and Drug Administration (FDA), the Food Standards Agency (FSA), and the European Food Safety Authority (EFSA) [17].

The Codex has adopted 0.2 mg kg^{-1} as the maximum level of arsenic in rice. This committee has the task of establishing international food safety and quality standards for consumers worldwide, which are widely used as a basis for national legislation.

To date, the European Union (EU) has not set maximum levels for arsenic in food. However, for water intended for human consumption, the value is 10 mg L^{-1} for total arsenic, with no distinction between the various chemical species of arsenic [18, 19]. Thus, extensive research should be conducted, aimed at more in-depth studies and attaining greater consistency in the data for the creation of relevant legislation regarding arsenic in foods [20–23].

2.4. Routes of entry into plants, animals, and humans

The most important route of exposure to arsenic is the ingestion of foods and beverages, which in most cases are contaminated from the use of irrigation water with a high concentration of the element. In general, the water is contaminated by dissolved minerals and contains different forms of arsenic [24]. Exposure to arsenic occurs via the oral route (ingestion), inhalation, dermal contact, and the parenteral route [25–27].

Arsenic occurs in both inorganic and organic forms, which exhibit large differences in their metabolism and toxicity. The high toxicity of arsenic is well known because arsenic compounds are readily absorbed by either inhalation or ingestion, the extent of absorption being dependent upon the solubility of the compound.

2.5. Metabolism or transformation in the living system

There is evidence that arsenic has a physiological role related to methionine metabolism [28]. However, the site of action of arsenic remains unknown. Studies with rats fed adequate amino acid–based diets have shown that arsenic deprivation had little effect on growth in rats. However, in rats fed suboptimal methionine, arsenic deprivation resulted in a significant reduction in body weight. It was shown that arsenic deprivation reduces the hepatic concentration of *S*-adenosylmethionine, indicating that arsenic maintains the metabolic pool of *S*-adenosylmethionine [28].

In addition, arsenic status affects DNA methylation in animal and cell culture models, resulting in an apparent hypomethylation of DNA [29]. This process is associated with an increased incidence of cancer. So, there is an amount of dietary arsenic that is harmful or beneficial to humans [30, 31].

2.6. Biological functions

In body, arsenic is present as arsenite and arsenate. Arsenic species interact strongly with sulfhydryl groups of organic molecules. It affects several enzymes, causing damage in several cell systems. Because of their similar properties, arsenate can substitute for phosphate and other phosphate intermediates in several biochemical reactions. At the cellular level, arsenate depletes adenosine triphosphate (ATP) in human erythrocytes, interrupting the production of energy [31, 32].

Although some research has indicated that arsenic is an essential nutrient for rats, chickens, and pigs, however, no studies have been published in the literature to determine the nutritional importance of arsenic in humans [32].

2.7. Mechanisms of toxicity of arsenic

The toxicity of arsenic is highly influenced by its oxidation state and solubility, as well as many other factors [33]. As(III) binds to thiol or sulfhydryl groups on proteins and can inactivate over 200 enzymes. As(V) can replace phosphate, which is involved in many biochemical pathways [34].

Mechanism by which arsenic exerts its toxic effect is due its ability to interact with sulfhydryl groups of proteins and enzymes and to substitute phosphorous in various biochemical reactions [34].

In humans, As(III) is methylated to two major metabolites via a non-enzymatic process to monomethylarsonic acid (MMA), which is further methylated enzymatically to dimethyl arsenic acid (DMA) before excretion in the urine [34].

Various hypotheses have been proposed to explain the carcinogenicity of inorganic arsenic. Nevertheless, the molecular mechanisms by which this arsenic induces cancer are still poorly understood [35, 36].

2.8. Incidence of (acute and chronic) toxicity

Arsenic can cause numerous human health effects. Several epidemiological studies have reported a strong association between arsenic exposure and increased risks of both carcinogenic and systemic health effects [25]. The severity of adverse health effects is related to the chemical form of arsenic and is also time and dose dependent [26, 37, 38]. Among the notable effects and diseases are skin lesions, neurotoxicity, cardiovascular diseases, abnormal glucose metabolism, diabetes, peripheral vascular diseases, coronary heart diseases, myocardial infarction, stroke, gangrene, kidney failure and liver failure, cancer of the internal organs, particularly the bladder and lung, skin pigmentation, keratoses, and skin cancer [24, 26, 34–37].

2.9. Comparative analysis of analytical techniques

Several techniques have been used for the detection of arsenic in foods. Among the combined techniques are the use of chromatography or inductively coupled plasma coupled with mass spectrometry (MS) [39], mass spectrometry–desorption electrospray ionization (MS-DESI) spectrometry, inductively coupled plasma–optical emission spectrometry (ICP-OES), and hydride generation–atomic absorption spectrometry (HG-AAS) [39, 40], which generally allow the chemical speciation of arsenic and other techniques such as capillary electrophoresis coupled to inductively coupled plasma and mass spectrometry (CE-ICP-MS) [41]. However, these techniques are expensive. In this regard, an interesting approach to determining arsenic species with detection by electrothermal atomic absorption spectrometry after cloud point extraction (ETAAS/CPE) was developed by Baig *et al.* [42] and Costa *et al.* [43].

3. Chromium

3.1. Chemistry of chromium

Cr is a naturally occurring element present in the earth's crust and is found in two oxidation states, namely hexavalent chromium (Cr(VI)) and trivalent chromium (Cr(III)). Hexavalent chromium (Cr(VI)) compounds are, in general, more toxic than Cr(III) compounds. Cr rarely occurs as a pure element. The most common ore of Cr is ferrochromite [44, 45].

3.2. Occurrence in the environment

Cr is an element available in the environment, found mostly in minerals, rocks, plants, soil, water, dust, and volcanic gases. Cr can be present as a contaminant in the environment from various natural and anthropogenic sources [46]. Cr released into the environment as anthropogenic activity occurs mainly from metallurgical and chemical industries such as tannery facilities, chromate production, stainless steel welding, and ferrochrome and chrome pigment production.

The health hazard associated with exposure to Cr depends on its oxidation state, ranging from the low toxicity of the Cr(III) form to the high toxicity of the Cr(VI) form [47]. Cr(III) is an essential trace mineral present in trace amounts in some foods, such as meat, whole grains, oleaginous plants, and legumes.

3.3. Dietary sources of chromium

Cr(III) is considered to be essential to mammals for the maintenance of glucose, protein, and lipid metabolism, whereas Cr(VI) is detrimental to human health even at relatively low concentration levels, because it can be involved in the pathogenesis of some diseases such as liver, kidney, lung, and gastrointestinal cancers [48–50]. Cr(III) is a stable and biologically active state of Cr, and it is found in many types of foods, including egg yolk, whole grains, cereals, coffee, nuts, green beans, broccoli, meat, beer yeast, and drinks produced from grapes. Cr is also available in many dietary supplements and is responsible for the proper functioning of the metabolism of carbohydrates and lipids. Table 1 shows the amounts of Cr that can be found in some foods.

Food	Amounts of chromium ($\mu g\ 100\ g^{-1}$)
Broccoli	11–22[b]
Dry garlic	60[a]
Mashed potato	1.5[a]
Whole wheat bread	4.4[a]
Champagne	1.1–3.6[c]
Red wine	0.7–9.0[c]
White wine	0.7–4.4[c]
Green grapes	0.3–2.1[c]
Red grapes	0.2–6.5[c]
Apple	0.8[a]
Grape juice	4.0[b]
Orange juice	1.0[a]

[a]Adapted source by NHI [51].

[b]Adapted source by Oliveira and Machine [52].

[c]Adapted source by Cabrera-Vique *et al.* [53].

Table 1. Quantity of chromium in some foods.

3.4. Routes of entry into plants, animals, and humans

Cr-containing compounds have been a major concern because of Cr release into the environment and the high risk of Cr-induced diseases in industrial workers occupationally exposed to Cr(VI) [54]. The route of human exposure to Cr is through skin and mainly through inhalation, and the lung is the target organ [55–57]. Non-occupational exposure occurs through ingestion of Cr- containing food and water. Cr content in foods varies greatly and depends on the processing and preparation of foods.

3.5. Metabolism or transformation in the living system

The main path for Cr(III) to get into the organism is through the digestive system. The mechanism of Cr intestinal absorption is not yet fully known, but it is known that Cr(VI) compounds are absorbed better than Cr(III) compounds. Absorbed Cr circulates in blood bound to the β-globulin plasma fraction and is transported to tissues bound to transferrin or other complexes at the physiological concentration [58]. Cr from blood is relatively quickly absorbed by bones, accumulating also in the spleen, liver, and kidneys. Cr is excreted especially by the urinary system.

3.6. Biological functions

Cr plays an important role in carbohydrate, lipid, and glucose metabolism [58–60]. Studies show evidence that Cr acts as a cofactor for insulin, and therefore, Cr activity in the organism is parallel to insulin functions. It is assumed that the activity of Cr is mediated by the anabolic action of insulin. Cr supplementation intensifies amino acid uptake by tissues such as these the binding of Cr to nucleic acids is stronger than in other metal ions [60]. Cr(III) seems to be involved in the structure and expression of genetic information in animals. Also, Cr protects RNA from heat denaturation and, among other functions, promotes the growth of the animals.

3.7. Mechanisms of toxicity of chromium

The toxicity of Cr compounds depends on its oxidation state and solubility [61–63]. Cr(VI) compounds are more toxic than Cr(III) compounds most likely due to the ease with which Cr(VI) can pass through cell membranes and its subsequent intracellular reduction to reactive intermediates [64–66]. As Cr(III) is poorly absorbed by any route, the reduction of Cr(VI) is considered as being a detoxification process. If Cr(VI) is reduced to Cr(III) extracellularlly, then Cr(III) is not readily transported into cells, and so toxicity is not observed. Under physiological conditions, Cr(VI) can be reduced Cr(III) by hydrogen peroxide (H_2O_2), glutathione (GSH) reductase, ascorbic acid, and GSH [66, 67].

3.8. Incidence of (acute and chronic) toxicity

Cr is of particular interest because its toxicity is highly dependent upon its chemical forms and concentration. Cr(VI) shows high toxicity and is related to clinical cases such as nasal irritation and ulceration, hypersensitivity reactions, and dermatitis through contact. The lethal dose is between 50 and 100 mg kg^{-1}, which is much lower than that of Cr(III), with a lethal dose between 1900 and 3300 mg kg^{-1} (both cases tested by oral ingestion in rats). Furthermore, Cr(VI) is classified as carcinogenic because it penetrates the cell membranes of living organisms [59]. Exposure to Cr(VI) can occur mostly through inhalation, skin contact, and ingestion. Cr(VI) inhalation, for example, besides causing severe irritation of the respiration system, is also carcinogenic. Although the WHO has established a limit for human consumption of 0.005 mg kg^{-1} body weight per day, no scientific studies have proved that Cr ingestion can cause disease. The potential effects of Cr(VI) vary mainly with the species, the amount absorbed into the bloodstream, and the route and duration of exposure [68–70]. Thus, Cr(VI) is found in most lists of high-toxicity elements for which strict control procedures apply. The difficulty in establishing a recommended dietary allowance (RDA) for Cr is mainly due to the limitations related to estimating the ingestion levels of this mineral, which range from the absence of data on the amount of Cr present in foods, due to analytical difficulties given the trace concentrations, to environmental contamination problems [70].

The ingestion of Cr(VI) is detrimental to human health even at relatively low concentration levels because it may be involved in the pathogenesis of some diseases, such as liver, kidney, lung, and gastrointestinal cancers. Following studies, many authors have suggested that chromium picolinate can cause DNA damage [60, 61], but there is no confirmation of carcinogenesis in animals [62]. There are reports of toxicity after supplementation, but the results of other investigations did not indicate hepatic alterations [64–70]. Based on this impasse, the US Agency of Toxic Substance and Disease Registration concluded that there is no conclusive evidence that supplementation causes liver damage, although it does have proven deleterious effects on the kidneys [67].

3.9. Comparative analysis of analytical techniques

The determination of Cr can be carried out through sensitive techniques that are able to quantify a few micrograms of this element. One of these techniques is graphite furnace–atomic absorption (GF-AAS), consolidated after 1981, and is able to detect Cr concentrations of around 0.2 μg kg^{-1} in food. More sensitive techniques developed later, such as ICP-MS and ICP-OES [71], are used for the determination of Cr. These techniques quantify only total Cr without promoting speciation, and separation techniques, such as chromatography, are required.

The official method for the analysis of Cr in food samples is ICP-OES using nitric acid and hydrogen peroxide to oxidize organic materials in food samples [71]. This technique has high sensitivity (of the order of 1 ng L^{-1}). However, this equipment is sophisticated and expensive with high operational costs. An option that combines higher sensitivity and lower cost is GF-AAS technique. This technique has important advantages such as a reduced amount of sample and high sensitivity, and the analysis can be carried out with minimal or no sample preparation [71].

4. Lead

4.1. Chemistry of lead

Pb is a heavy metal that has malleability, low melting point, low electrical conductivity, and high corrosion resistance. These properties allow its widespread use in the manufacture of blades and pipes of high flexibility and resistance in welds and coatings in the automotive industry; protective plates against ionizing radiation (e.g. X-rays); alloys; coating cables; and paints, dyes, and plastic additives [72]. Usually, inorganic Pb compounds are found as Pb(II) and rarely found as a pure element. Its most common mineral is galena or lead sulfide (PbS). The solubility of Pb compounds is enhanced at lower pH, suggesting that the increased mobility of the Pb is found in ecosystems under stress acidification [73].

4.2. Occurrence in the environment

Pb is a metal that occurs naturally, making up only about 0.0013% of the earth's crust. However, most Pb concentrations that are found in the environment are the result of human activities such as burning of fossil fuels and mining [74]. Pb can be found in the atmosphere in particulate form, being deposited in water systems, interfering with the characteristics of the water. In other cases, this metal may be found complexed with natural organic compounds [75]. In contact with the ground, Pb can remain for a long time and in various forms (such as insoluble and soluble complexes and colloids) and absorbed by plants, accumulating in the edible parts, causing contamination in humans and animals [74–78].

4.3. Dietary sources of lead

The WHO and the Expert Committee on Food Additives—"JECFA" initially established a provisional tolerable weekly intake (PTWI) for lead of 50 $\mu g\ kg^{-1}$ body weight for adults. However, after assessing the risk to health, the JECFA later reduced this value to 25 $\mu g\ kg^{-1}$ body weight, equivalent to 3.5 $\mu g\ kg^{-1}$ body weight per day (equal to 1.75 mg $week^{-1}$ or 1750 $\mu g\ week^{-1}$ for a person weighing 70 kg) [20, 21].

The Expert Committee noted, however, that some foods with high levels of Pb remain commercially available [22]. A reference value for Pb of 0.01 mg L^{-1} in drinking water was established by the WHO. The concentrations in drinking water are typically below 5 $\mu g\ L^{-1}$, although higher concentrations (above 100 mg L^{-1}) have been reported. The EPA regulations establish limits in the form of maximum contaminant levels (MCLs), and the value for Pb in drinking water is 0.015 mg L^{-1}, even though the EPA has also established a goal for zero Pb in this regard [23].

4.4. Routes of entry into plants, animals, and humans

The main routes of human exposure to Pb are by ingestion (food, water, and soil), inhalation, and skin [79]. The compounds of tetra-alkyl Pb (Pb tetra acetate, etc.), for example, are rapidly absorbed through the lungs, gastrointestinal tract, and also the skin. Usually, a high level of metal enters the body through the ingestion of contaminated cereals and vegetables [80]. Once absorbed, inside the body Pb is distributed by the blood reaching the soft tissue and then is deposited in the bones and other hard body parts. It is slowly excreted in urine and feces [79].

4.5. Metabolism or transformation in the living system

In the human body, Pb is not metabolized, but it forms complexes with macromolecules. Pb forms complexes (sulfur groups, –SH) through covalent bonds, causing the intoxication of humans [81]. Pb can disturb the metabolic functions in two ways: (1) it accumulates, thereby disrupting the function of vital organs and glands such as the heart, brain, kidney, bone, liver, so on and (2) it moves the vital nutritional minerals from their original location, hindering their biological function [82].

4.6. Biological functions

Pb is a toxic metal that would not have known beneficial effects to the body, and its accumulation over time in the bodies of animals and humans can cause severe illness [83].

4.7. Mechanisms of toxicity of lead

One of the main reasons by which Pb exerts toxic effect is its ability to substitute diverse cations (calcium, zinc, and magnesium) in their binding sites. Pb has a greater affinity than calcium and zinc ions to protein-binding site because of its larger ionic radius and greater electronegativity. For example, Pb interacts with oxygen and sulfur binds to sulfhydryl and amide groups of enzymes, altering their configuration and diminishing their activities, and competes with calcium in skeletal tissue and to interact with proteins [84].

In the blood, Pb is distributed to the remaining tissues, where it accumulates; the amount of metal accumulated depends mainly on the vasculature and metabolic characteristics of each tissue [85]. The half-life of Pb is 35 days in the blood and is about 2 years in the brain, and it can last for decades in bone.

Many investigators have demonstrated that Pb affects biomolecules and hence physiological systems. For example, calmodulin is a protein found primarily in the brain and heart. The binding of calcium ions of this protein allows the binding of this protein to cyclic nucleotide phosphodiesterase and adenylate cyclase with subsequent activation. Thus, this protein modulates the levels of AMP and cyclic GMP [86]. Pb is a more potent activator than calcium for calmodulin. According Kern, Pb modifies several signaling cascades and proteins that participate in the vesicular cycle [87]. The alterations caused by the abnormal protein operation in second messenger systems and exocytic processes greatly contribute to Pb neurotoxicity [87].

Pb affects various cellular organelles, for example, mitochondria and endoplasmic reticulum, in different ways. In the mitochondria Pb affects energy metabolism, while in the endoplasmic reticulum Pb increases the cytoplasmic concentration of calcium with a consequent reduction in ion concentration inside this organelle. Many signaling pathways that are within the endoplasmic reticulum are calcium dependent; because the amount is not appropriate, various processes are impaired. Besides, Pb binds to Ape1 nuclease, whose function is to detect and repair DNA damage, inhibiting its operation and allowing the accumulation of mutagenic damages [85–88].

4.8. Incidence of (acute and chronic) toxicity

Pb is one of the most common environmental contaminants. This element has no known physiological function in the organism, and its damaging effects can affect almost every organ and system in the body [89]. The main way in which Pb enters the body is through the respiratory route (occupational exposure), followed by the digestive route. Organic Pb compounds can penetrate in the body through skin contact and are rapidly absorbed [90, 91].

Exposure to Pb can result in a wide variety of biological effects, depending on the exposure level and duration. The major diseases related to Pb contamination are shown in Table 2. Pb is toxic to various organs and systems, and its effects may vary from enzyme inhibition and anemia to diseases of the nervous, immune, reproductive, and cardiovascular systems, impaired kidney function, and even death.

Studies have suggested an association between Pb exposure and lung cancer and, to a lesser extent, stomach cancer [90]. Pb is hypothesized to be a carcinogen and to enhance the genotoxic effects of other agents. Renal tumors developed in mice that had received high doses of certain Pb compounds and various other animal studies have shown increases in the yield or genotoxicity of known carcinogens. The US Environmental Protection Agency has determined that Pb is a probable human carcinogen [89–91].

Effects on health	Site of the body affected	Adverse effects
Neurological	Central nervous system, peripheral, and autonomic	Acute and chronic encephalopathy; peripheral neuropathy
Hematological	Blood	Anemia
Endocrine	Bone tissue and serum	Damage to the kidneys and development of cells, teeth, and bones. Possible damage to the thyroid.
Growth	Bone	Reduced growth
Reproductive	Male and female reproductive systems	Reduced fertility, high probability of miscarriages
Carcinogenic	Kidneys and cells—genomic DNA	Carcinogenic to animals and epigenetic involvement in the expression of the modified gene
Cardiovascular	Cardiovascular system	Likely increase in blood pressure, cardiac lesions, and abnormal electrocardiograms
Gastrointestinal	Gastrointestinal tract	Colic
Hepatic	Liver	Reduced functional capacity of the cytochrome P-450 to metabolize drugs

Table 2. Main health effects related to lead contamination [89–91].

4.9. Comparative analysis of analytical techniques

The determination of traces of Pb in various food samples is of great importance because Pb is recognized as a cumulative poison in humans and other animals [92]. The determination of Pb requires procedures that are sufficiently sensitive for detection at the pg L^{-1} level. Traditionally, GF-AAS has been applied in such cases, but the direct determination of Pb in complex matrices is usually difficult owing to matrix interference and separation procedures often being required before the sample analysis [93]. The ICP-MS technique is favored because of its low detection limits [93], although many researchers prefer AAS owing to its simpler and less expensive instrumentation. Lead hydride generation and its application to spectrometry analysis have been reviewed by Madrid and Cámara [94]. HG-AAS is a well-developed technique that can be used for the determination of volatile hydride-forming elements such as arsenic, selenium, antimony, and others at trace levels [95]. The advantages of HG-AAS over other atomic absorption spectrometric techniques such as the flame and graphite furnace methods are increase in atomization efficiency and higher selectivity because the analyte is removed from the matrix as a volatile compound and detection limits at the pg L^{-1} level or lower for the elements cited above. Considering these advantages, this technique could be applied for the determination of Pb, and it is possible to include this element in multi-element analysis schemes involving hydride generation.

The generation of Pb hydride was described by Carrijo et al. [96]. In this study, a flow injection–hydride generation–atomic absorption spectrometry (FI-HG-AAS) system was used for Pb determination. The main characteristics of the flow injection system, that is, high sampling rate and good accuracy, precision, and sensitivity, are maintained.

5. Mercury

5.1. Chemistry of mercury

Hg is a metal found in various chemical forms, which can be divided into the following categories: elemental or metallic Hg, inorganic Hg, mainly in the form of mercuric salts ($HgCl_2$ and HgS), and mercuric (Hg_2Cl_2) and organic Hg, for example, methylmercury and ethylmercury [68, 72].

Metallic mercury (Hg) is in the liquid state at room temperature and easily volatilizes into the atmosphere forming Hg vapors. Hg is a metal with widespread use, especially the production of scientific precision instruments, electrical industry, dentistry (production amalgams), the production of certain types of toys, mining, metal smelting, among others [97].

5.2. Occurrence in the environment

Hg is a metal found naturally in the earth's crust, occurring in air, soil, and water [98]. It rarely occurs free in nature and is found mainly in cinnabar ore (HgS). It can be found in metal form, as salts of Hg or organic Hg compounds. Once released, Hg remains in the environment among the circulating air, water, sediment, soil, and biota, which assumes various chemical forms.

Most emission to air occurs in the form of elemental Hg, which is very stable and can remain in the atmosphere for months or even years, enabling transport over long distances around the globe [98]. Most of the Hg released by human activities in air is by combustion of fossil fuels, mining, smelting, and combustion waste [99].

The Hg vapor in the atmosphere can be deposited or is converted into the soluble form, returning to the earth's surface in rainwater. From there, two important chemical changes may occur. The metal can be cast again and return the Hg vapor in the atmosphere or may be "methylated" by the microorganisms present in the water sediments, turning into methylmercury [98]. Furthermore, the Hg can also be released directly in the soil or in water by the application of agricultural fertilizer and disposal of industrial waste water [100].

Atmospheric emissions are the major source of environmental contamination, followed by water pollution and soil contamination, when there is improper disposal of effluents and waste [98].

5.3. Dietary sources of mercury

Usually, Hg contamination occurs by the presence of this metal in water, soil, air, or food, mostly in the form of methylmercury [100, 101]. The most important source of exposure through diet for the general population is the consumption of fish and other marine organisms. Hg is concentrated in the tissue of fish, becoming increasingly potent in predatory fish and mammals that feed on small fish. The larger carnivorous fish have higher concentrations than smaller ones [99]. The average daily intake of methylmercury (mainly from fish) that can cause demonstrable effects on the health of sensitive individuals is 300 mg day^{-1} or 4.3 μg Hg day^{-1} kg body weight^{-1} [102].

Industrial products can also be contaminated by Hg during the processing steps. Studies have shown the contamination of Hg in breast milk (4–15 μg kg^{-1}) [103], in tea (6 ng g^{-1}) [104], and in products for infant feeding (0.50 μg kg^{-1}) [105].

5.4. Routes of entry into plants, animals, and humans

The absorption of Hg by humans and animals can be by pulmonary route (inhalation), as well as by gastrointestinal or cutaneous route. In the case of pulmonary route, Hg after inhalation and the presence in the lung is distributed throughout the body, accumulating in various parts of body [106–108]. Soluble compounds are absorbed by mucous membranes following vapor inhalation and by the skin and the sebaceous glands. In the body, organic and inorganic Hg binds GSH [108]. It acts as inactivator because it readily binds to thiol groups of cellular enzymes and disrupts its function by inactivating the metabolism. The non-absorbed Hg is excreted in feces, and absorbed Hg forms are excreted via saliva and skin.

5.5. Metabolism or transformation in the living system

Hg and its organic compounds in low concentrations cause damage to human health. Their concentrations in surface and ground waters are below 0.5 mg L^{-1}. However, aquatic microorganisms convert the organic Hg into inorganic Hg compounds, which accumulate in the food chain. Methylmercury is the most relevant toxicant [109]. The gastrointestinal tract is the second way (after airway) through which Hg, already now in its organic form, enters the human body through the consumption of fish, shellfish, and other aquatic organisms.

5.6. Biological functions

Hg has no biological role. All Hg compounds are extremely toxic, particularly methylmercury [109].

5.7. Mechanisms of toxicity of mercury

The mechanism of toxicity of Hg is based on its chemical activity and biological features. The main mechanism of toxicity of Hg compounds involves their reactivity with sulfhydryl groups. Once in the cell, both Hg^{2+} and MeHg form covalent bonds with cysteine residues of proteins and deplete cellular antioxidants [110].

5.8. Incidence of (acute and chronic) toxicity

Metallic Hg and its organic compounds in very low concentrations cause damage to human health (such as neurotoxic, immunotoxic, and teratogenic properties) and can have high persistence and a high bioconcentration factor (BCF), accumulating in animals, fish, and the environment. Hg poisoning levels and the main symptoms and diseases related to acute and chronic poisoning by Hg are shown in Tables 3 and 4, respectively [111].

	24-hour urine
0.00–0.01mg	Non toxic
0.02–0.09 mg	Danger of poisoning
0.10–0.80 mg	Chronic intoxication
Above 1.00 mg	Acute intoxication
Above 2.00 mg	Subacute poisoning

Table 3. Mercury poisoning levels.

Acute intoxication	**Chronic intoxication**
Dark gray appearance in the mouth and pharynx	Digestive disorders
Severe pain	Nervous disorders
Vomiting (may even be bloody)	Cachexia
Bleeding gums	Stomatitis
Metallic taste in the mouth	Salivation

Acute intoxication	Chronic intoxication
Burning in the digestive tract	Bad breath
Severe or bloody diarrhea	Loss of appetite
Inflammation of the mouth (stomatitis)	Anemia
Tooth decay and/or loose teeth	Hypertension
Glossitis	Loosening of the teeth
Swelling of the gum mucosa	Central nervous system disorders
Kidney nephrosis	Mild kidney disorders
Serious liver problems	Possibility of chromosomal alteration
Can even cause sudden death (1 or 2 days)	–

Table 4. Main symptoms and diseases related to acute and chronic poisoning by mercury.

5.9. Comparative analysis of analytical techniques

The determination of Hg in food samples is critical to assess the degree of human exposure, and thus reliable analytical techniques with high sensitivity are required. However, in most situations, the determination of Hg species is not an easy task due to low concentrations in the samples and the characteristic volatility [112, 113]. The volatility of Hg requires special consideration when treating the sample. Food sample preparation using a microwave oven has been widely employed [114].

In the case of the quantification of methylmercury in fish samples and seafood, depending on the nature of the sample and the technique used, an additional pre-concentration step is required. Hg determination has been performed using cold vapor coupled to atomic absorption spectrometry (CV-AAS), cold vapor coupled to atomic fluorescence spectrometry (CV-AFS), inductively coupled plasma optical emission spectrometry (CV-ICP-OES), and inductively coupled plasma mass spectrometry (CV-ICP-MS). Hyphenated techniques involving gas or liquid chromatography separations with detection by element-specific detectors such as ICP-MS and atomic absorption/emission are the most commonly reported [114].

6. Conclusions

The information gathered herein highlights the risks associated with arsenic, Cr, Pb, and Hg contamination in foods. Therefore, measures should be taken to reduce exposure of the general population to these contaminants to minimize the risk of adverse health effects. The development of simple strategies suitable for obtaining quantitative information regarding some species of great interest should be encouraged.

In the food industry, analytical chemistry plays an important role, contributing new analytical procedures and instrumentation. Methods for the determination and monitoring of metals are still scarce. However, there are inherent difficulties associated with the types of samples involved. There are various types of food samples and a great variation in their compositions. This hinders the application of analytical techniques for the fast and accurate monitoring of metals in real samples. More sophisticated techniques are of interest in some fields of application, but these techniques have not yet reached the food industry. Thus, chemists need to direct their attention toward these trends with the aim of narrowing the gap between science and the food industry. These studies require an interdisciplinary approach to cover the various aspects involved and could achieve important advances in toxicology, chemistry, and food science.

Acknowledgements

The authors are grateful for financial support from the Brazilian Governmental Agencies [Conselho Nacional de Desenvolvimento Científico e Tecnológico (CNPq) and Coordenação de Aperfeiçoamento de Pessoal de Nível Superior (CAPES)], the MG State Government Agency [Fundação de Amparo à Pesquisa do Estado de Minas Gerais (FAPEMIG)], and the GO State Government Agency [Fundação de Amparo á Pesquisa do Estado de Goiás (FAPEG).

Author details

Luciana M. Coelho[1], Diego R. Pessoa[2], Kênia M. Oliveira[2], Priscila A.R. de Sousa[1], Lícia A. da Silva[1] and Nívia M.M. Coelho[2*]

*Address all correspondence to: nmmcoelho@ufu.br

1 Department of Chemistry, Federal University of Goiás, Catalão, GO, Brazil

2 Institute of Chemistry, Federal University of Uberlândia, Uberlândia, MG, Brazil

References

[1] Hogg M.F. Exposure assessment of chemicals from packaging materials in foods: A review. Trends Food Sci. Technol., 2007; 18: 219–230.

[2] Mamtani R., Stern P., Dawood I., Cheema S. Metals and disease: A global primary health care perspective. J. Toxicol., 2011; Article ID 319136, 01-11.

[3] Järup L. Hazards of heavy metal contamination. Br. Med. Bull., 2003; 68: 167–182.

[4] Hughes M.F., Beck B.D., Chen Y., Lewis A.S., Thomas, D.J. Arsenic exposure and toxicology: A historical perspective. Toxicol. Sci., 2011; 123: 305–332.

[5] Hindmarsh J., McCurdy R. Clinical and environmental aspects of arsenic toxicity. Crit. Rev. Clin. Lab. Sci., 1986; 23: 315–347.

[6] Templeton D., Ariese F., Cornelis R., Danielsson L., Muntau H., Leeuwen H., Lobinski R. Guidelines for terms related to chemical speciation and fractionation of elements. Definitions, structural aspects, and methodological approach. Pure Appl. Chem., 2000; 72: 1453–1470.

[7] FDA. Food and Drug Administration: Arsanilic Acid, 2008; 413–414, Washington, D. C..

[8] IPCS. Arsenic and Arsenic Compounds. 2nd ed. Geneva: World Health Organization, International Programme on Chemical Safety (Environmental Health Criteria), 2001; 224. http://whqlibdoc.who.int/ehc/WHO_EHC_224.pdf.

[9] EPA. National primary drinking water regulations: Arsenic and clarifications to compliance and new source contaminants monitoring. Proposed Rule [40 CFR Parts 141 and 142]. Fed. Regist., 2000; 65: 38888–38983.

[10] Alam Z., Rahman M. Accumulation of arsenic in rice plant from arsenic contaminated irrigation water and effect on nutrient content. Fate of Arsenic in the Environment, Bangladesh University of Engineering and Technology, Dhaka The United Nations University, Tokyo, 2003; 131–135.

[11] IARC. Some drinking-water disinfectants and contaminants, including arsenic. IARC Monogr Eval. Carcinog. Risks Hum., 2004; 84: 1–477.

[12] Carbonell-Barrachina A.A., Wu X., Ramírez-Gandolfo A., Norton G.J., Burló F., Deacon C., Meharg, A.A. Inorganic arsenic contents in rice-based infant foods from Spain, UK, China and USA. Environ. Pollut. 2012; 163: 77–83.

[13] Pasias I.N., Thomaidis N.S., Piperaki E.A. Determination of total arsenic, total inorganic arsenic and inorganic arsenic species in rice and rice flour by electrothermal atomic absorption spectrometry. Microchem. J., 2013; 108: 1–6.

[14] Food and Agriculture Organization/World Health Organization (FAO/WHO). Food Standards Programme Codex Committee on Contaminants in Foods. 5th session. Working document for information and use in discussions related to contaminants and toxins in the GSCTFF. CF/5 INF/1, 2011. Codex Alimentarius Commission, Netherlands, ftp://ftp.fao.org/codex/meetings/CCCF/cccf5/cf05_INF.pdf.

[15] Food Safety. Authority of Ireland. Toxicology Factsheet Series. Issue 1, 2009.

[16] Brown J.P. Risk assessment for arsenic in drinking water. In Howd R.A. and Fan A.M. (Eds.), Risk Assessment for Chemicals in Drinking Water. Hoboken, NJ: John Wiley & Sons, Inc., 2008; 213–266.

[17] WHO – Regional Office for Europe, Copenhagen. Air Quality Guidelines. 2nd edition. Denmark, 2000. http://www.euro.who.int/

[18] Shoji R., Yajima R., Yano Y. Arsenic speciation for the phytoremediation by the Chinese brake fern, *Pteris vittata*. J. Environ. Sci., 2008; 20: 1463–1468.

[19] ttp://www.fao.org/news/story/en/item/238802/icode/. Accessed 31 July 2015.

[20] World Health Organization (WHO). Safety evaluation of certain food additives and contaminants. Sixty-first meeting of the Joint FAO/WHO Expert Committee on Food Additives (JECFA), 2004; 563.

[21] World Health Organization (WHO). Summary and conclusions. Fifty-third meeting of the Joint FAO/WHO Expert Committee on Food Additives (JECFA), 1999; 21.

[22] SCOOP. Assessment of the dietary exposure to arsenic, cadmium, lead and mercury of the population of the EU Member States. Reports on tasks for scientific cooperation. Report of experts participating in Task 3.2.11. Directorate-General Health and Consumer Protection, European Commission, 2004.

[23] WHO. Guidelines for Drinking-Water Quality. 3rd ed. Geneva, Switzerland: World Health Organization, 2004.

[24] Guha-Mazumder D.N., Haque R., Ghose N., Santra B.K.A., Chakraborty D. Arsenic in drinking water and the prevalence of respiratory effects in West Bengal, India. Int. J. Epidemiol., 2000; 29: 1047–1052.

[25] Rahman M. Arsenic and contamination of drinking-water in Bangladesh: A public-health perspective. J. Health Populat. Nutr., 2002; 20: 193–197.

[26] Srivastava A.K., Hasan S.K., Srivastava R.C. Arsenicism in India: Dermal lesions and hair levels. Arch. Environ. Health, 2001; 56: 562–570.

[27] Liu C.W., Liang C.P., Huang F.M., Hsueh Y.M. Assessing the human health risks from exposure of inorganic arsenic through oyster (*Crassostrea gigas*) consumption in Taiwan. Sci. Total Environ., 2006; 361: 57–66.

[28] Ramirez T., Stopper H., Hock R., Herrera L.A. Prevention of aneuploidy by *S*-adenosyl-methionine in human cells treated with sodium arsenite. Mutat. Res., 2007; 617: 16–22.

[29] Chanda S., Dasgupta U.B., Guhamazumder D. DNA hypermethylation of promoter of gene p53 and p16 in arsenic-exposed people with and without malignancy. Toxicol. Sci., 2006; 89: 431–437.

[30] Danaee H., Nelson H.H., Liber H. Low dose exposure to sodium arsenite synergistically interacts with UV radiation to induce mutations and alter DNA repair in human cells. Mutagenesis, 2004; 19: 143–148.

[31] Snow E.T., Sykora P., Durham T.R., Klein C.B. Arsenic, mode of action at biologically plausible low doses: What are the implications for low dose cancer risk? Toxicol. Appl. Pharmacol., 2005; 207: S557–S564.

[32] WHO. Arsenic and Arsenic Compounds (Environmental Health Criteria 224). 2nd ed. Geneva: World Health Organization, International Programme on Chemical Safety, 2001.

[33] Hansen H.R.A., Raab A., Jaspars M. Sulfur containing arsenical mistaken for dimethylarsinous acid [DMA(III)] and identified as a natural metabolite in urine: Major implications for studies on arsenic metabolism and toxicity. Chem. Res. Toxicol., 2004; 17: 1086–1091.

[34] Shi H., Shi X., Liu K.J. Oxidative mechanism of arsenic toxicity and carcinogenesis. Mol. Cell. Biol., 2004; 255: 67–78.

[35] Chen C-J., Chuang Y-C., Lin T-M., Wu H-Y. Malignant neoplasms among residents of a blackfoot disease-endemic area in Taiwan: High-arsenic artesian well water and cancers. Cancer Res., 1985; 45: 5895–5899.

[36] Chen C-J., Kuo T-L., Wu M-M. Arsenic and cancers. Lancet, 1988; 331: 414–415.

[37] Otto D., Xia Y., Li Y., Wu K., He L., Telech J. Neurosensory effects of chronic human exposure to arsenic associated with body burden and environmental measures. Hum. Exp. Toxicol., 2007; 26: 169–177.

[38] Rintala E.M., Ekhom P., Koivisto P., Peltonen K., Venäläinem E. The intake of inorganic arsenic from long grain rice and rice-based baby food in Finland – Low safety margin warrants follow up. Food Chem., 2014; 150: 199–205.

[39] Abreu L.B., Augusti R., Schmidt L., Dressler V.L., Flores E.M.M., Nascentes C. Desorption electrospray ionization mass spectrometry (DESI-MS) applied to the speciation of arsenic compounds from fern leaves. Anal. Bioanal. Chem., 2013; 405: 7643–7651.

[40] Meharg A.A., Zhao F.J. Arsenic & Rice. Springer: Berlin, 2012.

[41] Hsieh M., Liu C., Chen J., Jiang S. Speciation analysis of arsenic and selenium compounds by CE-dynamic reaction cell-ICP-MS. Electrophoresis, 2010; 31: 2272–2278.

[42] Baig J.A., Kazi T.G., Shah A.Q., Arain M.B., Afridi H.I., Khan S. Evaluating the accumulation of arsenic in maize (*Zea mays* L.) plants from its growing media by cloud point extraction. Food Chem. Toxicol. 2010; 48: 3051–3057.

[43] Costa B.E.S., Coelho N.M.M., Coelho L.M.. Determination of arsenic species in rice samples using CPE and ETAAS. Food Chem., 2015; 178: 89–95.

[44] Jacobs J.A., Testa S.M. Overview of chromium(VI) in the environment: Background and history. In: J. Guertin, J.A. Jacobs, C.P. Avakian, editors. Chromium (VI) Handbook. Boca Raton, Fl: CRC Press; 2005, 1–22.

[45] Agency for Toxic Substances and Disease Registry (ATSDR). Toxicological Profile for Chromium. Atlanta, GA: U.S. Department of Health and Human Services, Public Health Service, 2012.

[46] U.S. EPA. Environmental Criteria and Assessment Office. Cincinnati, OH: United States Environmental Protection Agency; 1992.

[47] IARC. IARC Monographs on the Evaluation of Carcinogenic Risks to Humans, Vol. 49. Lyon, France: IARC Scientific Publications, IARC, 1990.

[48] Velma V., Vutukuru S.S., Tchounwou P.B. Ecotoxicology of hexavalent chromium in freshwater fish: A critical review. Rev. Environ. Health., 2009; 24: 129–145.

[49] Cohen M.D., Kargacin B., Klein C.B., Costa M. Mechanisms of chromium carcinogenicity and toxicity. Crit. Rev. Toxicol., 1993; 23: 255–281.

[50] Norseth T. The carcinogenicity of chromium. Environ. Health. Perspect., 1981; 40: 121–130.

[51] National Institutes of Health (NHI). Office of dietary supplements. Dietary supplement fact sheet: Chromium, 2005. http://ods.od.nih.gov/factsheets/chromium/. Accessed 31 July 2015.

[52] Dutra Oliveira J.E., Machine J.S.M.J. Ciências Nutricionais. São Paulo: Sarvier, 2003; 167–178.

[53] Cabrera-Vique C., Teissédre P.L., Cabanis M.T., Cabanis J.C. Determination and levels of chromium in French wine and grapes by graphite furnace atomic absorption spectrometry. J. Agri. Food Chem., 1997; 45: 1808–1811.

[54] Occupational Safety and Health Administration (OSHA). Federal Register. Washington, DC: Final Rule.

[55] Costa M. Toxicity and carcinogenicity of Cr(VI) in animal models and humans. Crit. Rev. Toxicol., 1997; 27: 431–442.

[56] Shelnutt S.R., Goad P., Belsito D.V. Dermatological toxicity of hexavalent chromium. Crit. Rev. Toxicol., 2007; 37: 375–387.

[57] WHO/IPCS. World Health Organization. Geneva, Switzerland, 1988.

[58] Ducros V. Chromium metabolism. A literature review. Biol. Trace Elem. Res., 1992; 32: 65–77.

[59] Dayan A.D., Paine A.J. Mechanisms of chromium toxicity, carcinogenicity and allergenicity: Review of the literature from 1985 to 2000. Hum. Exp. Toxicol., 2001; 20: 439–451.

[60] Speetjens J.K., Collins R.A., Vincent J.B., Woski S.A. The nutritional supplement chromium (III) tris. (picolinate) cleaves DNA. Chem. Res. Toxicol., 1999; 12: 483–487.

[61] Stearns D.M., Patierno S.R., Wetterhahn K.E. Chromium(III) picolinate produces chromosome damage in Chinese hamster ovary cells. FASEB, 1995; 9: 1643–1649.

[62] Fristedt B., Lindqvist B., Schutz A., Ovrum P. Survival in a case of acute oral chromic acid poisoning with acute renal failure treated by haemodialysis. Acta Med. Scand., 1965; 177: 153–159.

[63] Kaufman D.B., Dinicola W., Mcintosh R. Acute potassium dichromate poisoning. Treated by periodontal dialysis. Am. J. Dis. Child., 1970; 119: 374–376.

[64] Loubieres Y., De Lassence A., Bernier M., Veillard-Baron A., Schmitt J.M., Page B., Jardin F. Acute, fatal, oral chromic acid poisoning. J. Toxicol. Clin. Toxicol., 1999; 37: 333–336.

[65] Ivankovic S., Preussmann R. Absence of toxic and carcinogenic effects after administration of high doses of chromic oxide pigment in subacute and long-term feeding experiments in rats. Food Cosmet. Toxicol., 1975; 13: 347–351.

[66] Mackenzie R.D., Byerrum R.U., Decker C.F., Hoppert C.A., Langham R.F. Chronic toxicity studies. II. Hexavalent and trivalent chromium administered in drinking water to rats. AMA Arch. Industr. Health, 1958; 18: 232–234.

[67] Agency for Toxic Substances and Disease Registry (ATSDR). Toxicological Profile for Chromium, US Department for Health and Diseases, 1993. http://www.atsdr.cdc.gov/csem/chromium/docs/chromium.pdf. Accessed 15 July 2015.

[68] Holum J.R. Elements of General, Organic and Biological Chemistry. 9th ed. John Wily and Sons, 1995, New York, USA.

[69] Nriagu J.O., Nieboer E. Chromium in Natural and Human Environment. New York: Wiley, 1988.

[70] Lukaski H.C. Magnesium, zinc, and chromium nutriture and physical activity. Am. J. Clin. Nutr., 2000; 72: 585S–593S.

[71] Markiewicz B., Komorowicz I., Sajnóg A., Belter M., Barałkiewicz D. Chromium and its speciation in water samples by HPLC/ICP-MS – Technique establishing metrological traceability: A review since 2000. Talanta, 2015; 132: 814–828.

[72] Fergusson J.E., editor. The Heavy Elements: Chemistry, Environmental Impact and Health Effects. Oxford: Pergamon Press, 1990.

[73] Kabata-Pendia A., 3rd, editor. Trace Elements in Soils and Plants. Boca Raton, FL: CRC Press, 2001.

[74] IPCS. Inorganic Lead. Environmental Health Criteria 165. International Programme on Chemical Safety, Vol. 38. World Health Organization, 1995; 93–99, Geneva.

[75] Bradl H., editor. Heavy Metals in the Environment: Origin, Interaction and Remediation, Vol. 6. London: Academic Press, 2002.

[76] SCOOP. Assessment of the dietary exposure to arsenic, cadmium, lead and mercury of the population of the EU Member States. Reports on tasks for scientific coopera-

tion. Report of experts participating in Task 3.2.11. Directorate-General Health and Consumer Protection. European Commission, 2004.

[77] Moreira F.R., Moreira J.C. The effects of lead on the human body and its significance for health. Rev. Panam. Salud Pública, 2004; 15: 119–129.

[78] Copat C., Bella F., Castaing M., Fallico R., Sciacca S., Ferrante M. Heavy metals concentrations in fish from Sicily (Mediterranean Sea) and evaluation of possible health risks to consumers. Bull. Environ. Contam. Toxicol., 2012; 88: 78–83.

[79] Singh R., Gautam N., Mishra A., Gupta R. Heavy metals and living systems: An overview. Indian J. Pharmacol., 2011; 43: 246–253.

[80] Simeoni L.A., Brabarick K.A., Sabey B.R. Effect of a small-scale composting of sewage ludge on heavy metal availability to plants. J. Environ. Quality., 1984; 13: 264–268.

[81] Hertz-Picciotto I. The evidence that lead increases the risk for spontaneous absorption. Am. J. Ind. Med., 2000; 38: 300–309.

[82] Flora S.J.S., Saxena G., Gautam P., Kaur P., Gill K.D. Lead induced oxidative stress and alterations in biogenic amines in different rat brain regions and their response to combined administration of DMSA and MiADMSA. Chem. Biol. Interact., 2007; 170: 209–220.

[83] Papanikolaou N.C., Hatzidaki E.G., Belivanis S., Tzanakakis G.N., Aristidis M. Tsatsakis, Lead toxicity update. A brief review. Med. Sci. Monit., 2005; 11: RA329–RA336

[84] Aníbal G., Rosario V., Enrique S. Cellular mechanisms of lead neurotoxicity. Med. Sci. Monit., 2006; 12: RA57–RA65.

[85] Stevens F.C. Calmodulin: An introduction. Can. J. Biochem. Cell Biol., 1983; 61: 906–910.

[86] Agency for Toxic Substances and Disease Registry. Toxicological profile for lead. US Department of Health and Human Services, Public Health Service, 1999. Atlanta, U.S.A.

[87] Kern M., Wisniewski M., Cabell L., Audesirk G. Inorganic lead and calcium interact positively in activation of calmodulin. Neurotoxicology, 2000; 21: 353–363.

[88] Hermes-Lima M., Pereira B., Bechara E.J. Are free radicals involved in lead poisoning? Xenobiotica, 1991; 8: 1085–1090.

[89] Tchounwou P.B., Yedjou C.G., Patlolla A.K., Sutton D.J. Heavy metals toxicity and the environment, PubMed Commons, 2012; 101: 133–164.

[90] Rossman R.N. Mutagenesis and comutagenesis by lead compounds. Mutat. Res., 1992; 298: 97–103.

[91] Simons T. Lead-calcium interactions in cellular lead toxicity. Neurotoxicology. 1993; 14: 77–86.

[92] Harrison R.M., Laxen D.D.H. Lead Pollution. London: Chapman and Hall, 1981.

[93] Yan X.P., Adams F. Flow injection on-line separation and preconcentration with a knotted reactor for electrothermal atomic absorption spectrometric determination of lead in biological and environmental samples. J. Anal. At. Spectrom., 1997; 12: 459–464.

[94] Madrid Y., Cámara C. Lead hydride generation atomic absorption spectrometry: An alternative to electrothermal atomic absorption spectrometry. A review. Analyst., 1994; 119: 1647–1658.

[95] Dédina J., Tsalev D.L. Hydride Generation Atomic Absorption Spectrometry. Chichester: Wiley, 1995.

[96] Carrijo J.F.N., Brasil L.C., Coelho N.M.M. Determination of trace lead in waters by flow injection hydride generation atomic absorption spectrometry. J. Braz. Chem. Soc., 2005; 16: 520–525.

[97] Anderson A. Mercury pollution. Dry battery alarm in Japan. Nature, 1984; 309: 576–581.

[98] Ministry of Environment - Mercury. Available at http://www.mma.gov.br/seguranca-quimica/mercurio. Accessed July 31, 2015.

[99] Costa F.N., Korn M.G.A., Brito G.B., Ferlin S., Fostier A.H. Preliminary results of mercury levels in raw and cooked seafood and their public health impact. Food Chem., 2016; 192: 837–841.

[100] Patterson B., Ryan J., Dickley J. The toxicology of mercury. New Engl. J. Med., 2004; 350: 945–947.

[101] Kobal A.B. Glutathione level after long-term occupational elemental mercury exposure. Environ. Res., 2008; 107: 115–123.

[102] Jones D.W. Exposure or absorption and the crucial question of limits for mercury. Clin. Pract., 1999; 65: 42–46.

[103] Al-Saleh I., Shinwari N., Mashhour A. Heavy metal concentrations in the breast milk of Saudi women. Biol. Trace Elem. Res., 2003; 96: 1–3.

[104] Abdulla M., Chmielnicka J. New aspects on the distribution and metabolism of essential trace elements after dietary exposure to toxic metals. Biol. Trace Elem. Res., 1990; 23: 25–53.

[105] Martins C., Vasco E., Paixão E., Alvito P. Total mercury in infant food, occurrence and exposure assessment in Portugal. Food Addit. Contam. Part B Surveill. 2013; 6: 151–157.

[106] Valko M., Morris H., Cronin M.T.D. Metals, toxicity, and oxidative stress. Curr. Med. Chem., 2005; 12: 1161–1208.

[107] Clarkson T.W., Magos L. The toxicology of mercury and its chemical compounds. Crit. Rev. Toxicol., 2006; 36: 609–662.

[108] Rooney J.P.K. The role of thiols, dithiols, nutritional factors and interacting ligands in the toxicology of mercury. Toxicology., 2007; 234: 145–156.

[109] Zalups R.K., Koropatnik J., editors. Molecular Biology and Toxicology of Metals. London: Taylor & Francis, 2000.

[110] Clarkson T.W. Mechanisms of mercury disposition in the body. Am. J. Ind. Med., 2007; 50: 757–764.

[111] Brito Filho D. Human Toxicology and General. Atheneu, 2nd ed. Rio de Janeiro, 1988.

[112] Hamelink J.L., Landrum P.F., Harold B.L., William B.H., editors. Bioavailability: Physical, Chemical, and Biological Interactions. Boca Raton, FL: CRC Press Inc, 1994.

[113] Ferreira S.L.C., Lemos V.A., Silva L.O.B., Queiroz A.F.S., Souza A.S., Silva E. G.P., Santos W.N.L., Virgens C.F. Analytical strategies of sample preparation for the determination of mercury in food matrices – A review. Microchem. J., 2015; 121: 227–236.

[114] Capar S.G., Mindak W.R., Cheng J. Analysis of food for toxic elements. Anal. Bioanal. Chem., 2007; 389: 159–169.

7

Technologies for Detecting Botulinum Neurotoxins in Biological and Environmental Matrices

Luisa W. Cheng, Kirkwood M. Land, Christina Tam, David L. Brandon and Larry H. Stanker

Abstract

Biomonitoring of food and environmental matrices is critical for the rapid and sensitive diagnosis, treatment, and prevention of diseases caused by toxins. The U.S. Centers for Disease Control and Prevention (CDC) has noted that toxins from bacteria, fungi, algae, and plants present an ongoing public health threat, especially since some of these toxins could compromise security of the food supply. Botulinum neurotoxins (BoNTs), produced by *Clostridium* spp., are among those bacterial toxins that pose life-threatening danger to humans. BoNTs inhibit the release of acetylcholine at peripheral cholinergic nerve terminals and cause flaccid paralysis. BoNTs are grouped in seven serotypes and many subtypes within these groups. Rapid and accurate identification of these toxins in contaminated food as well as in environmental matrices can help direct treatment. Herein, we discuss current methods to detect BoNTs with a focus on how these technologies have been used to identify toxins in various food and environmental matrices. We also discuss the emergence of new serotypes and subtypes of BoNTs and the increasing number of cases of botulism in wildlife. Finally, we consider how environmental changes impact food safety for humans and present new challenges for detection technology.

Keywords: Botulism, Toxins, Food matrix, Environmental detection, Foodborne illness

1. Introduction

The U.S. Centers for Disease Control (CDC) have summarized the risks that biological toxins pose to human health [1]. Bacteria, fungi, parasites, and plants all produce toxins in the environment that can impact food safety. Furthermore, changes in the environment have caused emergence of new problems associated with toxins. One example is the production of toxins by *Clostridium botulinum*. This pathogen, which is a gram-positive, anaerobic spore-forming bacterium, produces botulinum neurotoxins (BoNTs). Humans are susceptible to the effects of these poisons, which are among the most toxic molecules known [1]. The parenteral lethal dosage for humans is 0.1–1 ng/kg, and the oral dose is 1 μg/kg. A single gram of BoNT released into the environment and subsequently inhaled could kill more than one million people [2, 3]. BoNTs exert their biological effects by blocking acetylcholine release by neurons. To date, BoNTs have been divided into seven serotypes, denoted as A through G, of which A, B, E, and F are known to be toxic to humans [4–6]. However, all of the botulinum serotypes are possibly toxic to people. In addition to the principal serotypes, at least 40 additional subtypes have been described based on differences in both primary peptide sequence and three-dimensional structure [4–6]. In this discussion, we review the basic biology of BoNTs and current methods to detect these molecules in biological and environmental matrices.

C. botulinum isolates are categorized into different groups [5, 6]. Members of Group I are referred to as "proteolytic" and produce toxin types A, B, or F. They are widely distributed in the environment and often found in various raw foods. BoNTs can cause symptoms at levels as low as 5 ng. Although the onset of symptoms typically takes 12–36 hours, the time course depends on the amount of toxin ingested [2, 3]. It can take much longer for symptoms to manifest. Initial symptoms include diarrhea and vomiting followed by neurological effects that include blurred vision, weakness, and difficulty in swallowing, talking, and breathing. Unless diagnosed early, mortality rates can be as high as 40% [1–4]. Timely response and current treatments have reduced mortality to less than 10%. The most common foods involved in outbreaks are improperly preserved meat or fish products, but a range of other foods have been implicated, such as cheeses (including vegetables preserved in oil and cheese). Because botulinum toxins are not heat stable, they can be inactivated at cooking temperatures.

Strains in Group II are classified as "non-proteolytic" [5, 6]. These *C. botulinum* strains synthesize neurotoxin B, E, or F. These bacteria can grow at temperatures <3°C are ubiquitous in the environment. Moreover, one can find Type E strains in aquatic habitats [1–4]. It is not known whether these strains can synthesize neurotoxins in refrigerated processed foods without visible spoilage. The endospores of strains in this group are not as resistant to heat as those strains in Group I. Neurotoxins synthesized by strains in Group II toxins have shown to be less potent than those of Group I; at least 0.1 μg of neurotoxin is required to cause symptoms of botulism. However, their other biological properties are similar. Foods involved in outbreaks of Group II botulism include cold-smoked fish and other preserved fish products.

Group III botulinum produces toxins of serotype C or D and is associated with avian and non-human mammalian botulism [5, 6]. Whole genome sequencing analysis indicates that strains of physiological group III are probably more related to *Clostridium haemolyticum* and *Clostridium novyi* than to *C. botulinum* serotypes belonging to Groups I and II. Group IV is rare and has not been well characterized. However, it does synthesize neurotoxin serotype G.

Bacteriophages contain the neurotoxin genes of *C. botulinum* serotypes C and D [5, 6]. The BoNT prophage replicates in the bacterium as a large plasmid, and strains containing the phage can become toxigenic via either type C or type D phages. The distinction between types C and D is not clear because chimeric sequences have been isolated from the environment. These toxin genes have been identified in avian isolates. They contain sections from both BoNT/C and BoNT/D genes and are referred to as type C/D [5, 6]. The chimeric toxin is more pathogenic to avians than either serotype C or D individually. In *C. botulinum* serotypes C and D, there is a small amount of a binary toxin, denoted as the C2 toxin. The genes encoding the C2 toxin have been localized to a plasmid. Structurally and functionally, the C2 toxin contains a translocation domain and an ADP-ribosylating domain that has been shown to target cellular actin. The occurrence of other chimeric botulinum toxin genes has yet to be determined [5, 6].

At the amino acid sequence level, BoNT serotypes can differ from one other by 34–64% [5, 6]. Significant genetic variation within each serotype has also been observed. In fact, 32 toxin subtypes with amino acid sequence differences of 2.6–32% have been identified thus far, and more will likely be identified in the future [5, 6]. This serotype and subtype diversity confound direct antibody and molecular-based assay designs. It is rare that one probe can bind to all serotypes. In *C. botulinum*, the neurotoxins are first synthesized as a large holotoxin (approximately 150 kDa). They are then processed by a trypsin-like protease in *C. botulinum* yielding two polypeptides (one approximately 100 kDa and the other approximately 50 kDa) that are still bound by a single disulfide [2, 3]. The neurotoxin structure mimics other known A–B dimeric toxins found in other bacterial pathogens. The ~100 kDa fragment is called the heavy chain (HC) and aids the binding of the neurotoxin to host cell receptors and its translocation from vesicles to the cytoplasm [2, 3]. The ~50 kDa fragment, called the light chain (LC), contains the enzymatically active domain of the neurotoxin. Recombinantly, expressed LC is routinely used for activity-based neurotoxin assays. Antibodies specific for the HC and LC are used for immunoassays for detecting neurotoxins as well as for neutralization.

2. Important factors to consider when developing toxin detection assays

The development of a robust assay for the detection of any pathogen or biological product of a pathogen (such as a toxin) requires consideration of several factors: sensitivity, specificity, matrix effects, and biological activity [8–10]. Each of these factors is briefly discussed below in the context of assay methods for *C. botulinum* toxins.

3. The mouse bioassay

In the laboratory, a rodent bioassay is considered the "gold standard" method for detecting BoNTs [8–10]. Despite much effort to replace the use of animals, it is still the most sensitive and reliable assay to model all aspects of BoNT intoxication: binding, translocation, enzymatic activity, and pathology. Alternatives to the mouse bioassay have been developed (discussed below) with shorter assay times and equal or greater sensitivity.

The mouse assay quantitatively determines the amount of BoNT required to kill all mice in a test group. This measurement is expressed as a minimal lethal dose (MLD). Although protocols may vary, mice are usually injected intraperitoneally with 0.5 mL of BoNT sample in a dilution series and then monitored over several days for signs of intoxication and death

[11, 12]. If enough sample is available for an assay, the specific neurotoxin can be identified using neutralization with antibodies specific to each of the neurotoxin serotypes (A–G). The toxin serotype is therefore revealed based on which antibody confers protection from death. The mouse bioassay is highly sensitive and useful for detection of different neurotoxins in different matrices. However, despite its versatility, the mouse assay has limitations that include: long assay times and the use of animals requiring specialized animal facilities, substantial costs, trained staff, and consideration of ethical issues, especially when death is used as an endpoint. There is also substantial variation in results observed among different research laboratories [8–12].

Alternative "refined" animal assays that do not use death as an endpoint, such as the mouse phrenic nerve hemi-diaphragm assay, have been evaluated [13, 14]. Despite being more sensitive and rapid compared to the use of whole animals, these assays usually require the use of specific equipment and personnel with specialized training. Furthermore, these alternative animal assays are not feasible with larger samples and those containing a complex matrix. However, a recent study described an *in vivo* assay using a toe-spread reflex model. This method was used to detect neurotoxins in simple buffer solutions, samples containing serum, and milk [15]. This new assay provides results more quickly than standard mouse bioassays. Whether these results can be translated into a user-friendly, deployable kit has yet to be determined.

4. DNA and other nucleic acid–based methods

Numerous nucleic acid methods have been developed for detecting clostridial DNA in biological and environmental matrices. The polymerase chain reaction (PCR) to identify the presence of *C. botulinum* DNA was originally used to detect the presence of bacterial spores in samples [16]. The method is capable of detecting the presence of as few as 100 spores per reaction mixture for serotypes A, E, and F and only 10 spores per reaction mixture for serotype B.

Multiplex PCR methods have also been developed to analyze unknowns for a battery of different targets such as different pathogens and/or associated gene products of these pathogens. Multiplexed assays employ different combinations or sets of PCR primers, each one specific for a gene of interest, to amplify multiple targets in one PCR tube. One such multiplex method could possibly discriminate among BoNT serotypes A, B, E, and F. As previously described, Peck et al. [16] developed a culture enrichment method that, when coupled with multiplex PCR, could identify strains of *C. botulinum* that were non-proteolytic (such as BoNT serotypes B, E, and F). This method was robust and rapid enough for use with food samples contaminated with *C. botulinum*.

Real-time PCR (RT-PCR) or quantitative PCR (qPCR) is also useful in studies of gene expression, specifically differential expression of genes under various environmental conditions or comparative studies of different organisms. For detection of clostridial DNA, RT-PCR methods examine expression of the NTNH (non-toxic, non-hemagglutinin) and numerous other genes in *C. botulinum* serotypes A, B, E, and F [18]. Pentaplex methods have been developed to simultaneously identify and discriminate among larger numbers of different serotypes, using a wider array of different genes [19]. This technology may prove to be efficient and cost-effective.

The GeneDisc Cycler is an apparatus to perform RT-PCR applications using the GeneDisc system. The GeneDisc is a disposable plastic reaction tray that is the size of a compact disc. This method has been designed for simultaneously testing for the BoNT/A, BoNT/B, BoNT/E, and BoNT/F genes. In 2011, Fach et al. evaluated the GeneDisc Cycler equipment with neurotoxin-producing clostridia and non-BoNT-producing bacteria isolated from various clinical, food, and environmental samples [20]. Results obtained using this "macroarray" were also compared to the mouse bioassay. The toxin genes were detected in all clostridial serotypes A, B, E, and F as well as in toxigenic *Clostridium baratii* Type F and toxigenic *Clostridium butyricum* Type E. No cross-reactivity was observed with bacteria not toxigenic to humans as well as *C. botulinum* Types C, D, and G. An evaluation of the GeneDisc array was performed in four European laboratories with BoNT-producing clostridia and 10 different samples that included food matrices and clinical isolates [20]. Results demonstrated the technology to be specific and reliable in the identification of *C. botulinum* cells containing genes encoding neurotoxins A, B, E, and F. Furthermore, contaminated food and fecal samples were successfully tested. This assay is highly sensitive, capable of detecting as low as 5–50 genome copies in each PCR assay. The GeneDisc Cycler can also be used for monitoring neurotoxin-producing clostridia in food samples, clinical samples, and environmental matrices. A similar study was carried out examining a focused microarray for detection of genes-encoding BoNTs [21].

Recently, Kolesnikov et al. [22] described a new method called "proteolytic PCR" in which PCR is used to assay the proteolytic activity of botulinum toxin. This technology starts with DNA–protein complexes attached to a solid phase. Proteolytic cleavage releases DNA into solution. The DNA can then serve as a template for PCR. This study described its use to detect botulinum toxin and tetanus toxin proteolytic activity [22].

5. Immunological and antibody-based assays

Enzyme-linked immunosorbent assay (ELISA) is a common assay used to detect BoNTs. This method utilizes anti-BoNT capture and detector antibodies arranged in a "sandwich" format. The detection formats are most commonly luminescent- or colorimetric-based. Prior generations of BoNT immunoassays were approximately 10 times less sensitive than the mouse bioassay described in the previous section. Although not as sensitive, ELISA methods are relatively fast, inexpensive, and simple to perform. They are also less subject to inhibitory matrix effects. An amplified ELISA for detecting toxins in food matrices has also been described [23]. Toxins for serotypes A, B, E, and F could be detected in liquids, solid food, and semisolid food. Assay performance was evaluated in a range of food matrices, such as broccoli, orange juice, bottled water, cola soft drinks, vanilla extract, oregano, potato salad, apple juice, meats, and dairy foods. The assay sensitivity varied for each botulinum serotype. The tests readily detected 2 ng/mL of serotypes A, B, E, and F in various foods tested. Recently, traditionally formatted, very sensitive sandwich ELISAs used high affinity mAbs against BoNT/A and BoNT/B to detect BoNT/A as low as 5 and 25 pg/mL in buffer and in a milk matrix, respectively; and BoNT/B at 100 fg/mL and 39 pg/mL in buffer and milk matrix, respectively [24–26].

These mAbs were used in an electrochemiluminesence (ECL) immunosorbent assay using the Sector 2400 Imager (Meso Scale Discovery [MSD], Rockville, MD, USA) instrument [27, 28]. Detection sensitivities for BoNT/A in this system were similar to traditional ELISAs in buffer but were markedly improved in liquid food matrices because of the reduced background signal. The higher sensitivity and reduced time required for these new immunosorbent methods make them potential alternatives to the mouse bioassay. Sharma et al. recently developed another ECL assay for simultaneous detection of several biothreat agents (including clostridial neurotoxins) in milk products, with limit of detection (LOD) of 40 pg/mL for BoNT/A complex [28]. The ECL assay was also successfully applied to screen *C. botulinum* serotype A outbreak strains. The study also showed that this sensitive ECL assay is rapid (it can be completed in less than 6 hours). The ECL assay also has potential for using as an *in vitro* screening method, complementing or replacing other immunoassays.

Cheng and Stanker [27] evaluated the performance of antitoxin mAbs using the same electrochemiluminescence immunoassay platform (Sector 2400 Imager, MSD). The ELISA and ECL methods were observed to be more sensitive than the mouse bioassay. In fact, the ECL assay was able to outperform ELISA in terms of detection sensitivity—including food matrices spiked with BoNT/A and in some food matrices spiked with BoNT/B. The ELISA and ECL methods are fast and simple alternatives to the mouse bioassay and can be used for detecting BoNTs in food matrices and serum samples.

One example of mAb development using a recombinant immunogen was the work of Liu et al. [29], who expressed the recombinant H(C) subunit of BoNT type A (rAH(C)). Two out of 56 mAbs were selected to establish a highly sensitive sandwich chemiluminescence enzyme immunoassay (CLEIA) with LOD for both rAH(C) and BoNT/A of 0.45 pg/mL. This CLEIA can be used to detect BoNT/A in matrices, such as milk and beef extract. This method is 20–40-fold more sensitive than the mouse bioassay and takes only 3 hours to complete, making it a useful method to detect and quantify BoNT/A.

The multiplex technology discussed above to detect nucleic acid has also been applied to the development of methods to analyze multiple epitopes on a single antigen and multiple targets in a single sample. This approach uses multiple mAbs as well as polyclonal antibodies to reduce false-positive and false-negative results. A commercialized system, Luminex xMAP technology, utilizes microsphere beads conjugated with antibodies. It employs paramagnetic beads instead of non-magnetic polystyrene beads and is very useful for the analysis of food matrices. The antibody-bead complexes detect multiple epitopes in a single sample. This technology was used to detect abrin, ricin, BoNTs, and staphylococcal enterotoxins in spiked food samples [30].

Zhang et al. [31] developed ELISA-based protein antibody microarrays to simultaneously detect six serotypes of BoNTs. Using numerous different food and other matrices, the microarray is capable of detecting BoNT serotypes A through F. Using engineered, high-affinity antibodies, these serotypes were detected to similar levels in various matrices and were comparable to detection in buffer.

Accurate and sensitive detection of contaminated food and other biological samples in the environment is critical. Brunt et al. [32] have developed an affinity column-based assay for detecting neurotoxin in food matrices—specifically serotypes A, B, E, and F. The detection limit for BoNT/A was reported as 0.5 ng, which is two-fold more sensitive than lateral flow methods (also see Section 6) [32]. For serotypes B, E, and F, the minimum detection limit ranged from 5 to 50 ng. Although not as sensitive as ELISA or mouse bioassay, rapid immunochromatographic methods generally require only 15–30 minutes to complete. They do not require enrichment steps and are amenable to use in the field.

Koh et al. have presented a new technology called SpinDx [33]. This method utilizes a centrifugal microfluidic platform to detect BoNTs based on a sedimentation immunoassay. A reagent mixture is prepared, consisting of capture beads conjugated with target-specific antibodies and fluorescent detection antibodies. The reagents are mixed with the sample and forced through a channel containing dense medium, a process that washes the sample and removes interfering substances. The beads that collect at the end of the channel are queried to determine the amount of antigen bound. SpinDx was used to quantify BoNTs with sensitivity that surpassed the mouse assay.

6. Lateral flow methods

The development of lateral flow methods for detecting toxins has also led to the commercial availability of numerous kits for sensitive and rapid testing. Lateral flow methods employ capture antibodies that are "printed" on nitrocellulose membranes in a process akin to inkjet printing technology. Detection antibodies are labeled with visible materials, such as colloidal gold or colored latex beads. The sample is added to a reagent pad containing labeled toxin-binding detection antibodies and is wicked across the membrane. Toxin is retained by the capture antibody, which also concentrates the labeled detection antibody. A positive reaction is revealed as a colorimetric change and is presented as a line on the device. In general, lateral flow methods are qualitative and simply determine the presence or absence of neurotoxin. Sharma et al. [34] compared several commercial lateral flow devices for their capacities to detect toxin in food samples. They were able to detect BoNT/A and BoNT/B as low as 10 ng/mL and BoNT/E at 20 ng/mL in various liquids, such as milk, soft drinks, and fruit juices. Ching et al. [35] used the same mAbs described in the ELISA section above [24–26] in lateral flow devices to achieve sensitivities of 0.5 and 1 ng/mL for BoNT/A in buffer and milk, respectively. Although simple lateral flow tests have lower sensitivities compared to other methods, they produce rapid results and are most useful for the rapid screening of samples suspected of frequent contamination at relatively high level of BoNT. They have many applications and are ideal for field use by non-technical personnel. Self-contained and not necessarily requiring additional reagents or equipment, they can be easily interpreted in the field.

An innovative approach for toxin detection has recently been developed that combines antibodies with the amplification power of PCR, immuno-PCR (I-PCR) [36]. In I-PCR, template DNA is conjugated to the antibody, replacing a secondary antibody conjugated to the detection enzyme. Upon binding of toxin by the antibody, the presence of toxin is revealed using PCR. Chao et al. [36] described an I-PCR method for detection of BoNT/A with femtogram (10^{-15} g) sensitivity. These investigators compared competitive and sandwich ELISA to the I-PCR method. The I-PCR method was 10^3–10^5 times more sensitive with LODs for the ELISA methods of about 50 fg. The use of I-PCR for highly sensitive detection of BoNT in food matrices or other biological and environmental backgrounds has yet to be reported (as of late 2015).

7. Mass spectrometry-based methods to detect toxins

Mass spectrometry (MS) has been used as a method to dissect components of botulinum toxin complexes [37–39]. The MS-based method, called ENDOPEP-MS, uses antibodies to concentrate and extract BoNT from test samples [38]. The concentrated toxins are then subjected to an endopeptidase activity–based assay to generate target cleavage products. Finally, MS is used to identify these products. This approach has been successful in identifying BoNT serotypes A, B, E, and F in various food and clinical matrices with greater sensitivity than the mouse bioassay.

Morineaux et al. [40] recently described a MS method that employs immunocapture enrichment by antibodies specific for BoNT/A-L chains. The enriched analyte is then analyzed by liquid chromatography–tandem mass spectrometry (LC–MS/MS) on a triple quadrupole mass spectrometer (QqQ) in multiple reaction monitoring (MRM) mode. Peptides from BoNT LC specific to the subtypes BoNT/A1–A3 and BoNT/A5–A8 could be identified. BoNT/A subtypes were correctly identified in culture supernatants, water, and orange juice samples with a LOD of 20–150 mouse lethal doses (LDs), but there was a lower sensitivity in serum samples.

Kalb et al. [41] have described the development of a quantitative enzymatic method for the detection of four BoNT serotypes using matrix-assisted laser desorption/ionization—time of flight (MALDI-TOF) MS. Factors that might affect the linearity and dynamic range for detection of BoNT cleavage products were carefully examined, including the concentration of the substrate and internal standard, the length of time for the cleavage reaction, and the components present in the reaction solution. Longer incubation time produced more sensitive results but was not capable of determining higher toxin concentrations, whereas a shorter incubation time was less sensitive. To address these limitations, a novel two-step analysis was developed [41]. By combining the results from a two-stage quantification, four or five orders of magnitude in dynamic range are observed for detection of BoNT serotypes A, B, D, and F. To minimize the number of cleavage reactions and analytical samples, the assay can be multiplexed using mixtures of different neurotoxin substrates. Numerous different research groups (including Kalb et al. [42], Björnstad et al. [43], and Hines et al. [37]) have used MS to dissect the components of the BoNT/G complex, revealing BoNT/G as well as other toxin protein components, namely NTNH, HA-17, HA-33, and HA-70. Overall, the use of MS can provide rapid and definitive results.

8. Enzymatic assays to detect toxins

Rapidly distinguishing between the presence of active versus inactive toxin is critical for effective medical intervention in toxicoses. As BoNTs are zinc metalloproteases, knowledge of the human targets for these enzymes has enabled development of enzyme-substrate assays. Activity assays have been developed using a wide variety of detection systems. Toxin samples can be treated with recombinant versions of host–target substrates (such as SNAP-25), and the cleavage products can be detected using immunoblotting. Alternatively, fluorogenic peptide substrates emit a signal when cleaved. One such system uses a peptide ("SNAPtide") with reverse-phase HPLC and a fluorescence detector to detect as low as 5 pg/mL of BoNT/A in skim milk [45]. Other peptide substrates (VAMPtide and SYNTAXtide) have been used for detection of their cognate BoNTs [46]. The levels of substrate cleavage correlate well with toxin activity.

9. Cell-based assays

Cell-based assays measure BoNT receptor binding, translocation, and enzymatic activity and can be *in vitro* alternatives to the mouse bioassay. Several neuronal and non-neuronal cell lines have been analyzed for use in neurotoxin assays. These include the following: BE(2)-M17 cells, chick embryo neuronal cells, neuroblastoma cells, and rat spinal cord cells [47–50]. In general, the endpoint of cell-based assays for BoNT/A is the proteolytic cleavage of its intracellular substrate, the vesicle-trafficking SNARE protein called SNAP-25. Recently, Hubbard et al. [51, 52] described the functional analysis of numerous different biological neurotoxins, including BoNTs, in networked cultures of stem cell–derived central nervous system neurons. The investigators demonstrate synaptic activity in cultured neurons of humans and rodents, suggesting that these could serve as comparable methods to animal studies. Hong et al. [53] have also developed a similar assay using a motor neuron-like continuous cell line. Pathe-Neuschäfer-Rube et al. [54] developed a N-terminal tagged luciferase-expressing neuronal cell line. Luciferase is released from these transfected cells during depolarization, which is blocked by botulinum toxin. Cell-based methods may prove to be equally sensitive, or better, than animal studies and may provide a new alternative for *in vivo* experiments. For example, for the first time, the U.S. Food and Drug Administration approved a cell-based assay developed by the biotechnology company Allergan, Inc. (Irvine, CA, USA) for its use as an alternative to the mouse bioassay. However, the details of the Allergan assay have not been published.

10. New antibody and biosensor technologies

Diamant et al. [55] have used an interesting approach for generating antibodies that have higher specificity against serotypes A, B, and E, and possess neutralizing capabilities. Mice were immunized with a "trivalent mixture" of recombinant fragments of neurotoxins A, B, and E. The method generated numerous different monoclonal antibodies against each serotype. Most of the monoclonal antibodies had higher ELISA titers compared to polyclonal antibodies and had specificities with five orders of magnitude greater specificity. These antibodies also protected against neurotoxin dosages of 10–50 LD_{50}. They also observed a neutralizing synergy when serotype-specific monoclonal antibodies were combined into an oligoclonal mixture.

Detection methods can also utilize highly sensitive antibodies to enrich or enhance sample preparation as well as amplify the signal. For example, an assay with a large immunosorbent surface area (ALISSA) [56, 57] utilizes an antibody to concentrate the neurotoxin onto the surface of a large bead. The "captured" toxin molecules are then used in an enzyme assay. Using food matrices, the LOD for ALISSA was observed as low as 50 fg/mL. This is far more sensitive than the mouse bioassay, immunoassay, or enzyme assay and suggests that it may be useful for detecting food contamination. Marconi et al. [58, 59] have also described the use of surface plasmon resonance (SPR) to examine synaptic vesicle capture by antibodies against BoNT substrates, such as SNAP25 and VAMP2. SPR could be used with cultured neurons in 96-well plates incubated with either BoNT/A or BoNT/B and may be an alternative to animal studies. Further development of label-free and optical biosensors for detecting botulinum toxin [61, 62] will provide additional technologies with possible impact on food safety.

11. Challenges for botulinum neurotoxin detection: new serotypes in the environment

Kull et al. [62] described the isolation of a novel *C. botulinum* strain associated with an outbreak of botulism in Germany. Genotyping of the isolate and subsequent comparison of its neurotoxin gene sequences with database sequences revealed it as a novel BoNT/A serotype. This novel isolate has been called BoNT/A8, and its neurotoxin gene is located within a HA-orfX+ locus. Unique among all other BoNT/A subtypes known so far, an arginine insertion was identified in the HC domain of the HC. Both the full-length neurotoxin and the recombinant LC of BoNT/A8 had lower endopeptidase activity compared to BoNT/A1. Reduced ganglioside binding and lower enzymatic activity may both contribute to lower biological activity of BoNT/A8 as determined using the phrenic nerve hemi-diaphragm assay. Nevertheless, the novel BoNT/A8 subtype caused severe botulism in a 63-year-old male. These findings reiterate that subtyping of BoNT is highly relevant to food safety, epidemiology, and clinical diagnostic and therapeutic practices. Hill et al. [63] carried out a detailed genetic analysis of *bont* genes and confirmed their location on chromosomes, phagemids, and plasmids, as well as variations among different genes. Close examination of sequences confirmed that horizontal gene transfer, site-specific insertions, and recombination events have contributed to the observed variation among different neurotoxins. Understanding the details of toxin gene sequences, protein sequences, and their function can pave the way for the development of novel therapeutics and tailor-made antitoxins. Ongoing development of diagnostics for new and emerging toxins is critical to food safety and human health.

12. Botulinum neurotoxin detection in the environment: role of climate change and algal blooms in avian botulism and the challenges of environmental matrices

Increased global temperature has been associated with increased algal blooms. The role of these algal blooms in disease is unclear. However, recently, a connection between algal blooms and botulism has been explored. Avian botulism is a disease that often occurs on a yearly cycle and results from the ingestion of neurotoxins by birds. This disease has become increasingly common in the U.S. Great Lakes [64], as have blooms of the green alga *Cladophora*, which can serve as a potential habitat for *C. botulinum*. The interactions between *Cladophora* and *C. botulinum* are unclear due to the complex food web associated with this disease. Investigators in several recently published studies [64–66] reported a high number of botulism cases in shoreline birds in Lake Michigan. This increased incidence was correlated with increasingly large accumulations of *Cladophora* in the water. Sadowsky et al. [65] examined algal mats that were collected from Lakes Michigan, Ontario, and Erie in 2011–2012 and then compared them with algal populations in sand and water. They found that 96% of *Cladophora* mats collected from the shorelines in 2012 contained *C. botulinum* Type E. Among the algae samples containing detectable *C. botulinum*, the large number of detected *C. botulinum* type E cells indicated that *Cladophora* mats are principal sources of this pathogen. Mouse toxin and antitoxin bioassays further confirmed the toxin in collected samples as serotype E. Further examination of *Cladophora*-associated *C. botulinum* may lead to a model system to study algal–clostridial interactions and result in lower bird mortality.

In a follow-up study using PCR, Sadowsky et al. [66] reported that algae mats from different shores of the Great Lakes contained the serotype E gene. Also, *C. botulinum* was found to be present in amounts of up to 15,000 cells per gram of dried algae, based on quantitation of gene copies encoding serotype E. Moreover, genes for serotypes A and B, which are associated with human diseases, were detected in several of the algal samples. Using mouse toxin assays and subsequent neutralization assays, it was confirmed that *Cladophora*-associated *C. botulinum* was serotype E. One might consider that with increased incidence of extreme drought and other environmental changes, algal blooms may happen more often in water-restricted areas, and *C. botulinum* growth may pose a threat to humans if toxin is produced in algal mats. Developing sensitive detection methods for toxins within algal matrices is urgent, as is monitoring other matrices that could provide an environment for botulinum toxin production. The increased avian botulism associated with increased algal blooms highlights the need to develop new technologies for detection of toxin in the environment, or a re-evaluation of current methods and their use in environmental matrices.

Vidal et al. [67] examined numerous environmental factors that influence the prevalence of the unusual mosaic BoNT serotype C/D. Between 1978 and 2008, 13 avian botulism outbreaks were observed, killing 20,000 birds. A significant association was found between the number of dead birds recorded in each botulism outbreak and the mean temperature in July (with average temperatures being higher than 26°C). The presence of *C. botulinum* type C/D in wetland sediments was detected by qPCR. Furthermore, low concentrations of chloride ions and high organic matter content were correlated with the presence of *C. botulinum*. The digestive tracts of dead birds found during botulism outbreaks were also analyzed; *C. botulinum* was present in almost 40% of the studied samples. Recently, Le Maréchal et al. [68] examined livers from dead birds suspected of having botulism and showed that this organ can serve as a reliable matrix for RT-PCR confirmation of disease. This finding may provide wildlife investigators with a faster method to confirm avian deaths due to botulism.

The presence of *C. botulinum* was detected in aquatic invertebrates and flesh-eating invertebrates collected around bird carcasses. Moreover, the presence of *C. botulinum* bacteria in the adult fly stage of some invertebrates raises the question of whether flies can transport *C. botulinum* from one carcass to another. The same investigators examined whether adult blowflies could play a significant role in botulism outbreaks by carrying *C. botulinum* between carcasses. A field experiment and subsequent laboratory tests determined that blowflies could transport *C. botulinum* Type C/D between carcasses [69]. These results confirm that adult flesh-eating flies could play a role in avian botulism outbreaks. An environmental monitoring protocol for botulinum-carrying flies has not yet been established. It is a matter for future research to determine whether these or other insects could serve as mechanical vectors for botulinum isolates that pose greater threats to humans than the avian isolates.

Probably, one of the greatest challenges is determining which environmental matrices should be collected and analyzed, and which ones would provide the most definitive information about potential threats to humans and animals. For instance, Anza et al. [70] examined the role of eutrophication and avian botulism outbreaks in wetlands receiving effluents from urban wastewater treatment plants. Numerous different avian pathogens, including clostridial pathogens, were present in wastewater and could pose a threat to birds living in wastewater wetlands. Methods to detect BoNTs in environmental matrices could be adapted from previous studies of food and clinical samples or may require new technologies. Future studies in this area are clearly warranted.

13. Future technologies to detect botulinum neurotoxins

The discussion herein has presented a general overview of methods currently being used to detect BoNTs. Many current methods to detect BoNTs in food and environmental matrices have been adapted from the clinical laboratory. New possibilities to consider, to name a few, could exploit the tools of nanotechnology, mHealth, and the use of mobile devices, the capability of miniaturization for even more sensitive and rapid detection of BoNTs. The application and practical use of these technologies might be valuable advancements to current methods to detect BoNTs.

14. Conclusions and recommendations

To maintain a safe food supply and to detect toxins in an ever-changing environment, an ongoing, concerted effort in assay development and validation is essential for human health and safety. Some areas for investigators to consider include the development of new antibodies and binding molecules specific to BoNT serotype F as well as new hybrid serotypes. The impact of different types of neurotoxin accessory proteins on the detection of BoNTs should also be examined. Furthermore, the impact of food processing conditions on the stability and bioavailability BoNTs is an area in need of further study. The development of new bioassays based on non-mammalian systems and cell cultures should also be supported as well as the advancement of new portable and field-deployable testing methods, including those based on miniaturization of current bench top instruments. These are only a few recommendations, but their development and use should help to further ensure food safety and animal and human health.

Acknowledgements

This work was funded by the U.S. Department of Agriculture, Agricultural Research Service (National Program 108, Project No. 5325-42000-048-00D). Larry H. Stanker also received funding from the U.S. Department of Homeland Security (Interagency Agreement No. 40768). Kirkwood M. Land was supported by the Department of Biological Sciences at the University of the Pacific.

Author details

Luisa W. Cheng[1*], Kirkwood M. Land[2], Christina Tam[1], David L. Brandon[1] and Larry H. Stanker[1]

*Address all correspondence to: luisa.cheng@ars.usda.gov

1 Foodborne Toxin Detection and Prevention Research Unit, Western Regional Research Center, Agricultural Research Service, U.S. Department of Agriculture, Albany, CA, USA

2 Department of Biological Sciences, University of the Pacific, Stockton, CA, USA

References

[1] http://www.cdc.gov/nczved/divisions/dfbmd/diseases/botulism/

[2] Rossetto, O., Pirazzini, M., and Montecucco, C. (2014). Botulinum neurotoxins: genetic, structural and mechanistic insights. Nat Rev Microbiol. 12:535–49. doi: 10.1038/nrmicro3295.

[3] Rossetto, O., Megighian, A., Scorzeto, M., and Montecucco, C. (2013). Botulinum neurotoxins. Toxicon 67:31–6. doi: 10.1016/j.toxicon.2013.01.017.

[4] Tighe, A.P., and Schiavo, G. (2013). Botulinum neurotoxins: mechanism of action. Toxicon 67:87–93. doi: 10.1016/j.toxicon.2012.11.011.

[5] Rummel, A. (2015). The long journey of botulinum neurotoxins into the synapse. Toxicon 107:9–24. doi: 10.1016/j.toxicon.2015.09.009.

[6] Hill, K.K., Xie, G., Foley, B.T., and Smith, T.J. (2015). Genetic diversity within the botulinum neurotoxin-producing bacteria and their neurotoxins. Toxicon 107(Pt A):2–8. doi: 10.1016/j.toxicon.2015.09.011.

[7] Hill, K.K., Smith, T.J., Helma, C.H., Ticknor, L.O., Foley, B.T., Svensson, R.T., Brown, J.L., Johnson, E.A., Smith, L.A., Okinaka, R.T., Jackson, P.J., and Marks, J.D. (2007). Genetic diversity among botulinum neurotoxin-producing clostridial strains. J Bacteriol. 189(3):818–32.

[8] Cheng, L.W., Land, K.M., and Stanker, L.H. (2012). Current methods for detecting the presence of botulinum neurotoxins in food and other biological matrices. In: Stephen Morse (Ed.), Bioterrorism. In Tech. ISBN 978-953-51-0205-2.

[9] Hakami, R.M., Ruthel, G., Stahl, A.M., and Bavari, S. (2010). Gaining ground: assays for therapeutics against botulinum neurotoxin. Trends Microbiol 18(4):164–72.

[10] Grate, J.W., Ozanich, R.M., Jr., Warner, M.G., Marks, J.D., and Bruckner-Lea, C.J. (2010). Advances in assays and analytical approaches for botulinum-toxin detection. Trends Anal Chem. 29(10):1137–56.

[11] Cheng, L.W., Onisko, B., Johnson, E.A., Reader, J.R., Griffey, S.M., Larson, A.E., Tepp, W.H., Stanker, L.H., Brandon, D.L., and Carter, J.M. (2008). Effects of purification on the bioavailability of botulinum neurotoxin type A. Toxicology 249(2–3):123–9.

[12] Cheng, L.W., and Henderson, T.D., 2nd (2011). Comparison of oral toxicological properties of botulinum neurotoxin serotypes A and B. Toxicon 58(1):62–7.

[13] Bigalke, H., and Rummel, A. (2015). Botulinum neurotoxins: qualitative and quantitative analysis using the mouse phrenic nerve hemidiaphragm assay (MPN). Toxins 12:4895–905. doi: 10.3390/toxins7124855.

[14] Rasetti-Escargueil, C., Jones, R.G., Liu, Y., and Sesardic, D. (2009). Measurement of botulinum types A, B and E neurotoxicity using the phrenic nerve-hemidiaphragm: improved precision with in-bred mice. Toxicon 53(5):503–11.

[15] Wilder-Kofie, T.D., Luquez, C., Adler, M., Dykes, J.K., Coleman, J.D., and Maslanka, S.E. (2011). An alternative in vivo method to refine the mouse bioassay for botulinum toxin detection. Comp Med. 61(3):235–42.

[16] Peck, M.W., Plowman, J., Aldus, C.F., Wyatt, G.M., Izurieta, W.P., Stringer, S.C., and Barker, G.C. (2010). Development and application of a new method for specific and

sensitive enumeration of spores of nonproteolytic *Clostridium botulinum* types B, E, and F in foods and food materials. Appl Environ Microbiol. 76(19):6607–14.

[17] De Medici, D., Anniballi, F., Wyatt, G.M., Lindström, M., Messelhausser, U., Aldus, C.F., Delibato, E., Korkeala, H., Peck, M.W., and Fenicia, L. (2009). Multiplex PCR for detection of botulinum neurotoxin-producing clostridia in clinical, food, and environmental samples. Appl Environ Microbiol. 75(20):6457–61.

[18] Fach, P., Micheau, P., Mazuet, C., Perelle, S., and Popoff, M. (2009). Development of real-time PCR tests for detecting botulinum neurotoxins A, B, E, F producing *Clostridium botulinum, Clostridium baratii* and *Clostridium butyricum.* J Appl Microbiol. 107(2):465–73.

[19] Kirchner, S., Kramer, K.M., Schulze, M., Pauly, D., Jacob, D., Gessler, F., Nitsche, A., Dorner, B.G., and Dorner, M.B. (2010). Pentaplexed quantitative real-time PCR assay for the simultaneous detection and quantification of botulinum neurotoxin- producing clostridia in food and clinical samples. Appl Environ Microbiol. 76(13):4387–95.

[20] Fach, P., Fenicia, L., Knutsson, R., Wielinga, P.R., Anniballi, F., Delibato, E., Auricchio, B., Woudstra, C., Agren, J., Segerman, B., de Medici, D., and van Rotterdam, B.J. (2011). An innovative molecular detection tool for tracking and tracing *Clostridium botulinum* types A, B, E, F and other botulinum neurotoxin producing *Clostridia* based on the GeneDisc cycler. Int J Food Microbiol. 145(Suppl. 1):S145–51.

[21] Raphael, B.H., Joseph, L.A., McCroskey, L.M., Luquez, C., and Maslanka, S.E. (2010). Detection and differentiation of *Clostridium botulinum* type A strains using a focused DNA microarray. Mol Cell Probes 24(3):146–53.

[22] Kolesnikov, A.V., Kozyr, A.V., Ryabko, A.K., and Shemyakin, I.G. (2016). Ultrasensitive detection of protease activity of anthrax and botulinum toxins by a new PCR-based assay. Pathog Dis. 74(1):pii:ftv112. doi: 10.1093/femspd/ftv112. Epub 2015 Nov 29.

[23] Sharma, S.K., Ferreira, J.L., Eblen, B.S., and Whiting, R.C. (2006). Detection of type A, B, E, and F *Clostridium botulinum* neurotoxins in foods by using an amplified enzyme-linked immunosorbent assay with digoxigenin-labeled antibodies. Appl Environ Microbiol. 72:1231–8.

[24] Stanker, L.H., Merrill, P., Scotcher, M.C., and Cheng, L.W. (2008). Development and partial characterization of high-affinity monoclonal antibodies for botulinum toxin type A and their use in analysis of milk by sandwich ELISA. J Immunol Methods 336(1):1–8.

[25] Scotcher, M.C., Cheng, L.W., and Stanker, L.H. (2010). Detection of botulinum neurotoxin serotype B at sub mouse LD(50) levels by a sandwich immunoassay and its application to toxin detection in milk. PLoS One 5(6):e11047.

[26] Stanker, L.H., Scotcher, M.C., Cheng, L., Ching, K., McGarvey, J., Hodge, D., and Hnasko, R. (2013). A monoclonal antibody based capture ELISA for botulinum neu-

rotoxin serotype B: toxin detection in food. Toxins (Basel) 5(11):2212–26. doi: 10.3390/toxins5112212.

[27] Cheng, L.W., and Stanker, L.H. (2013). Detection of botulinum neurotoxin serotypes A and B using a chemiluminescent versus electrochemiluminescent immunoassay in food and serum. J Agric Food Chem. 61:755–60. doi: 10.1021/jf3041963.

[28] Sachdeva, A., Singh, A.K., and Sharma, S.K. (2015). An electrochemiluminescence assay for the detection of bio threat agents in selected food matrices and in the screening of *Clostridium botulinum* outbreak strains associated with type A botulism. J Sci Food Agric. 94(4):707–12. doi: 10.1002/jsfa.6310.

[29] Liu, Z., Song, C., Li, Y., Liu, F., Zhang, K., Sun, Y., Li, H., Wei, Y., Xu, Z., Zhang, C., Yang, A., Xu, Z., Yang, K., and Jin, B. (2012). Development of highly sensitive chemiluminescence enzyme immunoassay based on the anti-recombinant H(C) subunit of botulinum neurotoxin type A monoclonal antibodies. Anal Chim Acta 20:23–30. doi: 10.1016/j.aca.2012.05.008.

[30] Garber, E.A., Venkateswaran, K.V., and O'Brien, T.W. (2010). Simultaneous multiplex detection and confirmation of the proteinaceous toxins abrin, ricin, botulinum toxins, and *Staphylococcus* enterotoxins A, B, and C in food. J Agric Food Chem. 58:6600–7. doi: 10.1021/jf100789n.

[31] Zhang, Y., Lou, J., Jenko, K.L., Marks, J.D., and Varnum, S.M. (2012). Simultaneous and sensitive detection of six serotypes of botulinum neurotoxin using enzyme-linked immunosorbent assay-based protein antibody microarrays. Anal Biochem. 430:185–92. doi: 10.1016/j.ab.2012.08.021.

[32] Brunt, J., Webb, M.D., and Peck, M.W. (2010). Rapid affinity immunochromatography column-based tests for sensitive detection of *Clostridium botulinum* neurotoxins and *Escherichia coli* O157. Appl Environ Microbiol. 76(13):4143–50.

[33] Koh, C.Y., Schaff, U.Y., Piccini, M.E., Stanker, L.H., Cheng, L.W., Ravichandran, E., Singh, B.R., Sommer, G.J., and Singh, A.K. (2015). Centrifugal microfluidic platform for ultrasensitive detection of botulinum toxin. Anal Chem. 87(2):922–8. doi: 10.1021/ac504054u.

[34] Sharma, S.K., Eblen, B.S., Bull, R.L., Burr, D.H., and Whiting, R.C. (2005). Evaluation of lateral-flow *Clostridium botulinum* neurotoxin detection kits for food analysis. Appl Environ Microbiol. 71:3935–41.

[35] Ching, K.H., Lin, A., McGarvey, J.A., Stanker, L.H., and Hnasko, R. (2012). Rapid and selective detection of botulinum neurotoxin serotype-A and -B with a single immunochromatographic test strip. J Immunol Methods 380:23–9. doi: 10.1016/j.jim.2012.03.008.

[36] Chao, H.Y., Wang, Y.C., Tang, S.S., and Liu, H.W. (2004). A highly sensitive immuno-polymerase chain reaction assay for *Clostridium botulinum* neurotoxin type A. Toxicon 43(1):27–34.

[37] Hines, H.B., Lebeda, F., Hale, M., and Brueggemann, E.E. (2005). Characterization of botulinum progenitor toxins by mass spectrometry. Appl Environ Microbiol. 71(8): 4478–86.

[38] Kalb, S.R., Goodnough, M.C., Malizio, C.J., Pirkle, J.L., and Barr, J.R. (2005). Detection of botulinum neurotoxin A in a spiked milk sample with subtype identification through toxin proteomics. Anal Chem. 77(19):6140–6.

[39] Kalb, S.R., Moura, H., Boyer, A.E., McWilliams, L.G., Pirkle, J.L., and Barr, J.R. (2006). The use of Endopep-MS for the detection of botulinum toxins A, B, E, and F in serum and stool samples. Anal Biochem. 351(1):84–92.

[40] Morineaux, V., Mazuet, C., Hilaire, D., Enche, J., and Popoff, M.R. (2015). Characterization of botulinum neurotoxin type A subtypes by immunocapture enrichment and liquid chromatography-tandem mass spectrometry. Anal Bioanal Chem. 407:5559–70.

[41] Wang, D., Baudys, J., Krilich, J., Smith, T.J., Barr, J.R., and Kalb, S.R. (2014). A two-stage multiplex method for quantitative analysis of botulinum neurotoxins type A, B, E, and F by MALDI-TOF mass spectrometry. Anal Chem. 86:10847–54. doi: 10.1021/ac502948v.

[42] Kalb, S.R., Baudys, J., and Barr, J.R. (2015). Detection of the HA-33 protein in botulinum neurotoxin type G complex by mass spectrometry. BMC Microbiol. 15:227. doi: 10.1186/s12866-015-0567-5.

[43] Björnstad, K., Tevell Åberg, A., Kalb, S.R., Wang, D., Barr, J.R., Bondesson, U., and Hedeland, M. (2014). Validation of the Endopep-MS method for qualitative detection of active botulinum neurotoxins in human and chicken serum. Anal Bioanal Chem. 406:7149–61. doi: 10.1007/s00216-014-8170-4.

[44] Feltrup, T.M., and Singh, B.R. (2012). Development of a fluorescence internal quenching correction factor to correct botulinum neurotoxin type A endopeptidase kinetics using SNAPtide. Anal Chem. 84:10549–53. doi: 10.1021/ac302997n.

[45] Shi, X., Garcia, G.E., Nambiar, M.P., and Gordon, R.K. (2008). Un-nicked BoNT/B activity in human SHSY-5Y neuronal cells. J Cell Biochem. 105:129–35. doi: 10.1002/jcb.21800.

[46] Wictome, M., Newton, K., Jameson, K., Hallis, B., Dunnigan, P., Mackay, E., Clarke, S., Taylor, R., Gaze, J., Foster, K., and Shone, C. (1999). Development of an in vitro bioassay for *Clostridium botulinum* type B neurotoxin in foods that is more sensitive than the mouse bioassay. Appl Environ Microbiol. 65(9):3787–92.

[47] Pellett, S., Tepp, W.H., Clancy, C.M., Borodic, G.E., and Johnson, E.A. (2007). A neuronal cell-based botulinum neurotoxin assay for highly sensitive and specific detection of neutralizing serum antibodies. FEBS Lett. 581(25):4803–8.

[48] Stahl, A.M., Ruthel, G., Torres-Melendez, E., Kenny, T.A., Panchal, R.G., and Bavari, S. (2007). Primary cultures of embryonic chicken neurons for sensitive cell-based assay of botulinum neurotoxin: implications for therapeutic discovery. J Biomol Screen. 12(3):370–7.

[49] Basavanna, U., Muruvanda, T., Brown, E.W., and Sharma, S.K. (2013). Development of a cell-based functional assay for the detection of *Clostridium botulinum* neurotoxin types A and E. Int J Microbiol. 2013:593219. doi: 10.1155/2013/593219.

[50] Basavanna, U., Muruvanda, T., Brown, E.W., and Sharma, S.K. (2014). Development of a cell-based functional assay for the detection of *Clostridium botulinum* neurotoxin types A and E. Int J Microbiol. 2013:593219. doi: 10.1155/2013/593219.

[51] Beske, P.H., Bradford, A.B., Grynovicki, J.O., Glotfelty, E.J., Hoffman, K.M., Hubbard, K.S., Tuznik, K.M., and McNutt, P.M. (2015). Botulinum and tetanus neurotoxin-induced blockade of synaptic transmission in networked cultures of human and rodent neurons. Toxicol Sci. 2015:pii:kfv254.

[52] Hubbard, K., Beske, P., Lyman, M., and McNutt, P. (2014). Functional evaluation of biological neurotoxins in networked cultures of stem cell-derived central nervous system neurons. J Vis Exp. 96. doi: 10.3791/52361.

[53] Hong, W.S., Pezzi, H.M., Schuster, A.R., Berry, S.M., Sung, K.E., and Beebe, D.J. (2016). Development of a highly sensitive cell-based assay for detecting botulinum neurotoxin type A through neural culture media optimization. J Biomol Screen. 21:65–73. doi: 10.1177/1087057115608103.

[54] Pathe-Neuschäfer-Rube, A., Neuschäfer-Rube, F., Genz, L., and Püchel, G.P. (2015). Botulinum neurotoxin dose-dependently inhibits release of neurosecretory vesicle-vargeted luciferase from neuronal cells. ALTEX 32(4):297–306. doi: http://dx.doi.org/10.14573/altex.1503061.

[55] Diamant, E., Lachmi, B.E., Keren, A., Barnea, A., Marcus, H., Cohen, S., David, A.B., and Zichel, R. (2014). Evaluating the synergistic neutralizing effect of anti-botulinum oligoclonal antibody preparations. PLoS One 9:e87089. doi: 10.1371/journal.pone.0087089.

[56] Bagramyan, K., and Kalkum, M. (2011). Ultrasensitive detection of botulinum neurotoxins and anthrax lethal factor in biological samples by ALISSA. Methods Mol Biol. 739:23–36. doi: 10.1007/978-1-61779-102-4_3.

[57] Bagramyan, K., Barash, J.R., Arnon, S.S., and Kalkum, M. (2008). Attomolar detection of botulinum toxin type A in complex biological matrices. PLoS One 3(4):e2041.

[58] Ferracci, G., Marconi, S., Mazuet, C., Jover, E., Blanchard, M.P., Seagar, M., Popoff, M., and Lévêque, C. (2011). A label-free biosensor assay for botulinum neurotoxin B in food and human serum. Anal Biochem. 410:281–8. doi: 10.1016/j.ab.2010.11.045.

[59] Marconi, S., Ferracci, G., Berthomieu, M., Kozaki, S., Miquelis, R., Boucraut, J., Seagar, M., and Lévêque, C. (2008). A protein chip membrane-capture assay for botulinum neurotoxin activity. Toxicol Appl Pharmacol. 233:439–46. doi: 10.1016/j.taap.2008.09.005.

[60] Ferracci, G., Marconi, S., Mazuet, C., Jover, E., Blanchard, M.P., Seagar, M., Popoff, M., and Leveque, C. (2010). A label-free biosensor assay for botulinum neurotoxin B in food and human serum. Anal Biochem. 410(2):281–8.

[61] Lévêque, C., Ferracci, G., Maulet, Y., Mazuet, C., Popoff, M.R., Blanchard, M.P., Seagar, M., and El Far, O. (2015). An optical biosensor assay for rapid dual detection of Botulinum neurotoxins A and E. Sci Rep. 5:17953. doi: 10.1038/srep17953.

[62] Kull, S., Schulz, K.M., Weisemann, J., Kirchner, S., Schreiber, T., Bollenbach, A., Dabrowski, P.W., Nitsche, A., Kalb, S.R., Dorner, M.B., Barr, J.R., Rummel, A., and Dorner, B.G. (2015). Isolation and functional characterization of the novel *Clostridium botulinum* neurotoxin A8 subtype. PLoS One 10:e0116381. doi: 10.1371/journal.pone.0116381.

[63] Hill, K.K., Xie, G., Foley, B.T., and Smith, T.J. (2015). Genetic diversity within the botulinum neurotoxin-producing bacteria and their neurotoxins. Toxicon 107:2–8. doi: 10.1016/j.toxicon.2015.09.011.

[64] Parks, N. (2008). Bird die-offs in the Great Lakes. Environ Sci Technol. 42:2209.

[65] Chun, C.L., Ochsner, U., Byappanahalli, M.N., Whitman, R.L., Tepp, W.H., Lin, G., Johnson, E.A., Peller, J., and Sadowsky, M.J. (2013). Association of toxin-producing *Clostridium botulinum* with the macroalga *Cladophora* in the Great Lakes. Environ Sci Technol. 47(6):2587–94. doi: 10.1021/es304743m.

[66] Lan Chun, C., Kahn, C.I., Borchert, A.J., Byappanahalli, M.N., Whitman, R.L., Peller, J., Pier, C., Lin, G., Johnson, E.A., and Sadowsky, M.J. (2015). Prevalence of toxin-producing *Clostridium botulinum* associated with the macroalga *Cladophora* in three Great Lakes: growth and management. Sci Total Environ. 511:523–9. doi: 10.1016/j.scitotenv.2014.12.080.

[67] Vidal, D., Anza, I., Taggart, M.A., Pérez-Ramírez, E., Crespo, E., Hofle, U., and Mateo, R. (2013). Environmental factors influencing the prevalence of a *Clostridium botulinum* type C/D mosaic strain in nonpermanent Mediterranean wetlands. Appl Environ Microbiol. 79(14):4264–71. doi: 10.1128/AEM.01191-13.

[68] Le Maréchal, C., Ballan, V., Rouxel, S., Bayon-Auboyer, M.H., Baudouard, M.A., Morvan, H., Houard, E., Poëzevara, T., Souillard, R., Woudstra, C., Le Bouquin, S., Fach, P., and Chemaly, M. (2015). Livers provide a reliable matrix for real-time PCR confirmation of avian botulism. Anaerobe 38:7–13. doi: 10.1016/j.anaerobe.2015.10.014.

[69] Anza, I., Vidal, D., and Mateo, R. (2014). New insight in the epidemiology of avian botulism outbreaks: necrophagous flies as vectors of *Clostridium botulinum* type C/D. Environ Microbiol Rep. 6(6):738–43. doi: 10.1111/1758-2229.12197.

8

Cholera – Epidemiology, Prevention and Control

Saulat Jahan

Abstract

Cholera is an important public health problem, causing substantial morbidity and mortality especially in the developing countries. It is an indicator of socioeconomic problems and is a global threat to public health. Worldwide, approximately 3–5 million cholera cases and 100,000–120,000 deaths due to cholera occur annually. Cholera is transmitted by drinking water or eating food, which is contaminated with the *Vibrio cholerae*. Approximately 5–10% of persons suffer from severe cholera and if untreated, 50% of severe cases are fatal. The frequency, severity, and duration of cholera infection vary and keep on changing in different parts of the world. Environmental factors such as climate variability, temperature, and rainfall play an important role in cholera transmission. Population density, urbanization, and overcrowding also influence cholera transmission. It is also closely associated with the social and behavioral aspects of individuals as well as communities. Each year, many cholera outbreaks are reported from different regions of the world. These outbreaks have negative impact on social and economic conditions of the affected countries. An integrated, multisectoral program, designed on evidence-based interventions, is required to prevent and control cholera.

Keywords: Cholera, foodborne diseases, outbreaks, public health, *Vibrio cholerae*

1. Introduction

Cholera is an acute diarrheal illness, caused by the toxigenic strains of the bacterium *Vibrio cholerae* serogroups O1 or O139 [1, 2]. It is one of the important public health problems in Asia and Africa and causes substantial morbidity and mortality [3]. Since centuries, cholera has been a subject of interest for epidemiologists. The studies regarding cholera helped in the development of new epidemiological methods which led to the understanding of not only cholera transmission but also helped in the development of the science of infectious disease epidemiology [4]. The purpose of this chapter is to discuss this important infectious disease, i.e., its historical aspects, clinical features, epidemiology, and the outbreaks caused by cholera.

Moreover, the preventive measures for cholera and methods of control of cholera outbreaks will also be discussed.

1.1. Historical aspects of cholera

For centuries, cholera remained one of the most horrific diseases [5]. It was first described by Hippocrates in the fifth century BC. Traditionally, the Ganges Delta region in Asia is considered the home of cholera. It is believed that cholera spread throughout the world from this region. Several epidemics occurred in Asia during the fifteenth and eighteenth centuries. Seven major pandemics of cholera have occurred since 1817 [3, 6].

Historians believe that the impact of cholera epidemics on the cultural evolution of Western Europe, was far reaching and it altered the social matrix of European culture. During the nineteenth century cholera was not only considered a terrifying disease, but was also a challenge to national identity and national economy [7].

John Snow performed pioneer work on cholera in London in the 1800s. He established an association between cholera transmission and contaminated water [4]. He discovered the method of prevention and control of cholera by tracing its source back to drinking water. Due to his work on the Soho cholera outbreak in 1854, John Snow has become a legend in epidemiology [5]. Later, in 1883, Robert Koch described the causative agent for cholera as a curved bacillus, *V. cholerae* [3].

In 1892, a major cholera outbreak occurred in Germany, killing 10,000 people. It was found to be caused by a defect in the design in the German waste removal system [7]. Seven cholera pandemics occurred during the nineteenth and twentieth centuries. The seventh pandemic began in Indonesia in 1961, reached West Africa in 1970 and the Americas in 1991 [3, 6].

In the United States of America, the risk of cholera is very low. However, in 2005, the floods caused by Hurricane Katrina, created the fear of a cholera epidemic for the first time in a century. The US Centers for Disease Control and Surveillance had 11 confirmed cases of people becoming ill from *Vibrio* sp. infection. However, only one of those cases had *V. cholerae*, which was not from the two epidemic-causing serogroups, i.e., O1 and O139 [5].

1.2. General burden of cholera on human population

Cholera is an indicator of a lack of social development and is a global threat to public health. With rapidly increasing population in developing world, the populations living in unsanitary conditions are increasing and the re-emergence of cholera has also been noted [8].

Every year, approximately 3–5 million cholera cases occur, worldwide. An estimated 100,000–120,000 deaths due to cholera occur every year. The number of cholera cases reported to the World Health Organization (WHO) continues to rise. In 2011, a total of 589,854 cholera cases including 7816 cholera deaths were reported from 58 countries, with a case-fatality rate of 1.3% [8, 9]. A total of 838,315 cases belonging to the period 2004–2008 had been notified to the WHO, as compared to 676,651 cases notified from 2000 to 2004 [1]. In 2006, 52 countries reported 236,896 cholera cases including 6311 deaths with a case fatality rate of 2.7% [10].

The above-mentioned numbers are an underestimation of the real picture as the true number of cholera cases is thought to be much higher. Many cases may not be reported due to limitations in health-care systems and the surveillance systems. There are inconsistencies in cholera case definitions and there is a lack of standard vocabulary. Notified cases are often diagnosed on clinical grounds, leading to a number of undiagnosed and unreported cases. Some countries notify only laboratory-confirmed cases. It is a challenge in the estimation of disease burden due to the dearth of microbiology laboratories capable of detecting *V. cholerae* O1 and O139 in the countries where the disease is prevalent. Fear of sanctions regarding travel and trade may also result in underreporting of cholera cases [10, 11].

2. Causative agent

The causative agent of cholera, *V. cholerae,* is a waterborne and foodborne gram-negative bacterium. *Vibrio cholerae* can cause global pandemics which makes it unique among the diarrheal pathogens [12]. The serotype O1 or O139 colonizes and multiplies within the human small intestine [13]. There are two biotypes of *V. cholerae* O1: Classical and El Tor. Each of these biotypes has two serotypes: Inaba and Ogawa. Both these biotypes cause similar symptoms, however, El Tor biotype mostly causes asymptomatic infection or causes a mild illness [9]. Until 1992, only *V. cholerae* serogroup O1 was responsible for causing epidemics. However, in 1992, a previously unrecognized serogroup of *V. cholerae,* labeled as O139, was found to be responsible for large outbreaks of cholera in India and Bangladesh [3]. Multiple antibiotic-resistant strains of *V. cholerae* have also emerged and the severity of the disease appears to be intensifying [1].

2.1. Mode of transmission

Cholera is transmitted by drinking water or eating food contaminated with the *V. cholerae.* Fecal contamination of water or foods may result in large epidemics. The disease may also be transmitted through eating contaminated raw or undercooked shellfish [9]. Before the late 1970s, it was believed that person-to-person transmission of cholera is the main mode of transmission. Now, *V. cholerae* is found in riverine, estuarine, and coastal waters throughout both temperate and tropical regions of the world. It is recognized as a component of coastal and estuarine microbial ecosystems [14].

2.2. Incubation period

The disease has a short incubation period of 18 hours to 5 days, and it can spread rapidly through a population [12].

3. Clinical features of cholera

The infection may be mild or asymptomatic in most of the cases, but sometimes it can be severe. Approximately 5–10% of persons suffer from severe cholera. In severe cases, the clinical features include profuse watery diarrhea, vomiting, increased heart rate, loss of skin elasticity, dry mucous membranes, low blood pressure, feeling of thirst, muscle cramps, metabolic acidosis, and restlessness or irritability. These patients can develop acute renal failure, severe electrolyte imbalances, and coma, leading to hypovolemic shock and death [3, 15]. If untreated, 50% of severe cases are fatal, while proper treatment and fluid replacement reduce mortality to less than 1% [3, 12].

3.1. Diagnosis

The confirmatory test for cholera is done by culture of a stool specimen or rectal swab. For transport of specimen, Cary Blair media is the most appropriate, and for isolation and identification of the organism, the selective thiosulfate–citrate–bile salts agar (TCBS) is the medium of choice. Commercially available rapid test kits should not be used for routine diagnosis as they cannot determine the subtypes and are not able to isolate the antimicrobial susceptibility. However, they are useful during epidemics [16].

3.2. Treatment

Rehydration by oral rehydration salts and, if required, intravenous fluids and electrolytes, is the cornerstone of cholera treatment. In addition, antibiotics may be needed for severe cases. Commonly used antibiotics include tetracycline, doxycycline, furazolidone, erythromycin, or ciprofloxacin [17].

4. Epidemiology of cholera

Cholera is said to be a 'forgotten disease' which mainly affects 'forgotten people' of the world. It comes into limelight when an extensive cholera outbreak occurs, although some neglected populations continually suffer from the recurrent episodes of cholera [10]. Cholera is the main cause of epidemic diarrhea in the developing countries. For the last four decades, there is an ongoing global pandemic in Asia, Africa, and Latin America [9].

4.1. Person distribution

In endemic regions, the majority of cases occur among children less than 5 years of age and in reproductive-age women [3]. However, various studies conducted in different parts of the world have reported varying age distribution for cholera patients. A study conducted in Pakistan during the period 2000–2001 found the mean age of patients infected with *V. cholerae* O139 as 40 years as compared to those infected with *V. cholerae* O1 strains where the mean age was 23 years [18]. Tamang et al (2005) conducted a study in a teaching hospital in Nepal between May 1, 2004 and October 31, 2004. They investigated 148 stool samples from patients with acute diarrhea and found out that 46 cases (31%) were positive for *V. cholerae* serogroup O1, biotype El Tor, serotype Ogawa. In their study, younger age group of less than 30 years was mostly affected. Males and females were equally affected [19]. During a cholera outbreak in southwestern Nigeria from November 20, 2005 to January 1, 2006, a total of 115 cases and 11 deaths were reported. Overall case fatality rate was 9.6% with a case fatality rate in males of 11.9% as compared to 7.1% in females. During this outbreak, the age group of 15 years and above was the most affected comprising of 68.3% cases and 90.9% deaths [20]. A study conducted in southeast of Iran examined 3,178 patients with watery diarrhea, referred to a teaching hospital over a period of 4 years. A total of 362 (11.4%) samples contained *V. cholerae* strains; 336 (92.8%) were *V. cholerae* O1 Ogawa strain. A majority of the cases (270) were referred from rural areas. In this study, the patients ranged in age from 1 to 65 years. Approximately one-fourth (26%) were children under the age of 5 years [21]. In Mexico, the distribution of cholera cases by gender was reported to be similar. Individuals aged 25–44 years had the greatest proportion of cases while individuals aged over 65 years had the highest rates of infection. The authors state that this distribution differs from the distribution of cholera cases in endemic regions. The authors interpret this difference because of the occupational exposure and eating habits of the older individuals [3]. Most of the studies have mentioned a pattern of disparities between richer and poorer people with a higher incidence of cholera in lower socioeconomic groups [22]. Malnourished children and people with low immunity, such as HIV cases, have a higher mortality risk from cholera infection as compared to the normal population [8].

4.2. Place distribution

The frequency, severity, and duration of cholera infections vary and keep on changing in different parts of the world [13]. Cholera is endemic in Africa, south and southeast Asia [23]. In contrast, cholera is almost eradicated from most of the developed countries [5]. Incidence is as low as 0–5 cases per year in the United States [9].

In some countries such as Bangladesh and India, cholera infections occur every year [13]. Cholera reached West Africa and Ghana during the seventh pandemic [24]. In East Africa, cholera epidemic was first reported in 1836 but no cases were reported across Africa between 1870 and 1970. Cholera returned to Africa in 1970 as a result of seventh cholera pandemic. In 1974, cholera cases were reported for the first time in Tanzania, and then reported each year since 1977 [12]. In 1970, the first cholera case was reported in Ghana, after which cholera has been endemic in Ghana. Between 1999 and 2005, Africa accounted for about 90% of the cholera cases and 96% of the cholera-related deaths worldwide [24]. In contrast, other regions such as parts of South America have historically had only sporadic epidemics [13].

During 2013, a total of 56,329 cases were reported from Africa, which shows a decrease of 52% as compared to 2012 when 117,570 cases were reported. Africa accounted for 43.6% of the total cases in 2013 as compared to 93–98% of the total cases during the period 2001–2009. In contrast, 11,576 cases were reported from Asia, representing an increase of 57% as compared to 2012 when 7367 cases were reported from Asia [11].

In 2013, a total of 26 countries reported deaths due to cholera; 17 of these countries belonged to Africa. The case fatality rate was <1% in 4 countries, 1–5% in 17 countries, and >5% in 5 countries. In 2013, a majority (65%) of the deaths were reported from the African continent. A total of 1366 deaths were reported with a case fatality rate of 2.43% [11].

Various studies have reported differences in incidence of cholera in rural and urban areas. Sepúlveda et al (2006) reported high incidence of cholera in rural and suburban areas due to lack of basic sanitation infrastructure [3]. Other studies have reported cholera predominance in the urban areas, due to overcrowding and unsanitary living conditions. In urban communities in Ghana, intermittent water supply along with indiscriminate sanitation practices increases the risk of cholera. Urban slum areas are also at high risk of cholera infection as a lack of safe drinking water, poor sanitation, high population density, and crowding are common features in these areas; all of which are the risk factors for cholera [12, 24].

Bompangue et al. (2008) explored geographic proximity of the cholera cases [25]. The number of cases was found to be statistically significantly higher in the presence of a lake, a main road, or a harbor. Proximity to surface water, high population density, and low educational status were also identified as the predictors of cholera in an endemic area of Bangladesh. Poverty, urbanization, and proximity to coastal areas were important geographic predictors of cholera in Mexico. Proximity and density of refuse dumps also play a significant role in cholera transmission [6]. Epidemiological studies have demonstrated an inverse relationship between diarrhea and access to tube well water, and positive associations with canal water compared with river or pond water [13].

4.3. Time distribution

Vibrio cholerae are found in seas, rivers, and ponds of coastal areas of the tropical countries. It is postulated that environmental factors such as climate variability, temperature, and salinity play an important role in cholera transmission. Cholera is also associated with rainfall patterns. Recurrence of epidemic cholera has also been related to climatic factors [6, 13, 14].

Even in endemic areas, the magnitude of cholera incidence varies from year to year. In rural Bangladesh, there is a premonsoon epidemic almost every year. Cholera is rare during the winter as the temperature is low and there is little rainfall. Gradually, the temperature gets hotter till the monsoon arrives. The incidence of cholera is low in the actual monsoon period as compared to premonsoon period. This lower incidence in the monsoon period is explained by the dilution effect reducing the amount of bacteria in the aquatic environment. However, postmonsoon epidemics are much larger than the premonsoon epidemics [22].

A regular seasonal cycle for cholera outbreaks related to the different strains: classical, El Tor and O139, is described for Bangladesh. El Tor is reported to be most incident from September to November. Some studies have reported two annual El Tor cholera peaks; smaller peak in April followed by a larger outbreak from September to December. *Vibrio cholerae* O139 has a similar seasonal pattern [13].

In Pakistan, the incidence of cholera increases from November to January and from April to May, while in India, cholera cases peak in April, May, and June. In South America, cholera peaks in summer months, January to February and in the rainy season. In eastern African countries such as Kenya, Somalia, Uganda, and Tanzania, cholera outbreaks occur following summer rainfall. From 1979 to 1983, two cholera peaks were reported from Tanzania; from October to December; and from March to May. Both of these peaks coincide with increased rainfall. In southern Tanzania, the highest incidence of cholera is reported in June and July, while during 2002 the northern parts of southern Africa reported maximum cholera cases from the last week of January to mid-March. In Mozambique, peak of cholera cases occurs from December to May [13].

Cholera outbreaks occur more commonly closer to the equator than at higher latitudes and do not follow a clear seasonal pattern near the equator. Although annual peaks are evident, it is difficult to determine distinct seasonal patterns in cholera outbreaks across regions. However, grouping countries by latitude range, rather than region, makes these seasonal peaks more obvious. Countries near the equator have higher and more constant temperatures, and have a greater incidence of cholera outbreaks [13].

The seasonality of cholera outbreaks may also be explained by secondary transmission. Several studies find that the severity of secondary transmission is associated with local environmental variables, predominantly water sources for household consumption. People who use contaminated surface water for drinking, cooking, and bathing are more likely to contract cholera than those who do not [13].

4.4. Social and behavioral epidemiology of cholera

Cholera transmission is closely associated with the social and behavioral aspects of individuals as well as communities. There are many demographic factors which predispose an individual or community to cholera infection. An increase in population density can result in overburdening existing water supply and sanitation systems. Thus, population density, urbanization, and overcrowding influence intensity of the cholera outbreak [24].

With increasing world population, urbanization is expected to increase. Most of this urban growth will occur in the developing countries. Because of economic issues, most of urban growth in the developing countries is the expansion of squatter settlements. These settlements comprise of households that lack access to safe drinking water and sanitation facilities. These conditions are of particular concern in the spread of infectious diseases which have been associated with conditions prevalent in urban squatter settlements [12].

There are obvious socioeconomic disparities in the occurrence of cholera. Even in the developing countries, the incidence of cholera is higher in socially deprived communities as compared to economically privileged class [13]. Among vulnerable groups living in areas at high risk for cholera, the case fatality rate is high. The secondary cases of cholera occur through fecal–oral transmission, and are related to poor hygiene, poor water, and sanitation environment [6, 22]. The water and sanitation environment as well as proper hygiene and educational level are associated with socioeconomic status of individuals as well as communities. Individuals with low socioeconomic status lack financial resources to install an appropriate sanitary system or obtain cleaner water sources. It is reported that cholera is more common in poorer households without safe water and proper sanitation system as compared to those having appropriate water and sanitation systems [22].

Surface water is contaminated with human excreta and sewage at place where rivers pass through overcrowded cities. In some African countries such as Ghana with multiethnic population, the traditional laws for protection of water bodies are no longer obeyed. Thus, waste dumps are a common sight at the banks of surface water bodies in urban communities. These urban inhabitants have to use such polluted water at the times of water shortage. The urban slums and squatter settlers have lack of accessibility and affordability to safe drinking water and sanitation. Sometimes, public utility providers are not bound legally to serve the urban slums because of technical and service regulations. In addition, many urban slums are located at low lying areas leading to higher vulnerability to floods, making them more prone to cholera infection. Once the disease is introduced in a population, these demographic and social variables further aggravate the situation [6, 24].

Rapid urbanization can lead to overcrowding, unprotected water sources, and improper disposal of solid waste, liquid waste, and human feces. All these conditions make the environment conducive for cholera transmission. In urban communities, cholera outbreaks have been attributed to poor waste management and sanitation systems [24].

The behavioral aspects of community regarding cholera are important because if cholera is recognized as a serious and life threatening condition by the community, it will be more receptive to community health education program. Human behavior is also important in the acceptance of vaccine program [2].

A humanitarian crisis in a country or region can result in cholera outbreaks. The risk of cholera epidemics also increases after large-scale natural disasters. Some areas of Democratic Republic of Congo have been the site of large scale disasters including invasion and occupation by foreign forces, civil war, population displacements, and earthquakes. These areas have also shown high incidence of cholera [26].

Outbreaks of cholera have negative impact on social and economic conditions of the country as it leads to widespread fear and travel sanctions. Moreover, within the cholera-affected countries, resources are diverted to the care of cholera patients. Certain countries such as Peru and Mexico suffered economically as a result of cholera outbreaks, because of a decrease in tourism and decreased exports by other countries [3].

5. Cholera outbreaks

Cholera outbreaks have always proved challenge for the resource limited health-care systems of countries that usually suffer from these outbreaks. During 2014, cholera outbreaks were reported from South Sudan while in 2013 Mexico reported cholera outbreak [27]. Since October 2010, there is an ongoing outbreak in Haiti and the Dominican Republic. During 2013, the cholera cases during this outbreak accounted for 47.3% of all reported cases [11]. In 2013, outbreaks were also reported from other Central African countries such as Angola, Burundi, and Congo [11]. Namibia reported 1557 cholera cases including 17 deaths while Congo reported 1624 cases and 221 deaths in an outbreak which started in November 2012 [11]. In Cuba, an outbreak started in July 2012, and a total of 181 cholera cases were reported in 2013 [11].

6. Prevention and control

6.1. Cholera prevention

For the prevention and control of cholera, it is important to understand the factors that are responsible for initiating and sustaining cholera infection in a community [28]. Measures for the prevention of cholera include provision of clean water and proper sanitation to the cholera-endemic communities.

6.1.1. Health education

Health education regarding personal hygiene and food hygiene should be provided to these communities. Media, community leaders, and religious leaders should participate in health education and social mobilization campaigns [29].

CDC has listed five basic cholera prevention messages. These include drinking and using safe water; washing hands with soap and water; using latrines or proper sanitation methods; proper cooking of food, covering it, and eating it hot; proper cleaning up of places used for bathing and washing clothes [30].

6.1.2. Vaccines

Currently, there are two oral cholera vaccines available: Dukoral and Shanchol. Dukoral is a killed whole cell vaccine including V. cholerae O1 serogroup and recombinant B subunit of cholera toxin. It can be given to children ≥2 years and to adults. For children 2–5 years of age, three doses, 1–6 weeks apart, are given orally, and booster dose is given after 6 months. For adults and children ≥6 years, 2 doses, 1–6 weeks apart, are given orally while booster dose is given after 2 years. The earliest onset of protection is 7 days after the second dose, and the protection at 6 months is 85–90% [31]. Shanchol is a killed bivalent (O1 and O139 serogroups) whole-cell vaccine suspension. It can be given to ≥1 year of age; 2 doses, 2 weeks apart, are given orally. The earliest onset of protection is 7–10 days after the second dose, and there is 65% protection for at least 5 years [31].

In cholera-endemic regions and those at high risk for outbreaks, cholera vaccines should be used along with other prevention and control strategies. In these regions, high risk population may be targeted for vaccination. The high risk groups include preschool and school-aged children, pregnant women, and HIV-infected individuals [23].

6.2. Cholera control

Cholera-endemic areas should prioritize cholera control measures [23]. Countries facing complex emergencies and displacement of internally displaced people (IDP) on a large scale or refugees to places where the provision of safe water and proper sanitation is compromised, and they are vulnerable to cholera outbreaks. In such situations, it is critical to depend on surveillance data to watch for an outbreak and to implement appropriate intervention measures [32]. Thus, strengthening of surveillance system and early warning system is vital in places at high risk of cholera outbreak [29].

The main strategies for cholera control include appropriate and prompt management of cholera cases; strengthening laboratories; training and capacity building of health-care workers; and availability of adequate medical supplies for management [3, 29]. In addition, access to safe water, proper sanitation, appropriate waste management; personal hygiene and food hygiene practices; improved communication and public information are needed for the control of cholera outbreaks.

6.2.1. Cholera vaccines

Oral Cholera Vaccine should always be used as an additional public health tool in complex emergencies and should not replace usually recommended control measures such as improved water supplies, adequate sanitation, and health education. Once a cholera outbreak has started, the vaccine is not recommended as it takes time to provide protection and is also not cost-effective [33]. Reyburn et al (2011) estimated that an organized mass vaccination campaign could prevent 34,900 (40%) cholera cases and 1695 deaths (40%) in Zimbabwe. However, the cost of the vaccines was an important barrier along with other logistic issues [34].

A well-organized, multisectoral approach is required to control cholera outbreaks. The effectiveness of public health interventions depends on an efficient surveillance system. There must be frequent and timely information-sharing at local as well as global level [11]. The administration of cholera vaccines may be considered for high risk population in high risk areas. Funds and resources should be provided to the deserving countries to improve cholera prevention and preparedness activities.

6.2.2. International travel and trade

Currently, there is no obligation of cholera vaccination for international travel. It is learned with experience that quarantine and restrictions on travel and trade are not very effective in controlling the spread of cholera. However, the travelers should be provided information regarding signs, symptoms, and prevention of cholera. The neighboring countries of cholera-affected areas should be advised to enhance their surveillance system for early detection and prompt response if any outbreak occurs [11].

Author details

Saulat Jahan

Address all correspondence to: saulatjahan@hotmail.com

Research and Information Unit, Public Health Administration, Qassim, Ministry of Health, Kingdom of Saudi Arabia

References

[1] Kanungo S, Sah BK, Lopez AL, Sung JS, Paisley AM, Sur D, et al. Cholera in India: An analysis of reports, 1997–2006. Bull World Health Organ. 2010 Mar;88(3):185–191.

[2] Schaetti C, Khatib AM, Ali SM, Hutubessy R, Chaignat C-L, Weiss MG. Social and cultural features of cholera and shigellosis in peri-urban and rural communities of Zanzibar. BMC Infect Dis. 2010;10:339.

[3] Sepúlveda J, Valdespino JL, García-García L. Cholera in Mexico: The paradoxical benefits of the last pandemic. Int J Infect Dis. 2006 Jan;10(1):4–13.

[4] Codeço CT, Coelho FC. Trends in cholera epidemiology. PLoS Med. 2006 Jan; 3(1):e42.

[5] 150 years of cholera epidemiology. Lancet. 2005 Sep 17;366(9490):957.

[6] Osei FB, Duker AA. Spatial dependency of V. cholera prevalence on open space refuse dumps in Kumasi, Ghana: A spatial statistical modelling. Int J Health Geogr. 2008;7:62.

[7] Matos TC. Choleric fictions: Epidemiology, medical authority and an enemy of the people. Modern Drama 2008; 51(3):353–368.

[8] WHO | Cholera [Internet]. Available from: http://www.who.int/mediacentre/factsheets/fs107/en/ [Accessed: 2015-07-15].

[9] Sources of Infection & Risk Factors | Cholera | CDC [Internet]. Available from: http://www.cdc.gov/cholera/infection-sources.html [Accessed: 2015-07-15].

[10] Deen JL, von Seidlein L, Sur D, Agtini M, Lucas MES, Lopez AL, et al. The high burden of cholera in children: Comparison of incidence from endemic areas in Asia and Africa. PLoS Negl Trop Dis. 2008;2(2):e173.

[11] WHO | Weekly Epidemiological Record: Cholera, 2013 [Internet]. Available from: http://www.who.int/wer/2014/wer8931.pdf?ua=1 [Accessed: 2015-07-15].

[12] Penrose K, de Castro MC, Werema J, Ryan ET. Informal urban settlements and cholera risk in Dares Salaam, Tanzania. PLoS Negl Trop Dis. 2010;4(3):e631.

[13] Emch M, Feldacker C, Islam MS, Ali M. Seasonality of cholera from 1974 to 2005: A review of global patterns. Int J Health Geogr. 2008;7:31.

[14] Constantin de Magny G, Murtugudde R, Sapiano MRP, Nizam A, Brown CW, Busalacchi AJ, et al. Environmental signatures associated with cholera epidemics. Proc Natl Acad Sci USA. 2008 Nov 18;105(46):17676–17681.

[15] Illness & Symptoms | Cholera | CDC [Internet]. Available from: http://www.cdc.gov/cholera/illness.html [Accessed: 2015-07-15].

[16] Diagnosis and Testing | Cholera | CDC [Internet]. Available from: http://www.cdc.gov/cholera/diagnosis.html [Accessed: 2015-07-15].

[17] Treatment | Cholera | CDC [Internet]. Available from: http://www.cdc.gov/cholera/treatment/index.html [Accessed: 2015-07-15].

[18] Jabeen K, Hasan R. Re-emergence of *Vibrio cholerae* O139 in Pakistan: Report from a tertiary care hospital. J Pak Med Assoc. 2003 Aug;53(8):335–338.

[19] Tamang MD, Sharma N, Makaju RK, Sarma AN, Koju R, Nepali N, et al. An outbreak of El Tor cholera in Kavre district, Nepal. Kathmandu Univ Med J (KUMJ). 2005 Jun; 3(2):138–142.

[20] Shittu OB, Akpan I, Popoola TOS, Oyedepo JA, Ogunshola EO. Epidemiological features of a GIS-supported investigation of cholera outbreak in Abeokuta, Nigeria. J. Public Health Epidemiol. 2010; 2(7):152–162.

[21] Khazaei HA, Rezaei N, Bagheri GR, Moin AA. The Epidemiology of *Vibrio Cholerae* in Zabol City, Southeast of Iran. Jpn J Infect Dis. 2005; 58(1):8–10.

[22] Emch M, Yunus M, Escamilla V, Feldacker C, Ali M. Local population and regional environmental drivers of cholera in Bangladesh. Environ Health. 2010;9:2.

[23] WHO Publication. Cholera vaccines: WHO position paper-Recommendations. Vaccine. 2010 Jul 5;28(30):4687–4688.

[24] Osei FB, Duker AA. Spatial and demographic patterns of cholera in Ashanti region - Ghana. Int J Health Geogr. 2008;7:44.

[25] Bompangue D, Giraudoux P, Handschumacher P, Piarroux M, Sudre B, Ekwanzala M, et al. Lakes as source of cholera outbreaks, Democratic Republic of Congo. Emerging Infect Dis. 2008 May;14(5):798–800.

[26] Bompangue D, Giraudoux P, Piarroux M, Mutombo G, Shamavu R, Sudre B, et al. Cholera epidemics, war and disasters around Goma and Lake Kivu: An eight-year survey. PLoS Negl Trop Dis. 2009;3(5):e436.

[27] WHO | Cholera [Internet]. WHO. Available from: http://www.who.int/csr/don/archive/disease/cholera/en/ [Accessed: 2015-07-15].

[28] Pascual M, Koelle K, Dobson AP. Hyperinfectivity in cholera: A new mechanism for an old epidemiological model? PLoS Med. 2006 Jun;3(6):e280.

[29] WHO | Prevention and control of cholera outbreaks: WHO policy and recommendations [Internet]. WHO. Available from: http://www.who.int/cholera/technical/prevention/control/en/index2.html [Accessed: 2015-07-15].

[30] Five basic cholera prevention messages| Cholera | CDC [Internet]. Available from: http://www.cdc.gov/cholera/five-messages.html [Accessed: 2015-07-15].

[31] WHO I Cholera: Oral Cholera Vaccines: Use of Oral Cholera Vaccines in Humanitarian Emergencies [Internet]. Available from: http://www.who.int/cholera/vaccines/OCV_in_humanitarian_emergencies_15Jan2014.pdf?ua=1 [Accessed: 2016-02-15].

[32] WHO I Cholera: Prevention and control [Internet]. WHO. Available from: http://www.who.int/topics/cholera/control/en/ [Accessed: 2015-07-15].

[33] WHO I Global task force on cholera control: Prevention and control of cholera outbreaks: WHO policy and recommendations: Cholera Vaccines. Available from: http://www.who.int/cholera/technical/prevention/control/en/index5.html [Accessed: 2015-07-15].

[34] Reyburn R, Deen JL, Grais RF, Bhattacharya SK, Sur D, Lopez AL, et al. The case for reactive mass oral cholera vaccinations. PLoS Negl Trop Dis. 2011;5(1):e952.

Permissions

All chapters in this book were first published in SPCFRD, by InTech Open; hereby published with permission under the Creative Commons Attribution License or equivalent. Every chapter published in this book has been scrutinized by our experts. Their significance has been extensively debated. The topics covered herein carry significant findings which will fuel the growth of the discipline. They may even be implemented as practical applications or may be referred to as a beginning point for another development.

The contributors of this book come from diverse backgrounds, making this book a truly international effort. This book will bring forth new frontiers with its revolutionizing research information and detailed analysis of the nascent developments around the world.

We would like to thank all the contributing authors for lending their expertise to make the book truly unique. They have played a crucial role in the development of this book. Without their invaluable contributions this book wouldn't have been possible. They have made vital efforts to compile up to date information on the varied aspects of this subject to make this book a valuable addition to the collection of many professionals and students.

This book was conceptualized with the vision of imparting up-to-date information and advanced data in this field. To ensure the same, a matchless editorial board was set up. Every individual on the board went through rigorous rounds of assessment to prove their worth. After which they invested a large part of their time researching and compiling the most relevant data for our readers.

The editorial board has been involved in producing this book since its inception. They have spent rigorous hours researching and exploring the diverse topics which have resulted in the successful publishing of this book. They have passed on their knowledge of decades through this book. To expedite this challenging task, the publisher supported the team at every step. A small team of assistant editors was also appointed to further simplify the editing procedure and attain best results for the readers.

Apart from the editorial board, the designing team has also invested a significant amount of their time in understanding the subject and creating the most relevant covers. They scrutinized every image to scout for the most suitable representation of the subject and create an appropriate cover for the book.

The publishing team has been an ardent support to the editorial, designing and production team. Their endless efforts to recruit the best for this project, has resulted in the accomplishment of this book. They are a veteran in the field of academics and their pool of knowledge is as vast as their experience in printing. Their expertise and guidance has proved useful at every step. Their uncompromising quality standards have made this book an exceptional effort. Their encouragement from time to time has been an inspiration for everyone.

The publisher and the editorial board hope that this book will prove to be a valuable piece of knowledge for researchers, students, practitioners and scholars across the globe.

List of Contributors

Bing Li and Lin Li
College of Light Industry and Food Sciences, South China University of Technology, Guangzhou, China

Mark E. Shirtliff
Department of Microbial Pathogenesis, School of Dentistry, University of Maryland, Baltimore, MD, USA

Zhenbo Xu
College of Light Industry and Food Sciences, South China University of Technology, Guangzhou, China
Department of Microbial Pathogenesis, School of Dentistry, University of Maryland, Baltimore, MD, USA

Brian M. Peters
Department of Clinical Pharmacy, College of Pharmacy, University of Tennessee Health Sciences Center, Memphis, TN, USA

Qinchun Rao and Yun-Hwa Peggy Hsieh
Department of Nutrition, Food and Exercise Sciences, Florida State University, Tallahassee, FL, USA

Antonio Valero, Magdevis-Yanet Rodríguez, Guiomar Denisse Posada-Izquierdo, Fernando Pérez-Rodríguez, Elena Carrasco and Rosa Maria García-Gimeno
Department of Food Science and Technology, University of Cordoba, Cordoba, Spain

Biljana Damjanović-Vratnica
Faculty of Metallurgy and Technology University of Montenegro Podgorica, Montenegro

Aslı Uçar, Mustafa Volkan Yilmaz and Funda Pınar Çakıroğlu
Ankara University Faculty of Health Sciences, Department of Nutrition and Dietetics, Ankara, Turkey

Luciana M. Coelho, Priscila A.R. de Sousa and Lícia A. da Silva
Department of Chemistry, Federal University of Goiás, Catalão, GO, Brazil

Diego R. Pessoa, Kênia M. Oliveira and Nívia M.M. Coelho
Institute of Chemistry, Federal University of Uberlândia, Uberlândia, MG, Brazil

Luisa W. Cheng, Christina Tam, David L. Brandon and Larry H. Stanker
Foodborne Toxin Detection and Prevention Research Unit, Western Regional Research Center, Agricultural Research Service, U.S. Department of Agriculture, Albany, CA, USA

Kirkwood M. Land
Department of Biological Sciences, University of the Pacific, Stockton, CA, USA

Saulat Jahan
Research and Information Unit, Public Health Administration, Qassim, Ministry of Health, Kingdom of Saudi Arabia

Index

www.ingramcontent.com/pod-product-compliance
Lightning Source LLC
LaVergne TN
LVHW080738210926
848485LV00003B/577